On Minds

On Minds

Reflections of a Cognitive Architect

Paul S. Rosenbloom

The MIT Press
Cambridge, Massachusetts
London, England

The MIT Press
Massachusetts Institute of Technology
77 Massachusetts Avenue
Cambridge, MA 02139
mitpress.mit.edu

The MIT Press would like to thank the anonymous peer reviewers who provided comments on drafts of this book. The generous work of academic experts is essential for establishing the authority and quality of our publications. We acknowledge with gratitude the contributions of these otherwise uncredited readers.

This book was set in Stone Serif and Stone Sans by Westchester Publishing Services. Printed and bound in the United States of America.

Library of Congress Cataloging-in-Publication Data

Names: Rosenbloom, Paul S. author
Title: On minds : reflections of a cognitive architect / Paul S. Rosenbloom.
Description: Cambridge, Massachusetts : The MIT Press, [2026] | Includes bibliographical references and index.
Identifiers: LCCN 2025048302 (print) | LCCN 2025048303 (ebook) | ISBN 9780262053259 paperback | ISBN 9780262053266 pdf | ISBN 9780262053273 epub
Subjects: LCSH: Artificial intelligence—Data processing | Cognition | Computer architecture | Rosenbloom, Paul S.
Classification: LCC Q336 (print) | LCC Q336 (ebook)
LC record available at https://lccn.loc.gov/2025048302
LC ebook record available at https://lccn.loc.gov/2025048303

EU Authorised Representative: Easy Access System Europe, Mustamäe tee 50, 10621 Tallinn, Estonia | Email: gpsr.requests@easproject.com

To Elaine and Michael, to whom I owe all.

Contents

Preface

This book is about *minds*—functional entities that can think—as explored in artificial intelligence (AI) and cognitive science, with the guiding assumption that it can be useful to ground such an exploration in what can be learned from both fields. If what you are looking for is an analysis of the current state and future prospects of deep learning and generative models—the current white hot and wildly successful approach(es) to AI—this book may not be the best place to start, although such topics do make an appearance where appropriate.

If instead what you seek is a deeper understanding of a broader take on AI and cognitive science from the perspective of a researcher who has spent most of the last half-century trying to understand what it means to put all of the pieces together into *cognitive architectures* that provide the fixed structures and processes underlying general humanlike intelligence, and who has spent a number of the past years reflecting on this experience—as well as on integrative approaches to minds more generally and on the combination of AI and cognitive science as a discipline—this book will hopefully both intrigue and inform you. Even if your interest is primarily in generative models or deep learning, this book may be of interest to the extent you care about building complete minds.

I've always had a tendency to reflect, whether on my own actions or on the content of my studies. As my life and career have progressed over the decades, this tendency has only grown, while being grounded in an ever-broader body of experiences and, hopefully, an increasingly mature perspective. Although most of my career has focused on developing and understanding a sequence of cognitive architectures—from XAPS to Soar to Sigma—a decade-long hiatus between the last pair, during which I explored new directions for the University of Southern California's Information Sciences Institute (USC/ISI), led to an extensive set of reflections on computing that ultimately ended up yielding a book.

On Computing: The Fourth Great Scientific Domain, which was published by the MIT Press in 2012, made the argument that computing was not only a scientific discipline but a *great scientific domain*, on a par with the physical, life, and social sciences. This involved reflections on the nature of computing, science, and engineering, and what it means for an intellectual discipline to be a great scientific domain. It also involved developing the *relational model* that was to provide insight into how the various disciplines within computing could be understood in terms of relationships between computing, itself, and the other three domains.

In the current book, the focus shifts from these earlier reflections on computing to more recent ones on minds, driven by attempts to better understand and explain my own work on Sigma, the community's work more broadly on cognitive architectures, and how this all relates to the even broader body of work in AI and cognitive science as a whole. Rather than yielding a single systematic story, as in the earlier book, this has yielded a set of three semi-independent stories that center on *definitions, dissections and mappings*, and *integrations* and that serve as the themes of the three parts of this book. Still, a key feature of telling these three stories in the same context is the value added that results from their interactions.

The relational model from the earlier work plays only a peripheral role in the overall story here, but the definitional work there—on computing, science, engineering, great scientific domains, and interdomain relationships—developed into the methodology of *essential analysis* that is at the heart of part I's focus on definitions. In guiding how to rethink the definitions of key terms in AI and cognitive science, it has proven fruitful in exploring multiple topics critical to the study of minds. Four of these explorations are included in this part, on *architectures*, *symbols*, *intelligence*, and *theories*. The first two are relatively short and straightforward while still taking these topics in new directions. The latter two are longer and more complex, reflecting their more challenging nature but, hopefully, also correspondingly rewarding.

As an undergraduate, a nascent interest in philosophy led me to enroll in a course on ethics that I hoped would provide insight into ethical systems. However, the course turned out instead to center on what seemed like endless arguments concerning the meanings of ethics-related terms such as *ought*. Not seeing how this addressed the fundamental questions about ethical systems that had led me to take the course, this experience largely distanced me from an interest in philosophy for a number of years. Although the focus in part I is similarly on definitions, the idea here is to leverage essential analysis to cut through existing Gordian knots rather than further entangling them. In the process, the aim is to provide fresh insights into key aspects of the philosophical basis of the field. As a consequence, this part is likely to be of more interest to those focused on understanding minds than to those wanting to leverage them for more practical ends.

Part II, concerning dissections and mappings, grew out of several failed attempts in 2019 to publish explanations of how Sigma was able to support many of the capabilities found across AI and cognitive science without including distinct modules within the architecture specifically built for them—satisfying the core Sigma desideratum of *functional elegance*. Reconsidering this effort more generically, as an attempt to understand the space of technologies relevant to minds—whether symbolic or neural (although with more of a slant toward the former given my own background and how rapidly the latter is advancing)—evolved into the methodology of *dichotomic analysis*, of how contrasts such as asymmetric versus symmetric processing have shaped, and can shape, how we understand this overall space.

A cross product of dichotomies effectively dissects the overall space of technologies into regions with uniform properties, with differences across regions providing contrasts, to ultimately yield a structured view of the space and of how the individual technologies both relate to and complement each other. Dichotomic analysis also yields insights into how combinations of technologies across regions yield integrated systems such as cognitive architectures, and how extensively such systems can span the full space. When combined with relevant evaluation criteria—such as accuracy, efficiency, and generality—and an understanding of how these criteria fracture across dichotomies, dichotomic analysis further provides a way to compare the trade-offs embodied in both individual technologies and systems that combine multiple of them.

Altogether, this supports structured mappings of technologies, and of their associated criterial trade-offs, that span both the entire discipline and systems within it, making part II ideally of interest to anyone working in the field who seeks a better understanding of the discipline, its space of available technologies, the systems that integrate together these technologies, and the criterial trade-offs involved in all of this. To facilitate following this analysis, appendix A provides informal characterizations of the various technologies mentioned in it.

Part III is a *personal intellectual memoir* that was begun as I was recovering from a serious illness and that reflects on my career—and life—as a whole.[1] Here I characterize myself as an *architectural explorer*. Much of this concerns my personal grand challenge of developing cognitive architectures that aid in understanding the human mind and in understanding and building artificial minds, plus work on the *common model of cognition*, an attempt at developing a community consensus concerning what must be in any architecture that is to support humanlike cognition. However, given the broad notion of architecture developed in part I, this notion of architectural exploration turns out to be one way to view much of the rest of my research career. As a result, part III also includes material related to the dichotomic analysis of part II as well as material from the earlier book on the relational model and the idea that computing is the fourth great scientific domain. It does not, however, include material related to part I, related to essential analysis, as that developed after the memoir was completed.

Interwoven into part III are personal details that influenced my path throughout this career, or have some entertainment value, or simply reveal key aspects of my life. This part in toto thus reflects *integrations* across disciplines, technologies, architectures—which themselves focus on integrations—and aspects of my life, while wrapping up with a set of maxims in appendix B that have been extracted from this experience and that may be of use to others. The result is likely the most broadly accessible part of this book and is appropriate for anyone interested in such a story.

The core of the book kicks off in chapter 1, which goes into more depth on the book's overall structure before introducing a *reflective method* that generalizes across essential analysis, dichotomic analysis, and the relational model. This provides one form of value added that came about from exploring the commonality lurking behind the three semi-independent stories told in this book, although one that readers less interested in methodology may want to either skip or skim. Chapter 1 also provides some background on the Soar and Sigma cognitive architectures, as well as the common model of cognition, which will be drawn upon in various ways throughout parts I (chapters 2–5), II (chapters 6–12), and III (chapters 13–19).[2] Chapter 20 wraps things up with final thoughts.

Acknowledgments

It may seem odd to mention, but the role of a serious illness and the resulting long, slow, yet still incomplete recovery—and my consequent retirement—yielded the time and inclination to spend such a large fraction of what productive time I have had over the past few years on reflections. Early versions of parts II and III were begun during a final sabbatical semester and a stretch on long-term disability, both of which were funded by the University of Southern California (USC), so I would like to thank the university for this support. I am also thankful to have received helpful comments on parts of this work from a number of people, including Stevan Harnad, Randall Hill, Claire Jansa, Julia Kim, John Laird, Christian Lebiere, Paul Middlebrooks, Paul Allen Newell, Andrew Nuxoll, Judith Pfeffer, Andrea Stocco, Volkan Ustun, and Sashank Varma.

Reaching out more broadly, the work on cognitive architectures upon which these reflections were based came about originally as a collaboration with John Laird and Allen Newell and were facilitated by the collegial and open-minded environment around the Carnegie Mellon University Computer Science Department in the late 1970s and early 1980s. The lion's share of the funding for my research on cognitive architectures came from the Defense Advanced Research Projects Agency (DARPA) in the early years and the US Army in later years, but also with a mixture of other federal, foundation, institute, and industrial support over the decades. The leadership, colleagues, and environments at USC's Information Sciences Institute (ISI) and Institute for Creative Technologies (ICT) also played a major role in support of my research on cognitive architectures during my many years associated with them.

Reaching out even a bit more, the artificial intelligence (AI) and cognitive science communities were both supportive of the early work on Soar. They have also been supportive of the work on the common model of cognition. However, with the development of Sigma and the continued evolution of the AI community, support for Sigma shifted more away from AI in general toward the artificial general intelligence (AGI) community, which has also been supportive of my more recent work that has taken on a more reflective cast. The cognitive science community has remained generally in support, in one form or another, throughout this time.

1 Introduction

I have pursued research in artificial intelligence (AI) and cognitive science for most of the last half-century.[1] Both of these disciplines study *minds*—functional entities that can think. In natural systems they are implemented by organic brains, whereas in artificial systems they are implemented by computers. AI focuses on building artificial minds and thus cares most about how such minds can be built. Cognitive science focuses on modeling natural—preferably human—minds and, thus, cares most for understanding such minds. From their combined perspective, the field of minds can be taken generically to concern the understanding and construction of natural and artificial minds.

Although some early views of cognitive science viewed it this broadly—for example, Simon and Kaplan (1993) described it as "the study of intelligence and its computational processes in men (and animals), in computers, and in the abstract"—and there have been recent attempts to resurrect this view of the field (e.g., Rosenbloom & Forbus, 2019), cognitive science has instead cleaved most closely to the study of human minds, and thus, to cognitive psychology, rather than reflecting this entire field.

In addition to cognitive science—which will be assumed to also act as a stand in for cognitive psychology—and AI, the study of minds also clearly plays an important role in neuroscience and robotics. Neuroscience concerns the structure and function of brains, and thus, with respect to minds, cares most—in the form of cognitive neuroscience—about how they arise from brains. Robotics concerns building and controlling artificial bodies, and therefore, with respect to minds, cares most about how they control such bodies. Philosophy of mind is also relevant, with much of what can be found in this book considerable as part of a sufficiently broad construal of this discipline. Other disciplines also touch significantly on the topic of minds, but my own experience and expertise—and thus, this book—are centered on AI and cognitive science, with these other disciplines playing a role mostly in how they interact with these two.

During my time in the field, many trends have come and gone. Some of the most significant in AI, for example, have centered on search (e.g., Nilsson, 1971), logic (e.g., Genesereth & Nilsson, 1987), knowledge (e.g., Hayes-Roth et al., 1983), probabilities (Pearl, 1988), and machine learning (e.g., Murphy, 2012). The latest trend, deep learning (Goodfellow et al., 2016)—including its more recent variant of generative models and their use in large language models (LLMs) (e.g., Brown et al., 2020)—is a form of machine learning, but one that has likely had more impact than all of the previous ones combined.

My own work has included aspects of all of these trends, except for logic and generative models, but it has not been fixated on any one of them individually. Instead, it has centered on *cognitive architectures*—that is, hypotheses concerning the fixed structures and processes that yield a mind, whether natural or artificial (Newell, 1990; Chong et al., 2007; Langley et al., 2009; Samsonovich, 2010; Ye et al., 2018; Kotseruba & Tsotsos, 2020; Rosenbloom, 2022a).

Cognitive architectures must be integrative, in striving ultimately to accommodate the full range of capabilities required of a mind. They also must be general, in applying across many, if not all, problems and domains. A breadth of investigations across a sequence of three such architectures—*XAPS* (Rosenbloom, 2006) → *Soar* (Laird et al., 1987) → *Sigma* (Rosenbloom et al., 2016a)—across so many years has more recently inspired an epoch of reflection about both cognitive architectures and the disciplines of AI and cognitive science within which such work occurs.

A communal piece of reflection on cognitive architectures that has formed an important additional basis for the reflections here is work on the *common model of cognition* (Laird et al., 2017; Rosenbloom et al., 2022), an ongoing effort to develop a community consensus concerning what must be in any architecture capable of supporting *humanlike cognition*—that is, human cognition plus any other forms of cognition that are similar enough to it to fit comfortably within a common abstraction.

Figure 1.1 summarizes the three levels of subject matter upon which the reflections in this book are based, starting with cognitive architectures as the

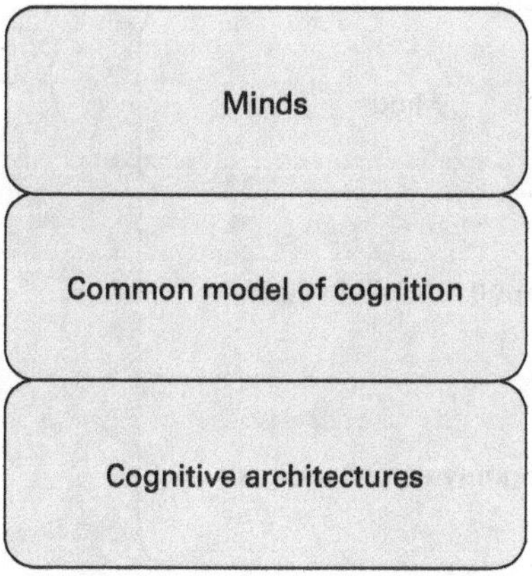

Figure 1.1
Three levels of subject matter upon which the reflections in this book are largely based.

most concrete or specific level. Abstracting and broadening from there yields the common model for humanlike minds. The final level comprises the generic notion of minds, whether natural or artificial, and whether humanlike or otherwise.

Ideally, reflections of any sort should not only provide insight into the past and present but also induce new perspectives that help to frame the future. The intention here is to capture, structure, and present reflections and perspectives that do just this. The three parts of this book are organized around reflections centered on: (I) definitions; (II) dissections and mappings; and (III) integrations.

Each of the first two parts uses a singular form of analysis that is apt for its style of reflection—*essential analysis* for definitions and *dichotomic analysis* for dissections and mappings. A third form of analysis shows up briefly in part III—*relational analysis*—but only as part of a summary of earlier reflections on the broader subjects of computing and science (Rosenbloom, 2012a).

Definitions in general ascribe meaning to terms. Sometimes this is straightforward. For example, a thermometer is by definition an instrument that measures temperature. There may be many different ways of measuring temperature as well as many different scales usable for such measurements, but none of this complexity belies the simplicity and noncontroversial nature of the definition. In other cases, attempting to define a term may lead to unending controversy. This may be because the term has been used colloquially

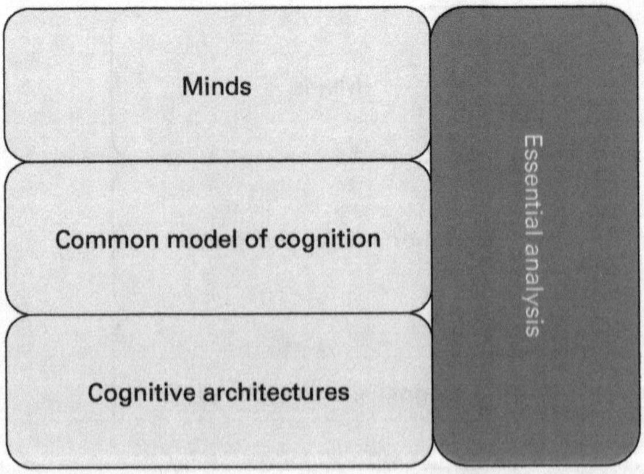

Figure 1.2
Part I: Applying essential analysis across the three levels of subject matter to yield new definitions, definitional spaces of variations, and additional implications.

for an extended period of time before any serious attempt has been made to give it a precise definition. Or it may be because it refers to a phenomenon that remains poorly understood. The term *consciousness*, for example, is difficult to define due to both of these causes, as a colloquial term for a poorly understood phenomenon.

Part I includes reflections on the meaning of four key terms—*architectures*, *symbols*, *intelligence*, and *theories*—that arise in the study of the three levels identified in figure 1.1, according to the method of essential analysis that yields new definitions, spaces of variations around these definitions, and additional implications of these definitions and spaces (figure 1.2). There has not been a systematic process by which these four terms have been selected as being the most critical ones, either with respect to their importance for the field or the difficulty in defining them. However, in the context of my own work at these three levels, each of these four terms has proven itself to be important enough and difficult enough to define that I was driven to reflect seriously on their meanings and to write up the results for others to consider. The reflections on intelligence and theories proved to be involved, yielding outsized chapters in comparison to those on symbols and architectures.

The notion of a cognitive architecture can trace it roots back through computer architecture to building architecture. As characterized in chapter 2, architectures in general concern *fixity*, inducing decompositions of systems into fixed architectures plus variable content. A cognitive architecture in these terms amounts to the fixed structures and processes that yield a mind, with its accompanying variable content being the skills and knowledge that are acquired, stored, and processed by the architecture. The reflections in this chapter yield implications into such topics as architectural constraint, multi-layered architectures, theoretical computational architectures, and architectural exploration.

A particular notion of symbols—the *Physical Symbol Systems Hypothesis* (e.g., Newell & Simon, 1976; Newell, 1980)—has been central to much of the study of minds since the earliest days of AI and cognitive science. However, the recent ascendance of neural-network-based approaches has raised major questions about both the nature and importance of symbols for minds. Chapter 3 takes a fresh look at the question of how symbols should be defined in the context of minds—as *atoms* or *placeholders*—and whether there might be a hybrid rethinking of the classical approach that brings simplicity, flexibility, and coherence to how the term is used, to include not only traditional symbol systems but also neural systems, probabilistic systems, and blends of these. Although intended to support the full scope of AI and cognitive science, the core ideas for this hybrid approach arose more specifically from work on the Sigma cognitive architecture (sections 1.2.2 and 15.4) and the common model of cognition (section 1.2.3 and chapter 16).

Intelligence is at the heart of what minds are about, with both AI and artificial general intelligence (AGI) including it in the very names of their fields. Yet there is little consensus about what it is. This is bad enough when trying to define a single notion of intelligence, but it gets even more complicated when acknowledging that there may be many kinds and levels of intelligence across different natural species and artificial systems as well as across and within individuals of any of these species or systems.

Chapter 4 provides some background on this overall quandary before attempting to bring at least partial clarity to it through identifying three notable levels of intelligence—*immediate*, *cumulative*, and *full-spectrum*—that span many of the existing definitions while also yielding insight into theories of multiple intelligence and a potential pathway toward a simpler, more essential definition of intelligence. This chapter also proposes building out an *intelligence space* around these levels, via approximations to them, that spans many of the variations of intelligence. From this all we can also begin to derive additional implications for topics such as the intelligence singularity, generative AI, Sigma, ethics, intellectual property, and free will.

Theories clearly matter when considering minds, whether they concern how natural systems work or how artificial systems might be constructed. But they are obviously also critical across all of science and engineering and, in fact, beyond this to most human intellectual endeavors.[2] In some of these areas, the nature of theories may seem straightforward and, thus, handleable in standard fashions, but in areas such as cognitive architectures, their nature is both subtle and complex. This has led to the reflections in chapter 5, which focuses on the notion of *aboutness* (and *metasystems*) but with a rich space of possibilities around this notion. The approach here has been to start by considering theories as broadly as possible—including what they are, how they are used, and how they are evaluated—while keeping in mind the nature of cognitive architectures, and then to explore the implications of these general results more specifically for cognitive architectures.

Part II uses dichotomic analysis to dissect both the overall field and integrated systems that are on the path to exhibiting intelligence at human level or above (figure 1.3). Dissection provides an opportunity to understand a complex entity by cutting it up in a manner that exposes the pieces out of which it is composed and how those pieces fit together.

Rather than dissecting along traditional module/capability boundaries, dissection occurs here across three core dichotomies—*(a)symmetric*, *(un)controlled*, and *(meta)data*. This *dichotomic analysis* yields multidimensional maps that organize the field's core technologies and help in understanding the similarities and differences among them. With the introduction of relevant evaluation criteria, these maps also yield insight into the trade-offs embodied by both these technologies and the integrated systems they combine to form.

Each dichotomy cuts the field in two, so dissecting it along all three dichotomies yields a map with eight regions, all of which are distinct from the others while sharing properties with those that align along any of the dimensions. Smaller maps based on just a subset of these dichotomies are also possible and, in fact, can be useful in a variety of ways.

As shown in figure 1.3, dichotomic maps come in two flavors. *Disciplinary maps* span the study of minds, providing an overall space in which the field's underlying technologies can be situated, their similarities and differences can be assessed along each of the dimensions, and the distances

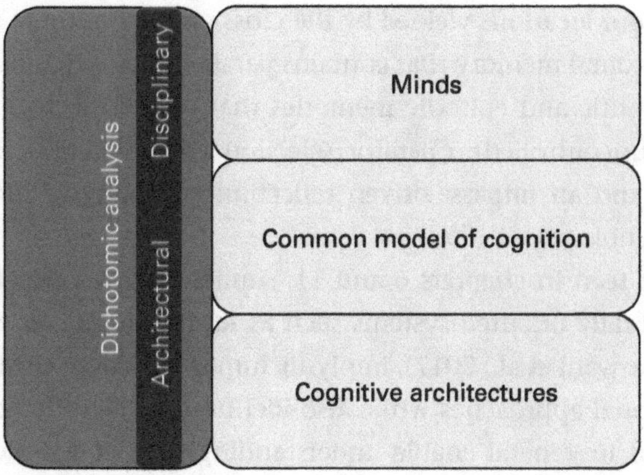

Figure 1.3

Part II: Applying dichotomic analysis across the three levels of subject matter to structure the space of relevant technologies and to understand how they differ from and complement each other, the trade-offs they involve, and how they may combine into integrated systems.

among them can be measured. Such maps can, for example, enable decomposing traditional, relatively uninformative, coarse-grained distinctions, such as the one between rules and neural networks, into combinations of finer grained, more meaningful similarities and differences.

In this case, both rules and neural networks are asymmetric and uncontrolled but differ in whether they focus on symbolic data versus quantitative metadata—that is, data about data. They are thus not opposites, as they are often portrayed. Opposition here would involve differing along all three dimensions, whereas these two technologies differ along only one dimension, making them simply two variations in a much larger space. When considering the entire map, technologies such as nonmonotonic logic, problem space search, and planning are more truly in opposition to neural networks than are rules.

Architectural maps are similar to disciplinary maps and are structured along the same three dimensions. However, as will be seen in chapter 6, they concern the technologies implicated in individual systems, such as cognitive architectures and the common model of cognition, rather than to the field as a whole. Consider, for example, the cross product between the first two dichotomies, (a)symmetric and (un)controlled. The latest incarnation of the Soar cognitive architecture (Laird, 2012) employs distinct technologies at each of the four locations yielded by the cross product of these two dimensions: a procedural memory that contains parallel rules (asymmetric uncontrolled); semantic and episodic memories that contain facts and episodes (symmetric uncontrolled); operator selection and execution (asymmetric controlled); and an impasse-driven reflection capability (symmetric controlled) (Rosenbloom, 2019a).

As will be seen in chapters 6 and 11, similar analyses are also possible for more neurally oriented systems such as AlphaZero (Silver et al., 2018) and LLMs (Vaswani et al., 2017), implying important relationships with the more traditional approaches while also identifying key differences. Architectural maps in general enable understanding such connections and, in combination with the evaluation criteria mentioned earlier, enable comparisons of entire integrated systems. They may also guide the development of new systems based on motivated combinations of technologies.

In addition to defining the dichotomies and laying out these maps, part II also yields multiple further results. It reduces a potpourri of preexisting dichotomies from across the field down to the core three—such as by mapping both the distinctions of rules versus logic and heteroassociative versus autoassociative networks onto (a)symmetric—helping to reduce the cacophony of ways researchers have previously attempted to divide up the field. The evaluation criteria that are introduced in chapter 7—which, at a high level, are *accuracy*, *efficiency*, *generality*, *simplicity*, *improvability*, *usability*, and *operability*—also help to explain the trade-offs induced by the various dichotomies and why it can make sense to support both ends of each rather than arguing—sometimes endlessly—for one end versus the other. This further inspires explorations into how these dichotomies might be spanned so as to yield the best of both worlds.

The integrative reflections (part III) take a rather different form, as a *personal intellectual memoir* that highlights my intellectual journey as an *architectural explorer*, starting with work on cognitive architectures and the common model of cognition, through the broader study of the mind, to the understanding of computing and science. These latter bits, concerning computing and science, resulted from an earlier period of reflection—summarized in Rosenbloom (2012a) and shown graphically in figure 1.4—that was inspired by a decade away from work on cognitive architectures exploring new directions for the University of Southern California (USC) Information Sciences Institute (ISI). This included early forms of essential analysis focused on the

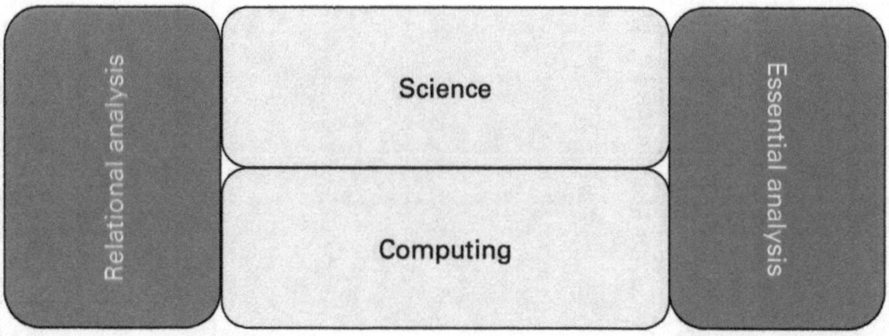

Figure 1.4
The structure of the material in *On Computing: The Fourth Great Scientific Domain* (Rosenbloom, 2012a) as two levels of subject matter plus two forms of analysis.

definition of *computing* and on the nature of *great scientific domains*, such as the physical, life, and social sciences, to make the case that computing was not only a science but in fact the fourth such domain—the *computing sciences*. In the process it also reinterpreted the meanings of both *science* itself and its companion, *engineering*. Beyond these early applications of essential analysis, this work introduced *relational analysis*, although called the *relational model* there, that identified how the various disciplines across the computing sciences—as well as potentially all human intellectual endeavors—could be understood in terms of a combination of the domains implicated in those disciplines and the relationships that hold among them.

In combination, the reflections to be covered in this book can be seen in figure 1.5 as a composition of figures 1.1–1.4. This composite structure assumes that minds are a special case of computing, matching my own perspective as well as the dominant view in AI and much of cognitive science.

As will be discussed further in part III, figure 1.5 can be seen as an extension to figures 13.6 and 19.1, which were earlier developed as maps of my career. This extension has been driven by a combination of what has been learned while formulating the material in part I, which postdates the memoir, and the process of melding the material in this book into a coherent whole.

Part III also includes a smattering of material on other topics I have worked on during roughly the past half-century that don't quite fit this overall narrative, as well as background on me as a human being, that together help round out my character and history and how this combination has impacted my intellectual journey. This part concludes with a set of maxims inspired

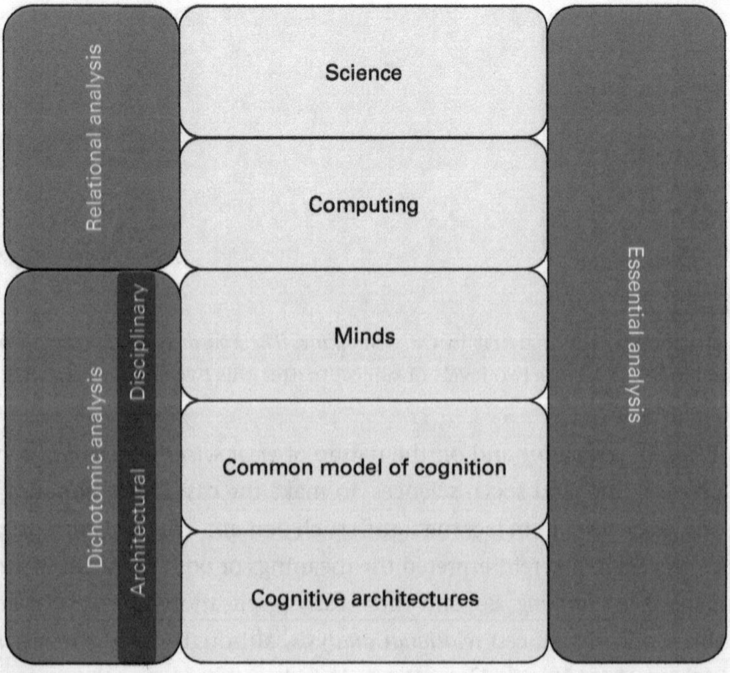

Figure 1.5
The five levels of subject matter plus three forms of analysis included in this book. This overall structure can also be viewed as a map of my career, although a somewhat different map is used to drive the memoir in part III.

by this journey that may be of use to others, either directly by application to them or by indirectly inspiring reflections in them that yield more individually relevant alternatives.

Digging more deeply into figure 1.5, parts I and II of this book can be seen to focus on two distinct forms of analysis that bracket the bottom three levels of subject matter. Part III focuses on a single trajectory through multiple levels of subject matter—most particularly the lower ones—while also traversing all three kinds of analyses that bracket the full set of five levels. Together these three parts yield a complementary set of perspectives in which minds are conceptually at the heart—as they are in figure 1.5—while extending to both more specific and more general levels of subject matter and to multiple forms of analyses across these levels. This complementary combination of perspectives, plus how they mesh together into a coherent whole in figure 1.5, is part of the value added from bringing all of this material together in one place in this book. Taken as a whole, the hope is that this all provides significant food for thought for readers with a variety of interests around the study of minds.

The subtitle of this book is *Reflections of a Cognitive Architect.* All three of the coming parts focus on reflections that typically are either grounded in my own work on cognitive architectures or have implications for cognitive architectures more broadly. The remainder of this chapter leads into to this by providing necessary background on reflective methodology and cognitive architectures.

Section 1.1 reviews the three methodologies included in figure 1.5—essential, dichotomic, and relational analysis—and which will be employed in parts I–III, before introducing the *reflective method* as a generalization over them. In doing so, this section not only explains the methods that will be used in the remainder of the book but also adds its own (methodological) contribution.

Section 1.2 provides introductory material on cognitive architectures, with a focus on two in which I have played central roles—Soar and Sigma—plus the common model of cognition. In contrast to section 1.1., no new contributions are introduced here.

1.1 Methodology

In Rosenbloom (2012a), I referred to reflections on computing as a form of *metascience*—that is, science about science. This might more conventionally be considered as philosophy of science, but it was further argued that an appropriately broad definition of science would include philosophy as part of it. Most of the reflections found in parts I–III may also be considered as instances of metascience, although here typically concerning minds rather than computing. In these terms, the five levels of subject matter in figure 1.5 can be seen as parts of science, whereas the three forms of analysis at the heart of this book can be seen as parts of metascience.

This section lays out some of the groundwork for this metascientific activity by pursuing a form of *meta²science*—that is, science about metascience. In particular, it breaks down into two pieces. Section 1.1.1 reviews the three analytical methods applied in this book—essential, dichotomic, and relational analysis. One form of added value that has arisen from juxtaposing these three previously distinct methods in this manner has been the identification of a more general method—the *reflective method*—that effectively subsumes all three of these more singular forms. Much of the detail in this section is necessary to establish and relate the four methods, but for readers more interested in the results of the various reflections than in the methodology used for these reflections, focusing on just the specifications of the methods without worrying about much of the rest of the commentary should be more than enough to understand the reflections in the coming chapters. This new method is explored in some detail in section 1.1.2, along with how each of the earlier methods can be viewed as a specialization of it.

1.1.1 A Review of Essential, Dichotomic, and Relational Analysis

Essential analysis grew out of early work—reviewed in chapter 18 of part III, on (re)defining the notions of computing, science, engineering, great scientific domains, and interdomain relationships—before being crystalized as a means of (re)defining the notions of architectures, symbols, intelligence, and theories in part I. As articulated in Rosenbloom (2023b), it comprises a sequence of four steps.

1. Strip out many of the elaborations that are normally part of a topic's definition, and which are often a source of dissonance among researchers and communities, to yield its *essence.*
2. Use what has been stripped out, and possibly more, in specifying a *definitional space* of variations on the topic.
3. Populate this space with *exemplars* that flesh it out.
4. Derive novel *implications* from the results of the first three steps.

For example, in chapter 3 the notion of a symbol is first stripped down to the essence of an atom or placeholder in step 1. In step 2, a definitional space is based on aspects of traditional definitions of the notion that are thus stripped out—composition, designation, and interpretation—plus the addition of an option for a symbol to be hybrid. In step 3, a few exemplars are introduced in examing physical symbol systems, hybrid symbol systems, and neural-network systems. Finally, in step 4, implications are derived, including proposing replacing the physical symbol systems hypothesis with the hybrid symbol systems hypothesis.

Dichotomic analysis, as applied in part II, is a four-phase method that reflects on dichotomies, evaluation criteria, technologies relevant to AI and cognitive science, and systems that combine such technologies. The first phase (chapter 6) employs a manual, informal variant of *conceptual clustering* (e.g., Michalski, 1980) to group many of the existing dichotomies into three categories, with an essential, or core, dichotomy identified for each of these categories. The second phase (chapter 7) explores criteria for evaluating the technologies of interest and that may thus be useful in understanding trade-offs across the individual dichotomies and, beyond this, to cross products among them. In the third phase, each of the essential dichotomies is explored in some detail (chapters 8–10), including partitioning the space of AI technologies across its poles; identifying a list of existing dichotomies that map onto it; and exploring the dichotomy's relationship to learning, the criterial trade-offs the dichotomy involves, and how it might be spanned to achieve the best of both worlds. The fourth phase explores interactions among the three essential dichotomies along with the implications of these interactions (chapter 11).

A set of maps proves to be crucial to the latter two phases of dichotomic analysis. Locations in these maps are identified by the poles of the dichotomies that define their dimensions; so, for example, a 1D map has two locations, a 2D map has four, and a 3D map has eight. One variant of these maps localizes AI technologies along these dimensions, providing insight into their similarities and differences and into how they may be combined in integrated systems. A second variant identifies the *criterial advantages* characteristic of the various locations—such as which differentially advantage accuracy or efficiency—for use in evaluating technologies, dichotomic trade-offs, and integrated systems. To support the development of this second variant, a *criterial advantage (CA)* language is developed that facilitates succinct representations of the potential advantages characteristic of the different locations in these maps.

Relational analysis, as reviewed in chapter 18 of part III,[3] focuses on topic areas within computing, but also potentially extending more broadly across all human intellectual enquiry. It interprets these topic areas in terms of the great scientific domains—*physical*, *life*, *social*, and *computing*—that are relevant to them and the kinds of binary relationships—*implementation* and *interaction*—that hold between these domains. Much of computing can then be summarized in a single domain-that-interacts-with-computing × relationship-between-computing-and-the-other-domain map populated with topic areas.

To go beyond this map, a *metascience expression (ME)* language facilitates characterizing topics that involve more than just one relationship between two domains, as well as topics that are completely outside of computing.[4]

1.1.2 The Reflective Method

The reflective method was developed from reflections on essential, dichotomic, and relational analysis. Since these three methods are themselves metascientific—that is, they enable engaging in science about science—work on the reflective method can be viewed as part of meta^{2}science, or science about metascience. This is how it will primarily be viewed in this section. However, the result of this meta^{2}science process is a new analytical method that is a generalization of the previous three, and which, thus, could itself be used directly in pursuit of metascience, either in lieu of any of the existing methods or in areas not covered by them. It turns out it is also possible to view the development of the reflective method as a recursive application of itself. A bit more will be said about this later in this section, but the focus will initially be on how it can serve as a generalization across the three previously identified metascientific methods.

The reflective method is based on the following six steps:

1. Identify *topics* of interest.
2. Amass *exemplars* of these topics.
3. Identify the *essences* of these topics.
4. Identify *structures* within and/or across these topics.
5. Assort *exemplars* of these topics across these structures.
6. Derive *additional implications* from the previous steps.

Although the steps appear ordered here, they need not proceed precisely in the order listed, nor must they be unitized in the sense of occurring at just one point in the process. One example concerns step 2, of amassing exemplars. Some of this may occur earlier to enable conceptual clustering (e.g., Michalski, 1980) for topic identification in step 1, whereas some may also occur later in support of step 5. Another example is step 6, where some implications may be derived after any of the earlier steps. Still, the numbering here provides a roughly useful way to consider the ordering of the steps, as well as providing simple labels that can be used for them.

Given this caveat, step 1 kicks things off by identifying the topic(s) over which reflection is to occur. The title of Rosenbloom (2012a) was *On*

Computing, whereas the title of this book is *On Minds*, indicating unsurprisingly that the two books are about computing and minds, respectively. However, other related topics are also implicated in each. In the earlier book, this included science, engineering, the notion of a great scientific domain, and the relational structure of interdisciplinary fields. These are summarized in chapter 18 and form the basis for the discussion of relational analysis in the current section. New topics related to mind are also in focus here, such as architectures, symbols, intelligence, and theories in part I, and categories of dichotomies and AI technologies in part II.

Step 2 simply amasses exemplars of the topics identified. These may be well-known exemplars within the field of interest or, particularly if this amassing happens later in the overall process where essences and structures are available to suggest exemplars, ones that are less well known or even only implicit. This latter phenomenon is notable in the dichotomic analysis found in part II. Either way, the exemplars amassed may provide anything from a sparse illustrative set to a dense, and even possibly exhaustive, list of possibilities.

Step 3 is the linchpin of the reflective method, and a key part of the value its application can add, in providing the posssibility of perspective change. It ultimately centers on judgment calls as to what is essential for a topic versus optional with respect to it. This judgment is typically grounded in analyzing any exemplars already amassed and stripping out what does not seem essential, all from the perspective of how the essence is to be used.

In an eloquent guest essay in the *New York Times*, the Italian physicist Carlo Rovelli (2023) stated, "History has many examples of scientific discoveries that came about through the difficult and subtle art of changing perspective." He followed this up with "I think that the answer is to grope for a delicate balance—a balance between how much of our previously accrued learning we take with us and how much we leave at home, freeing ourselves to reconsider what we think we know." This overall captures the spirit of the reflective method, and of step 3 in particular—finding a balance between what to keep and what to leave behind when understanding the essence of a topic, in service of a new perspective on it and the implications that can result from such a perspective.

Consider, for example, the identification of the essence of a symbol as an *atom*, or *placeholder*, in chapter 4. This essence is much more minimal than traditional definitions of the term, in service of a unified notion of hybrid symbol systems that subsumes both traditional (and probabilistic) symbol systems and neural networks. When focused solely on traditional symbol systems, aspects of symbols that are essential to constructing fully operational systems—such as composition and designation—are naturally included as part of their definition. Yet, when considering neural networks, such aspects play a much more limited role. The judgment has therefore been that, from the broader perspective of hybrid symbol systems, such aspects should no longer be considered definitional, making them optional rather than essential.

When the exemplars are definitions, as is the case with essential analysis, determining an essence typically involves converting what were previously considered mandatory strictures into optional aspects. A side benefit of this process can be to enable neutral discussions of where the optional aspects are relevant without necessarily introducing additional questions that are more contentious than useful. For example, with respect to symbols it becomes possible to talk about where now-optional aspects of traditional symbols may be relevant in neural networks while avoiding the question of whether this necessarily means that such systems should be considered "merely" as traditional symbol systems.

Although some essences might be simple, such as the notion of an atom (or placeholder) for symbols (chapter 3) and aboutness (or metasystems) for theories (chapter 5), others can be more complex. Cognitive architectures, for example, are intended to serve as the essence of a mind, even if arrived at via more scientific than reflective/metascientific methods, and may end up being quite complex in the process. At an intermediate point of complexity, intelligence (chapter 4) is first defined as a nested, three-level structure, with *immediate intelligence* nested within *cumulative intelligence*, which is in turn nested within *full-spectrum intelligence*, before an attempt is made at a somewhat more minimal essence.

The *least common core* of a topic might be considered as what is common among the exemplars of the topic, in analogy with a least common denominator for a set of fractions. An essence may be a least common core, but there is also freedom to go beyond this if there is a sufficient reason to determine that the essence should be more minimal. In the process, the topic may absorb other similar but traditionally differentiated topics that then become part of its overall space of variations. For example, the definition that emerges for the notion of a theory in chapter 5 strips out the common notion of *lawfulness* to include learned bodies of knowledge that may be incoherent, such as the massive sets of learned parameters found in LLMs (e.g., Brown et al., 2020); or unprincipled, such as the large numbers of instances acquired via instance-based learning (e.g., Aha et al., 1991).

Step 4 is a second potential source of added value from application of the reflective method. If the topics identified in step 1 can be related to each other in some systematic manner, we can hope to see useful across-topic structures identified. This yields what can be referred to as *topical maps*—or simply *maps*—whose purpose is to support deeper insight into how the topics are related. If there is no systematic relationship among the topics—that is, they are simply a hodgepodge or related in an unsystematic manner—there should still be the possibility of useful within-topic structures being identified. This in turn yields what can be called *topical spaces* whose purpose is to provide deeper insight into the individual topics themselves. Topical maps are central to dichotomic and relational analysis in parts II and III, whereas topical spaces are central to essential analysis in part I, where although there are a number of points of contact among the four topics—architectures, symbols, intelligence, and theories—these contacts are unsystematic.

Step 5 involves assorting exemplars of the topics according to the structures identified in step 4. This is key to a deeper understanding of the exemplars and of the similarities and differences that the structures can reveal among them. Although not as deep a form of value added as may result from steps 3 or 4, the insight that can be provided here into individual exemplars and the relationships among them can itself be of significant value. For example, in part I, there is a significant focus on localizing cognitive architectures within multiple topical spaces, and in thus understanding how they relate to other exemplars of the respective topics.

Step 6 is a catchall for any further implications that can be derived from the results of the prior steps. For example, in chapter 4 (part I), further implications from the analysis of intelligence are considered with respect to the intelligence singularity, generative AI, Sigma, ethics, intellectual property, and free will. Similarly, with the addition of evaluation criteria for AI technologies and integrated systems in part II, further implications can be considered with respect to the comparative advantages across these technologies and systems.

Each of the three analytical methods reviewed in section 1.1.1 can be seen at the top level as specializations of this more general reflective method. However, all but essential analysis also require one or more additional specializations of this method in support of this top-level one. For this reason, only part I will be structured in terms of the relevant specialization of the reflective method, with part II and the relevant portion of part III structured directly in terms of dichotomic and relational analysis, respectively. Yet, it is still worthwhile to understand all three of the analytical methods in terms of the reflective method.

Let's start with essential analysis, which is both the simplest of the three and the first to be employed in the coming pages. The six steps in the reflective method, as specialized for essential analysis, are:

1. Identify a topic to be (re)considered.
2. Amass exemplars and existing definitions of the topic.
3. Identify an essential definition of the topic.
4. Identify a definitional space for the topic in terms of the presence/absence of now optional/nondefinitional aspects, and possibly more.
5. Assort exemplars of the topic within its definitional space.
6. Derive any additional novel implications from the earlier steps.

As reviewed in section 1.1.1, essential analysis was originally defined in terms of a sequence of four steps. These map onto the six here as follows. The first step there, of identifying the essence of a topic, maps onto step 3 here—steps 1 and 2 here were implicit in the earlier specification. The second step there maps onto step 4 here, with the definitional space providing the topical space for essential analysis. The third step there, of populating the definitional space with exemplars, maps onto step 5 here, although part of step 2 here effectively occurred out of sequence as part of this third step there, where exemplars would be amassed in service of assorting them across the definitional space. The fourth step there maps to step 6 here.

At the top level, dichotomic analysis concerns categories of AI technologies. However, it also depends on two additional specializations of the reflective method, one concerned with categories of dichotomies and the other concerned with evaluation criteria. Going through the four phases of dichotomic analysis identified in section 1.1.1 will result in exploring these two supporting specializations first, before considering the top-level specialization.

To begin, the first phase—of identifying dichotomic categories along with a core dichotomy for each—involves a specialization of the reflective method that proceeds as follows:

1. Identify categories of dichotomies from across AI and related disciplines.
2. Amass exemplars of these categories.
3. Identify a core dichotomy for each category.
4. Identify the list of core dichotomies.
5. Assort dichotomies across the list of categories.
6. Reduce the cacophony of dichotomies across the field.

Steps 1, 2, and 5 here, irrespective of their nominal order, combine to yield a populated list of categories of dichotomies from across AI and related disciplines. Step 3 identifies a core dichotomy for each category, and step 4 simply yields a list of these core dichotomies. Step 6, based on steps 4 and 5, yields a dramatically simplified view of the space of dichotomies found across AI and related fields.

The second phase of dichotomic analysis focuses on the criteria to be used in evaluating AI technologies and integrated systems. It depends on a rather simplified specialization of the reflective method, amounting to identifying evaluation criteria (step 1); structuring them as a hierarchical list (step 4), although with bits of steps 2, 3, and 5 involved in this; and deriving the evaluation criteria for use in dichotomic analysis from the hierarchical list of possibilities identified in step 4 plus the criterial advantage language (step 6).

The third phase of dichotomic analysis focuses on developing a deeper understanding of the individual core dichotomies. Much of this falls under the same specialization of the reflective method that underlies the first phase. It populates the core dichotomies with exemplars (steps 2 and 5), considers their relationship to learning (steps 2 and 6), and considers how to span these dichotomies (step 6). However, two components of this phase are more appropriately considered as parts of other specializations of the reflective method. The component centered on within-dichotomy criterial trade-offs fits best as an additional aspect of step 6 from the second-phase specialization that is focused on criteria. The component centered on partitioning AI technologies across the core dichotomies' poles fits as part of the to-be-discussed-next top-level specialization at the heart of the fourth phase, where interactions among the core dichotomies come to the fore.

The top-level specialization leverages dichotomies to understand both the space of AI technologies and the structure of integrated systems, with the previous two specializations playing supporting roles in it. This specialization proceeds as follows:

1. Identify categories of AI technologies with similar properties.
2. Amass exemplars of these categories.
3. Identify essences as combinations of poles of core dichotomies.
4. Identify dichotomic maps based on cross products of core dichotomies.
5. Assort AI technologies across the categories laid out in the maps.
6. Derive similarities and differences among AI technologies; the compositional structure and completeness of integrated systems; and criterial trade-offs among technologies and integrated systems.

The first supporting specialization, in identifying the core dichotomies, plays a critical role in supporting all six of these steps. The second plays a more limited, but still essential, role in support of step 6 here. Due to the overall complexity of mapping dichotomic analysis onto the interactions among these three distinct specializations of the reflective method, part II is structured directly in terms of the four phases of the former rather than the three specializations of the latter.

The complexity of relational analysis falls between essential analysis and dichotomic analysis, implicating only a top-level specialization of the reflective method plus one supporting specialization. The latter is in fact just essential analysis applied to a range of existing and novel topics of importance in relational analysis—such as computing, science, engineering, and great scientific domains—as well as the types of relationships that can exist among great scientific domains. Implications of this include the broadening of the notions of science and engineering to understanding and shaping; the proposal that the computing sciences is a great scientific domain on a par with the physical, life, and social sciences; and the development of the ME language based on the great scientific domains and the relationships among them.

It is actually possible to either view this specialization directly as part of relational analysis or simply as a parallel application of essential analysis to the subject matter levels of computing and science. The description in this section aligns with the former, whereas figure 1.4 aligns with the latter. The truth is that it is a bit of both. Relational analysis could not proceed without this application of essential analysis, and the early work on this topic clearly assumed that the latter was in service of the former. However, as discussed in chapter 18, a perspective change led to understanding that identifying computing as the fourth great scientific domain ended up being important in its own right and, in fact, likely of more importance than the notion of relational analysis itself. The title of the resulting book thus focused on this rather than on relational analysis. Still, some aspects of this application of essential analysis—such as identifying the two key relationships among great scientific domains—is best considered as subservient to relational analysis. Thus, a hybrid approach, of considering some parts of this application of essential analysis as subservient to relational analysis and others as parallel to it, may be the most appropriate way to think about what is going on here.

This all being said, the top-level specialization for relational analysis proceeds as follows:

1. Identify categories of (sub)disciplines in the computing sciences (and possibly beyond).
2. Amass exemplars of these categories.
3. Identify ME expressions for these categories.
4. Identify the structure of the computing sciences (and possibly beyond) in terms of the ME expressions for its (sub)disciplines.
5. Assort exemplars across the structure of (sub)disciplines.
6. Derive a better understanding of computing as an interdisciplinary domain; the similarities and differences among its (sub)disciplines; and the nature of complex topics that involve multiple domains, relationships, and/or (sub)disciplines.

In step 4, the structure of *dyadic computing,* which relates computing to one other great scientific domain, comes across clearly in table 18.1. For more complex forms of computing, described in terms of arbitrary ME expressions, the structure may be less directly apparent, but it is embodied in the nature of the individual expressions and the similarities and differences among multiple of them. These forms of structure are critical for step 6, where, for example, attempts have been made at deriving a better understanding of such complex topics as computing-systems hierarchies, sociotechnical systems, mixed reality, the use of computing in/for science, and organizations that study computing.

Between essential, dichotomic, and relational analysis we have now seen five distinct specializations of the reflective method, with the one for essential analysis showing up a second time in the context of relational analysis. Viewing these specializations as exemplars of metascientific analysis leads to a recursive application of the reflective method that identifies itself as the essence of metascientific analysis, at least in the form evidenced by these exemplars. It is not necessary to lay out the details of this explicitly, as it is just the general method seen earlier, albeit specialized to the topic of metascientific analysis. The first three steps here combine to yield the reflective method as the essence of interest. Although not much emphasis has yet been placed on steps 4–6, that could change with further developments.

The reflective method in general provides a counterpoint to the more traditional *divide-and-conquer* methodology that presumes you already have a sufficient characterization of the topic and thus can immediately proceed to break it up into at least semi-independent pieces (Simon, 1969) before considering how to approach each of these pieces in turn. The reflective method instead starts with an attitude that can be considered either more humble or more grandiose, depending on your perspective, in questioning at the beginning what the topic really should be before proceeding with further analyses. From a related perspective, in Rosenbloom (2012a) I referred to myself as a *lumper* (or unifier) rather than a *splitter* (or decomposer), terminology that stems from mid-nineteenth century biology (Enderby, 2009). The reflective method effectively refines this identification by noting that what is actually key is to lump (step 3) prior to splitting (step 4), rather than just to lump, with the lumping ideally then yielding new approaches to splitting plus other useful forms of analysis (step 6).

It can also be useful to contrast the reflective method with the traditional *scientific method* that is based on observations, hypotheses, and experiments (figure 1.6[5]). The latter is focused on science, whereas the former is focused on metascience. However, if metascience is itself part of science, then it might be expected that it would use the same method(s) as other sciences. Indeed, there is some commonality between the two methods. For example, step 1 of the reflective method amounts to stating the question whereas step 2 corresponds to observation. Step 3, and in some cases step 4, then roughly correspond to hypothesis generation.

The biggest difference is the absence of classical experiments that test the hypotheses in the reflective method. In some sense these experiments are replaced by steps 5 and 6, which explore the consequences of the hypothesis. Yet, while these consequences may provide some feedback, they fall far short of experiments that can invalidate hypotheses. Still, as discussed in Rosenbloom (2012a), science in its broadest sense, as improving our understanding of the world around us, must proceed with respect to each topic via

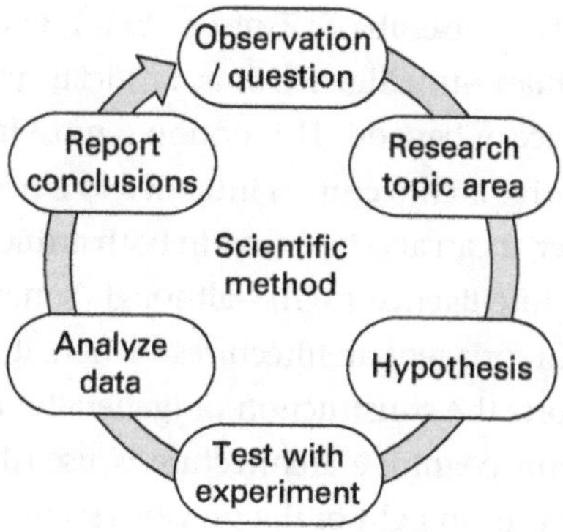

Figure 1.6
The scientific method.

the strongest methods available—as defined by the level of validity they can assure—for that topic. The basic argument here is that the reflective method is a strong method for areas of metascience in which experimentation is not a standard option.

The development of the common model of cognition provides an interesting hybrid case study, involving aspects of both science and metascience, and thus a blending of the reflective and scientific methods. With respect to the reflective method, step 1 amounts to asking what a consensus might be concerning what must necessarily be part of a humanlike cognitive architecture. Step 2 includes both observations on human minds and exemplars of existing cognitive architectures. Its essence/hypothesis—that is, the common model itself—is then generated in step 3 through a combination of scientific and metascientific reasoning. Like cognitive architectures themselves, this is a much more complex essence than the definitional ones found in part I, but it is still an essence. The common model then also structures a topic space for cognitive architectures in which each of its modules can be further specified in a variety of ways that are consistent with its overall assumptions. Implications are then derived, some of which have been amenable to experimental testing (e.g., Stocco et al., 2021).

1.2 Cognitive Architectures

Cognitive architectures are central to many attempts to capture key aspects of what is necessary for general intelligence (Langley et al., 2009; Goertzel,

2014; Laird et al., 2017; Kotseruba & Tsotsos, 2020), whether as models of natural—typically human—intelligence or as artificial systems that embody human-level intelligence or beyond. This notion is most familiar in cognitive science, where a cognitive architecture is intended to embody a theory of the human mind. However, it can also be found in both artificial intelligence (AI) and artificial general intelligence (AGI)—although sometimes under other names, such as AI, AGI, or agent architectures—where it amounts to a fixed framework that supports the construction of (generally) intelligent systems. For simplicity, the term cognitive architecture is used here as the generic across all of these variants. In light of the previous section, a cognitive architecture can be viewed as the essence of a mind, even though work on cognitive architectures is typically more science than metascience.

The purpose of this section is to provide general background on three efforts centered on cognitive architectures that underlie many of the reflections to come in this book—the Soar cognitive architecture (Laird et al., 1987; Rosenbloom et al., 1993; Laird, 2008; Laird, 2012), the Sigma cognitive architecture (Rosenbloom et al., 2016a), and the common model of cognition (Laird et al., 2017; Rosenbloom et al., 2022). Soar and Sigma are two attempts at articulating a theory of the essence of humanlike minds, each of which I was closely associated with for over a decade. Soar has been under active development since the late 1970s and is one of the two longest lived and most thoroughly developed and applied architectures, with ACT-R (Anderson et al., 2004) being the other. Sigma is much newer and, concomitantly, less well developed and applied, having been under active development from the late 2000s to the late 2010s.

The common model is even newer, having been under development just since the mid-2010s. It is also not a cognitive architecture in the traditional sense. It is instead intended to be an abstract, on-paper design that reflects a community consensus rather than a detailed, working system that reflects the perspective of a single research group. Still, in a more generic sense that is consistent with the essential analysis of architecture in chapter 2, it can be considered a cognitive architecture that provides a hypothesis about the fixed structures and processes necessary for humanlike cognition. Alternatively, it can be viewed as a *meta-architecture*, capturing as it does the essence across a range of existing cognitive architectures. These existing architectures serve as the exemplars while judgments about what should be part of the essence are based on community consensus.

References to these three architectures will be found in many places in the remainder of this book. Additional context on each can also be found in part III, chapters 15–16.

1.2.1 Soar

An explanation of Soar at this point is best partitioned according to what can be characterized as the *early version* (Laird et al., 1987; Rosenbloom et al., 1993) versus the *late version* (Laird, 2008; Laird, 2012). The early version was notable for, for the first time, bringing together problem-solving (in the form of *state-space* or *problem-space search*), knowledge (in the form of *rules*), decision-making (in the form of a *preference-based decision procedure*), and learning (in the form of the *acquisition of new rules from experience*). A diagram of its main components can be seen in figure 1.7.

At the center of the early version was a *working memory*, which represented the current cognitive state of the system in terms of symbolic identifiers, attributes, and values, for example, encoding that some object i2 has color red and shape square. Operating on this working memory was a *long-term memory* comprising condition-action rules; for example, if an object is red and square, it should be picked up. Each rule independently checked

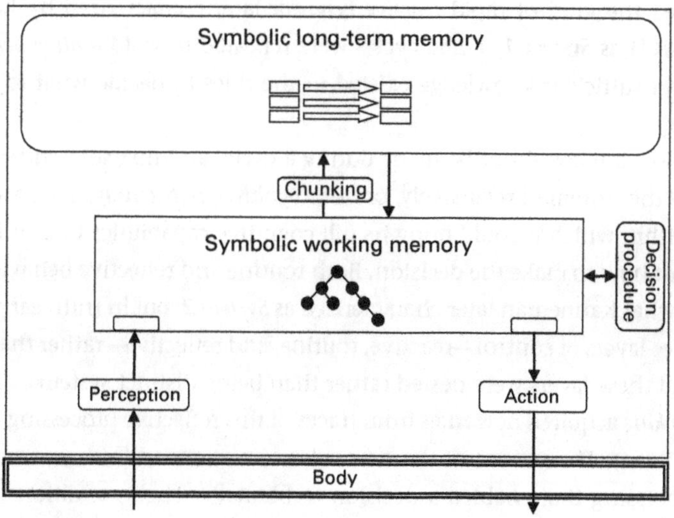

Figure 1.7
A diagram of the early version of Soar.

whether, and in what ways, its conditions matched to working memory and executed its actions to modify working memory for all appropriate matches. This led to parallel firing of rules, a process that continued until quiescence, that is, until there were no more ways any of the rules could fire. All long-term knowledge was stored in the form of such rules. Although rules naturally encoded procedural knowledge, for declarative knowledge, facts were stored as actions with the corresponding conditions determining when to retrieve them into working memory.

Upon quiescence, more persistent changes were made to working memory. This could involve a decision procedure selecting an internal operator to execute based on symbolic preferences generated by rules, or it could involve the application of the already-selected operator, by operator-application rules, to modify the state of the system represented in working memory. Operator-application rules latched their results in working memory whereas the remaining rules operated more like a justification-based truth maintenance system, or JTMS (Doyle, 1979), in that the results of their actions were automatically retracted when their conditions were no longer met.

This cycle of rule-firing and decision-making was intended to map to the ~50 msec *cognitive cycle* found in humans (Card et al., 1983; Anderson, 2007; Kieras & Meyer, 1997; Laird, 2012; Rosenbloom, 2012d; Laird et al., 2017). It yielded the kind of rapid *reactive behavior* later characterized by Kahneman (2011) as *System 1*. These cycles were repeated to yield *routine behavior* whenever sufficient knowledge existed in the rules to decide what to do at each step.

If a decision could not be made during a cycle, an impasse occurred that enabled the system to recursively, or *metacognitively/reflectively*, generate subgoals within which it could bring its full cognitive capabilities to bear in figuring out how to make the decision. Both routine and reflective behavior fall within what Kahneman later characterized as *System 2*, but in truth early Soar had three layers of control—reactive, routine, and reflective—rather than just two, and these layers were nested rather than being distinct systems.

Chunking acquired new rules from traces of this reflective processing (Laird et al., 1986a). The actions of the new rules corresponded to those results of this processing that enabled a decision to be made. Their conditions corresponded to those aspects of the pre-impasse situation that led to these results. Chunking of this sort could straightforwardly reduce the time to perform a task by converting extended problem-solving episodes into rules that directly yielded the corresponding results. Given the flexibility of Soar and the types of impasses it experienced, it could also yield many other forms of learning that previously required distinct learning mechanisms (Steier et al., 1987).

Perception and *action* with respect to the world outside of Soar were not well-defined aspects of the architecture, but in effect it included APIs that enabled perception to generate specific structures in working memory and for external action to be triggered by other specific structures in working memory.

All in all, this early version of Soar was conceptually a simple architecture that elegantly combined one representation for long-term knowledge (rules), one way to make decisions (based on symbolic preferences), one way to learn (chunking), and one way to reflect on its own behavior (impasse-driven subgoaling). From this conjunction of capabilities plus appropriate knowledge, a wide range of humanlike intelligent behavior was demonstrated in Soar, including problem-solving and planning, learning, knowledge-intensive behavior, natural language processing, and robotic control. Its most significant application was to the development of intelligent automated pilots and commanders for large-scale military simulations (Tambe et al., 1995a; Tambe et al., 1995b; Hill et al., 1997; Jones et al., 1999).

Although Soar was originally hypothesized as an architecture for general intelligence, and many forms of intelligent behavior were exhibited, over the years a number of limitations became apparent that could not be adequately resolved with just this set of architectural capabilities, even with further incremental development (Laird & Rosenbloom, 1996). The late version of Soar (figure 1.8[6]), with which I was not involved, dealt with many of these limitations by dramatically expanding the set of architectural capabilities found in the early version.

The most critical additions were two new *long-term declarative memories*—a *semantic memory* for facts and an *episodic memory* for experiences—with associated learning mechanisms for each. A second form of procedural learning—*reinforcement learning* (*RL*; Sutton & Barto, 2018)—was also added, although instead of acquiring new rules it acquires numeric preferences from experience. The dichotomy of symbolic data versus quantitative metadata has already been mentioned briefly in discussing disciplinary maps. Here it shows up in the sense that numeric preferences provide quantitative metadata associated with symbolic operators. A second form of quantitative metadata was also added, in the form of *activation levels* for elements in both

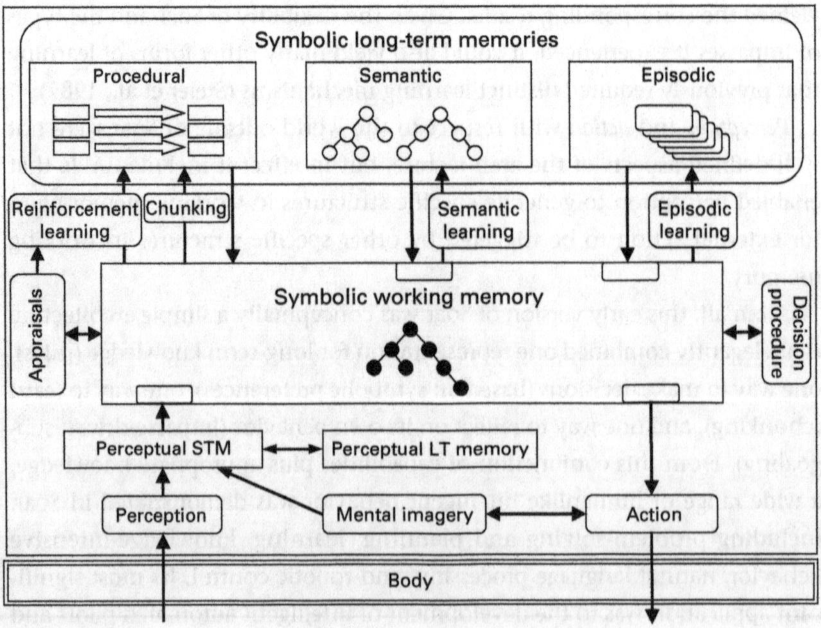

Figure 1.8
A diagram of the late version of Soar.

working memory and declarative memory, introducing a hybrid aspect to the whole architecture. Beyond this, the late version also has a *spatial-visual system* capable of working with mental imagery as well as other more minor changes. The most significant application area for this late version of Soar has been in *interactive task learning,* where an agent learns to perform tasks from a combination of direct experimentation with the task and in-context instruction from a human tutor (Laird et al., 2017).

1.2.2 Sigma

Sigma (Rosenbloom et al., 2016a) can be considered as a spinoff of the early version of Soar that had only one architectural module for each major capability. Its development has primarily been driven by four desiderata:

- *Grand unification* concerns spanning not only traditional cognitive capabilities but also key noncognitive aspects, such as perception, motor control, and affect.
- *Generic cognition* concerns spanning both natural and artificial cognition at a suitable level of abstraction.

- *Functional elegance* concerns yielding broad cognitive (and noncognitive) functionality—ultimately all that is necessary for humanlike intelligence—from a simple and theoretically elegant base.
- *Sufficient efficiency* concerns executing quickly enough for real-time applications involving virtual humans and intelligent agents/robots, as well as for large-scale experiments in modeling human cognition.

Its description is best partitioned into two levels—a *graphical architecture* and a cognitive architecture—rather than two versions.

Sigma's graphical architecture is based on a version of factor graphs (Kschischang et al., 2001), a general form of probabilistic graphical model (Koller & Friedman, 2009) that has been extended to handle Sigma's needs (Rosenbloom et al., 2016b). Factor graphs are bipartite, linking *variable* and *factor* nodes.

Variable nodes represent bindings of variables. In support of a first-order aspect, a variable node in Sigma maintains quantitative values over all possible elements in the domain. This can, for example, represent the full set of possible bindings of a variable in a rule condition, with a 1/0 value for each that designates whether it is a correct binding. Or it can represent a probability distribution over the possible bindings of the variable. Or it can represent a continuous 1D mental image with 1/0 indicating whether locations are occupied (Rosenbloom, 2011a; Rosenbloom, 2012c). Or it can represent a distributed vector where the quantitative values represent activations over the elements of the domain (Ustun et al., 2014).

The domain of possible elements for a variable is natively represented as a portion of the real number line, which is then automatically partitioned into segments over which the quantitative value does not change, yielding a vector of segments with constant values. More generally, segments may have their values defined via linear—rather than constant—functions over their domain elements, with segmentation then being required whenever a different linear function is implicated.

The native continuous nature of a domain can be specialized to be discrete if partitions distinguish unit intervals that correspond to the integers—1, 2, 3, and so on. A domain can be further specialized to be symbolic if labels are assigned to its unit segments in a manner that the architecture can understand. In terms of how symbols are defined in chapter 3, as *atoms* or *placeholders*, all segments of a domain are technically symbols, whether continuous, discrete, or symbolic. However, Sigma's more traditional use of the term assumes designation of labels by segments, yielding what is essentially a classical symbol table.

Altogether, Sigma's approach here enables representing both numeric and symbolic information and, for numeric information, both continuous and discrete information. Furthermore, all three forms of representation are enabled via a uniformly continuous substrate. It is, however, important to note that this trifold nature of Sigma's variable domains is distinct from the inherently numeric nature of its variable ranges. These ranges are also inherently continuous, while possibly being specialized to discrete values such as 0 versus 1 to, for example, represent false versus true or absent versus present. In the terms of the common model of cognition (section 1.3), the domains of these variables amount to the data whereas the ranges embody quantitative metadata about the data.

In addition to representing single variables, variable nodes in Sigma may also represent a set of variables, yielding an *n*-dimensional domain that can capture interactions among the variables. A combination of one segment along each dimension then yields an *n*-dimensional region that has either a single real value or a linear function over its variables. In the terms of chapter 3, each such region can be seen as one form of symbol structure, composed of the segments/symbols along each of its dimensions. More complex symbol structures can be built up via relations across such regions.

A complete set of regions for a set of variables can be viewed as a *region tensor*. For example, the pseudo example of a 2D region tensor in table 1.1 shows a symbolic dimension for *concept type*, an integral dimension for *weight*, and either a constant value or a linear function of weight for each 2D region. One or more dimensions of a region tensor may encode a (joint) distribution—which may or may not be normalized—that is effectively conditioned on the remaining, nondistributional, dimensions. Nondistributional dimensions may, for example, represent anything from combinations of variable bindings for a rule to a mental image.

Function nodes in general represent *n*-dimensional functions, structured as region tensors, over the *n* variables connected to them. *Messages*, which are at the heart of the sum-product algorithm (Kschischang et al., 2001) that has been adapted for use in solving Sigma's extended factor graphs, are also represented as region tensors. They are transmitted among connected variable and factor nodes in the overall graph. The algorithm proceeds by forwarding messages from variable nodes along links to factor nodes that

Table 1.1
Pseudo example of a Sigma region tensor

P(weight | concept)	Walker	Table	. . .
[1,10>	.01*w*	.001*w*	. . .
[10,20>	.2–.01*w*	"	. . .
[20,50>	0	.025–.00025*w*	. . .
[50,100>	"	"	. . .

are the products of the messages coming in from all of the other links, and forwarding messages from factor nodes along links to variable nodes that are the products of the messages coming in from all of the other links and the function stored at the node, while summing out all of the variables not included in the target variable node.

Learning updates function at a subset of the factor nodes via gradient descent, based on comparisons of the nodes' existing functions with the products of their incoming messages (Rosenbloom et al., 2013), gradually shifting the stored functions to correspond to ongoing experience. This can be viewed as a variant on the *free energy principle* (Friston, 2010)—the existing functions yield predictions, while the products of incoming messages represent current reality.

This single learning mechanism can support various forms of symbolic, probabilistic, and neural learning (Rosenbloom et al., 2013; Rosenbloom, 2012b; Rosenbloom, 2014; Rosenbloom et al., 2017). What it does not yet support, however, is the acquisition of new rules from experience, such as is provided by Soar's chunking mechanism or, in fact, any form of structure learning.

Like learning, decisions are made at a subset of the factor nodes based on the product of the incoming messages. In particular, a decision is made jointly over the combination of *selectional* dimensions in a region tensor, given the context established by its nonselectional dimensions. The joint distribution over the selectional dimensions acts much like Soar's numeric preferences, enabling operator selection, as in Soar. However, decisions in Sigma are not limited to this one special case of operator selection; they can, for example, also occur in deciding how an operator is to alter a state.

The overall graphical processing cycle (bottom of figure 1.9[7]) is divided into two phases—the *elaboration phase* comprises message passing and

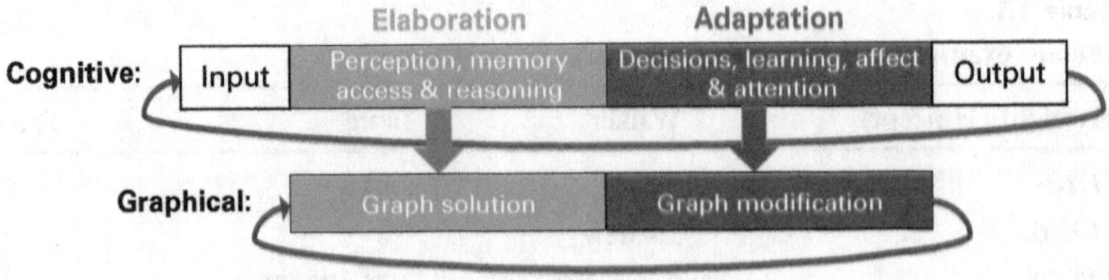

Figure 1.9
Sigma's two architectural layers—graphical and cognitive—and the cycles of processing that they support.

sum-product computations while the *adaptation phase* modifies the graph via such processes as learning and decision-making. This dual phase approach is itself akin to what was found in the early version of Soar. The results of decisions in the adaptation phase are latched—as with Soar's operator-application rules—whereas, also in Soar fashion, processing in the elaboration phase is retractable in a JTMS style.

Sigma's cognitive architecture is implemented on top of its graphical architecture. As illustrated in figure 1.10[8], different regions of the overall graph are specialized to the different capabilities required by the cognitive architecture. Long-term memory comprises regions in which learning occurs, whereas working memory comprises regions in which decision-making occurs plus additional regions in which results are retractable. At this architectural level, region tensors are viewed as continuous-valued, multi-attribute predicates—such as *object(concept, weight)* in table 1.1—whose attributes correspond to variable domains that can be continuous, discrete, or symbolic. The more persistent aspects of Sigma's working memory, such as those that correspond to operator selection and application in Soar, correspond to predicates for which decisions are made. More transient portions of working memory comprise predicates whose values never make it into factor functions.

Conceptually, long-term memory consists of *conditionals*, a generalization of rules that not only supports the traditional asymmetric patterns of *conditions* and *actions* but also symmetric patterns, called *condacts*, that blend the two to directly support the bidirectional message-passing necessary for probabilistic graphical models. Conditionals also have associated functions whose dimensions correspond to variables found in its patterns. Sigma's long-term memory has turned out to be functionally elegant enough to support the kind of rule-based procedural memory that Soar started with

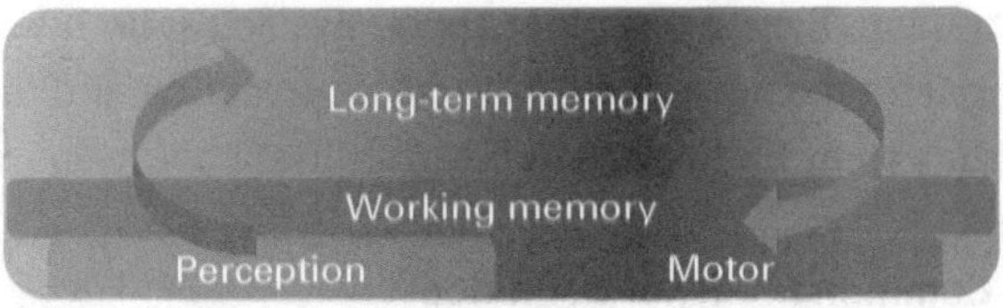

Figure 1.10
The Sigma cognitive architecture, emphasizing how key memory and processing components arise as variations on the underlying structure of extended factor graphs.

plus the two kinds of long-term declarative memories later added to it—for semantic and episodic knowledge—plus probabilistic reasoning, neural networks, and forms of perception and motor control.

As with the graphical processing cycle, the cognitive processing cycle can be divided into elaboration and adaptation phases—bracketed by input and output—each of which is implemented by the corresponding phase of the graphical processing cycle (top of figure 1.9). However, in cognitive terms this processing includes such things as perception, memory access, and reasoning during the elaboration phase and decisions, learning, affect, and attention during the adaptation phase.

Leveraging combinations of Sigma's existing mechanisms plus limited extensions to the architecture—such as including the ability to restrict messages to flow in only one direction along particular links in support of asymmetric technologies (chapter 8)—and appropriate skills and knowledge enabled by the architecture, the traditional symbolic capabilities of the early Soar versions have been extended without additional modules to include not only rules (Rosenbloom, 2010a), but also sum-product networks (Joshi et al., 2018), mental imagery (Rosenbloom, 2011a; Rosenbloom, 2012c), simultaneous localization and mapping (SLAM; Chen et al., 2011), object and speech recognition (Joshi et al., 2014; Joshi et al., 2016), distributed vector and holographic representations (Ustun et al., 2014), and neural networks (Rosenbloom et al., 2016b) and deep learning (Rosenbloom et al., 2017). Sigma has also been extended in recent years to incorporate aspects of theory of mind (Pynadath et al., 2013; Pynadath et al., 2014), attention and emotion/affect (Rosenbloom et al., 2015), and motivation (Rosenbloom & Ustun, 2019).

In contrast to Soar, Sigma is a deeply hybrid architecture in which tensor regions can be viewed as symbol structures—as discussed further in chapter 3—with functional values over these regions providing quantitative metadata for them. Region tensors are pervasive across both working memory and long-term memory, as well as across perception and motor control (at least to the extent that these latter capabilities have been explored).

1.2.3 The Common Model of Cognition

Over the years, many different cognitive architectures have been investigated to one extent or another.[9] A recent paper surveys eighty-four different architectures, forty-nine of which were still under active development (Kotseruba & Tsotsos, 2020). It also mentions the existence of 111 additional architectures, for a total of 195. Some of these bear a similarity to architectures like Soar and Sigma, but others differ in striking ways, particularly when considering AI and AGI architectures that make no attempt to align with human cognition. Understanding, evaluating, and comparing these architectures can be challenging, as each is typically a complex software system that further requires content in the form of skills and knowledge to be added into the mix before it can actually behave.

The common model of cognition is an attempt to abstract away from individual cognitive architectures to find a community consensus that may be incomplete[10]—it is not expected to be an implemented architecture itself—but that captures what is agreed upon in a common language (Laird et al., 2017; Rosenbloom et al., 2022). The common model focuses on *humanlike* cognitive architectures; that is, architectures intended either as models of human cognition or that are similar enough to models of human cognition—whether natural or artificial—as to be considered in the same category (figure 1.11[11]). This acknowledges that there may be multiple distinct ways in general of achieving human-level, or even higher levels of, cognition—making the existence of a common model over all such approaches open to question—while explicitly assuming that there must be substantive commonality for humanlike minds.

If successful, the common model should provide an evolving statement of where the field of (humanlike) cognitive architectures considers itself to be, while improving communication among researchers within this field and with others outside of the field. It should also help in comparing and contrasting distinct architectures by providing an interlingua to which they can all be related. It may also help guide the development of new architectures and of applications that either require models of human minds or working instances of artificial minds.

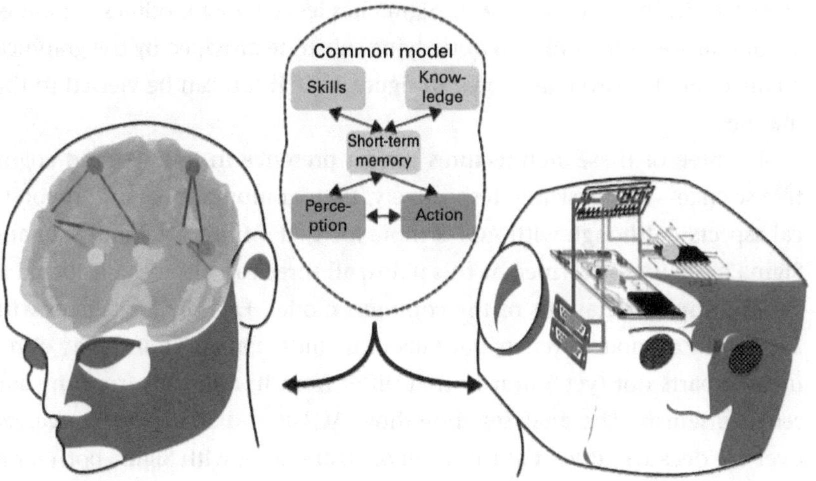

Figure 1.11
The common model of cognition as an abstract consensus over natural human minds and artificial humanlike minds.

As shown in figure 1.11, the common model includes a central, short-term working memory; long-term procedural (skills) and declarative (knowledge) memories; and perception and action modules. All communication goes through working memory except for the direct connection between perception and action. Accompanying this overall structure are sixteen assumptions that underlie the common model, such as that there is: parallelism both within and across the modules, sequential behavior that arises from a cognitive cycle that operates at ~50 msec in humans, a hybrid of symbolic data and quantitative metadata, and learning that occurs incrementally as a side effect of performance.

The initial work on the common model focused on generalizing across key capabilities found in ACT-R, Soar, and Sigma, while taking other architectures and results from AI and cognitive science less directly into account. In a sense, how the common model is shown in figure 1.11 can be seen as part way between the early (figure 1.7) and late (figure 1.8) versions of Soar, being most like the early version but with a declarative memory that abstracts over the later version's semantic and episodic memories. ACT-R is typically diagrammed differently, with procedural memory at its center, surrounded by working-memory buffers that then connect to the other modules, but it can be viewed in this manner as well. Sigma has less distinct modules, grounded as they all are in the uniform underlying substrate provided by the graphical architecture. However, as shown in figure 1.10, it too can be viewed in this manner.

All three of these architectures have a presence in both AI and cognitive science—emphasizing, respectively, their computational and theoretical aspects—although with ACT-R more focused on the latter and Soar and Sigma more on the former. At this point, all three can also be considered as variant implementations of the common model. Each largely agrees with the common model's tenets but fails to include parts of it and may either include parts not (yet?) in it or that differ from it, and thus from the current consensus. The analyses show how ACT-R and Soar have converged over the decades, at least at this level of abstraction, with Sigma both more incomplete—not surprising given its relative newness—and more in disagreement with the consensus, particularly in terms of what it considers part of the architecture versus stemming at least in part from knowledge and skills above the architecture.

By now, more than a dozen cognitive architectures have been mapped onto the common model or been described as common model architectures, including many suggestions for how it might be improved (e.g., Choi, 2020; Gobet & Lane, 2020; Kieras, 2020; Rosenbloom et al., 2024; Laird et al., 2025). The common model has also been used as a guide in various applications (e.g., Ketola et al., 2020; Mohan, 2021) and served as a means of integrating and building bridges across disciplines and paradigms (e.g., Meireles et al., 2018). Most excitingly, the common model has been tested as a model of high-level brain architecture. This work mapped the common model's components onto functional circuits in the brain and used data from the Human Connectome Project to compare it with more standard hub-and-spoke and hierarchical proposals from neuroscience (Stocco et al., 2021). The result has been an overwhelming statistical dominance for the common model over these other approaches.

I Definitions: Essential Analysis

2 On Architectures

Much has already been said about cognitive architectures in chapter 1, including descriptions of two architectures—Soar and Sigma—and the more abstract common model of cognition. The goal of this chapter is to step back from such models and their descriptions, to apply essential analysis, as reformulated in section 1.1.2 as a specialization of the reflective method, to the general notion of an architecture itself and to consider some of the resulting implications for architectures broadly and for cognitive architectures more specifically.[1]

One existing definition of a cognitive architecture is as a *hypothesis concerning the fixed structures and processes that together yield a mind, whether natural or artificial* (Rosenbloom et al., 2016a). Explicit in this definition is that a cognitive architecture should be *fixed* and that it may concern *natural or artificial minds*. Implicit in it is that a cognitive architecture should support nonfixed—that is, variable—content and be *theoretical* ("hypothesis concerning") and *computational* ("structures and processes"). Computer architectures—the direct inspiration for cognitive architectures—share all of these traits except for not concerning minds and not in general being theoretical, in the sense of being *about* something else (see chapter 5 for more on theories and the notion of aboutness). Reaching further back in the inspirational chain, building architectures further differ in not being computational.

In his classic work on *Unified Theories of Cognition*, Newell (1990) defined an architecture as a *symbol system* (Newell & Simon, 1976). This shares the concern with fixity and the notion that an architecture in general need not be theoretical—the symbols themselves possess aboutness, but the system that processes them need not—but it then narrows the definition down to a form of universal computation. In the process it rules out building architectures and any other noncomputational structures—such as noncomputational scientific theories—or even many special purpose computational devices, while still admitting typical computer architectures and symbolic cognitive architectures. By having identified minds with symbol systems (Newell & Simon, 1976)—an idea that remains controversial to date—Newell had no need to add that cognitive architectures are about minds, whereas under the version in Rosenbloom et al. (2016a), they are explicitly concerned with understanding and/or building minds.

Further discussion of symbol systems, the issues around them, and their relationship to both cognitive architectures and minds can be found in chapter 3, but do not otherwise play a role here. The essential analysis at the heart of this chapter instead begins with step 1 of its specialization of the reflective method—identifying *architecture* as the topic—before proceeding to step 2 in section 2.1, where how this concept has developed over the sequence of three types of architectural exemplars mentioned earlier—building, computer, and cognitive architectures—is explored. A key part of this process is identifying relevant *distinctions* that these three types of architectures have introduced, as this provides critical input for identifying the definitional space in step 4. Section 2.2 then proceeds with step 3, of identifying an essential definition, and step 4, of identifying the definitional space.

Step 5, of populating the definitional space, has not been an explicit focus here, although it does at least include the three types of architectures just mentioned and other variations that will arise in the flow of this chapter. The remainder of this chapter focuses on step 6, with discussions of four issues of interest that bear on both architectures in general and cognitive architectures in particular: architectural constraint (section 2.3), multilayered architectures (section 2.4), theoretical computational architectures (section 2.5), and architectural exploration (section 2.6).

2.1 Historical Context

The word *architecture* stems from the Greek ἀρχιτέκτων (*arkhitekton*), where it referred to a master builder or director of works ("Architect," 2025; "Architecture," 2024a). In modern usage it includes "both the process and the product of sketching, conceiving, planning, designing, and constructing buildings or other structures" ("Architecture," 2024a), with its practice oriented toward both practical and expressive requirements, and thus serving both utilitarian and aesthetic ends ("Architecture," 2024b). Distinctions that arise in such architectures that resonate in later forms of architecture include:

1. The *fixed* nature of the architecture versus the *variable* nature of its contents. In French, this is succinctly *immobilière* (real estate) versus *mobilière* (furniture).
2. *Design* versus *implementation.* A design is part of the process that specifies what is to be implemented, whereas an implementation is the heart of the resulting product.
3. (*Function* versus *structure*) versus *form.* Function concerns how the product is used whereas structure concerns how this function is provided. Both relate to utility whereas form in this context concerns the aesthetics of how these are presented.
4. *Simple* versus *complex.* Although this may impact utility, the greater concern is with form/aesthetics, such as Scandinavian or Japanese versus Victorian or rococo.

Computer architecture (Bell, and Newell, 1971; Hennessy & Patterson, 2012) is a more recent conceptualization inspired by the notion of building architecture. It induces a partitioning of a computational system—that is, a system that transforms information (Rosenbloom, 2012a)—into a fixed architecture versus variable programs and data. The design is a description, whereas the implementation actively computes. The earliest computers each had their own *instruction set*—a specification of the instructions the computer could execute—that was tied to the hardware they came with. However, it was not long before this was separated from the implementation, with the architecturally defined instruction set specifying what was to be implemented, and multiple hardware implementations being developed to span sizes and generations of computers. The "IBM System/360" (2024) provided a notable early example, where a single instruction set was implemented in a variety of ways that yielded both different trade-offs and different generations. A general language—Instruction Set Processor (ISP)—was even developed to enable specifying instruction sets across the field of computer architecture (Bell & Newell, 1971).

The function of a computer architecture is to support computation via an instruction set. The structure comprises the hardware (and possibly firmware) components that yield this plus their organization. Form is often deprecated, although some manufacturers do take it seriously. There is an analog of simple versus complex, in the debate between reduced versus complex instruction set computers—that is, reduced instruction set computer (RISC) versus complex instruction set computer (CISC)—although the focus in this debate is more on utility than aesthetics.

A computer architecture induces one additional distinction of note:

5. *Transformer* versus *container*. A transformational architecture transforms its variable content rather than simply containing it.

A computational architecture transforms information, with a computer architecture further specializing this to be based on hardware (and possibly firmware) structures. This contrasts with building architectures, which act as containers. Parts of computer architectures, such as memories may also appropriately be considered as containers, in contrast to central processing units (CPUs) and graphics processing units (GPUs) that transform the information retrieved from memory.

The notion of a *cognitive architecture* is even newer, being a form of computational architecture that originated in analogy to computer architecture, with Allen Newell having played formative roles in both areas (Bell & Newell, 1971; Newell, 1973a). Its function is to provide a (fixed) *mind* that can transform (variable) *knowledge and skills*. Its structure comprises the mechanisms and their organization that together yield a mind. In line with this, it was intended to sit at the base of the *cognitive band* in a timescale analysis of human cognition, sitting just above the *biological band* where neural processing took place (Newell, 1990). The bottom of the cognitive band supports the ~50 msec cycle time that was mentioned several times in section 1.2.

The question of form for cognitive architectures arises primarily in terms of the simple versus complex distinction. A simple, RISC-like architecture focuses on achieving intelligence from the interactions among a small number of very general mechanisms (e.g., Hutter, 2005; Rosenbloom et al., 2016a; Vaswani et al., 2017; Silver et al., 2021), whereas a complex, CISC-like architecture includes a wider range of more complex mechanisms (e.g., Laird, 2012; Goertzel et al., 2014). This distinction may be thought of as between a physics (i.e., beauty) mindset (e.g., Wilczek, 2015)—although even there it can be controversial (Hossenfelder, 2018)—versus a biological (i.e., evolution-as-a-tinkerer) one, with the computer science (i.e., modular) perspective possibly sitting in between. Both utilitarian and aesthetic aspects are implicated here. The introduction of cognitive architecture to this overall history also induces additional distinctions, with the one of most importance here being:

6. *Theoretical* versus *atheoretical*. This amounts to whether an architecture is *about* something else. Architectures in cognitive science tend to be theoretical—being about human minds—whereas A(G)I architectures, along with building and computer architectures, need not be so.

A theoretical architecture—or *theory*—with or without accompanying variable content, represents—or is about—key aspects of a phenomenon or domain of interest. Architectures represent complex phenomena when used in *understanding* them (Rosenbloom, 2012a), as is typical in the sciences. Much more on the general notion of a theory—including what it means for a theory to be beautiful—along with its implications for cognitive architectures can be found in chapter 5.

Transformational architectures may also be used to generate (or *shape*) complex phenomena, as is typical in engineering and other professions (Rosenbloom, 2012a). Such architectures may also represent something but need not do so. A computer architecture, for example, is central to engineering all kinds of computational systems, with its foremost purpose being atheoretical—to aid in the development of such systems rather than to represent or to help in understanding anything. The architecture in such a case could conceivably be considered as a model of computation, as with a Turing machine (Turing, 1937), but this is deep in the background in most applications.

Cognitive architectures, according to this analysis, can be both computational and theoretical. They do, however, cede being computational when limited to a design on paper rather than an implemented and executable body of code. In the other direction, they cede being theoretical when they are purely engineering artifacts that are not directly about anything. Yet, even when not about human intelligence, they can potentially be about a more abstract domain of intelligent phenomena and thus still be theoretical in this sense.

If intelligence is considered as one aspect of computation—as discussed in chapter 1—the theoretical side of cognitive architectures could be considered as existing within the general area of *theory of computation*, the discipline within computer science concerned with understanding computing itself as a complex body of phenomena, including considering such questions as what can be computed and what can be computed tractably, including the million-dollar question of "P=NP?" ("P versus NP problem," 2024). Here a sample question might then be, "What can be intelligent?," although using cognitive architectures in answering this question may be a less formal approach than is the standard in theory of computation.

An *inspired* architecture, such as a biologically inspired cognitive architecture (BICA) (Stocco et al., 2010), sits at a hybrid position with respect to theoretical versus atheoretical, via a combination of understanding and shaping. At the top level, the goal may be atheoretical, to support the development of useful AI systems with no concern as to what might be represented in the process. However, understanding of an existing system becomes an instrumental subgoal in guiding the design of the atheoretical architecture. In some cases the existence of this subsidiary understanding is considered purely heuristic, and fine to ignore or dismiss when convenient, while in others it is considered a virtue to be extolled.

2.2 Defining Architecture

From these three overall classes of architectures—building, computer, and cognitive—an essence can be identified that centers on a *fixed framework* that enables and delimits a *variational space*, with each of the three classes of architectures having an essence that narrows this down to its subject area. Such a definition focuses on distinction 1, with the other five helping to scope the space of architectural types, and thus the definitional space for architectures, while not themselves being definitional. Simon (1974) discussed the importance of identifying *invariants* because of "their power to strip away the complexity and diversity of a whole range of phenomena and to reveal the simplicity and order underneath." In this sense, an architecture is a set of invariants; however, it also typically goes beyond this to consider the interactions among them and what a complete set of them might be.

The notion of a variational space for an architecture might simply appear to be a different way of characterizing its definitional space; however, the two are, in truth, quite different. The definitional space is populated by alternative architectures, whereas the variational space is populated by *variable content* that augments an architecture to yield a *system*. This can succinctly be semi-formalized via the equation *system = architecture + variable content*.[2] This harks back to earlier syntactically similar equations, such as *algorithms + data structures = programs* (Wirth, 1976) and *algorithm = logic + control* (Kowalski, 1979); however, each of these equations makes a distinct point. In the spirit of the second equation, the familiar equation *system(s)=program(s)+data* can be seen as a specialization of the equation proposed here to computational systems, where the program is the fixed, architectural component and the data yields the variable content.

More broadly, the systems of interest here may be conceptual, mathematical, computational, physical, and so forth. An *application* is then simply a system that yields value. In general, different kinds of architectures enable different forms of variable content and thus yield different types of systems and applications. As we have just seen, computer architectures enable programs and data while cognitive architectures enable skills and knowledge. In addition, modular architectures enable module definitions and API (i.e., application programming interface) specifications; hierarchical and graphical architectures enable node and link definitions; tables and maps enable a structure of locations and entries at these locations; and buildings enable furniture, appliances, and ornaments. In tightly specified theories or models, in which nearly the entire system is architecture, the variable content may simply be parameter values.

A classic slogan in AI, although it seems to have originated hundreds of years ago with Francis Bacon as *scientia potentia est*, is that *knowledge is power* (Feigenbaum, 1992). We could also add that *applications are payoff*. In architectural terms, knowledge is a form of variable content enabled by cognitive architectures, whereas applications are complete systems that result from combining appropriate architectures and variable content. In analogy, we can now go a step further to say that *architecture is essence*. In other words, a system's architecture captures its core nature.

As a final definitional note here, there can be a slippery slope from architecture to *toolkit*, that is, something that provides many options that can be leveraged as appropriate. A toolkit nominally provides fixity, yet the proliferation of options weakens this notion. For theoretical frameworks with many options, the variety of options is assumed necessary to match the scope of what is being modeled, and so there remains comfort in considering such as architectures. Whether an atheoretical framework is an architecture versus "just" a toolkit can, however, be contentious. When there are few options, "architecture" generally seems the appropriate label. When there are many, the result may be more a toolkit than an architecture. However, a functionally elegant approach that yields its options from interactions among a small set of common mechanisms yields what might be considered a hybrid, providing an architecture at one layer that engenders a toolkit at a higher layer (unless the higher layer itself has a theoretical stance).

2.3 Architectural Constraint

By providing a level of fixity while delimiting the range of variable content, an architecture inherently provides constraint on a body of phenomena, whether the goal is to model the phenomena or to produce it. Architectures, and thus the constraint they provide, can be nominal. For example, the concept of modules and connections can by itself comprise an architecture, as can the concept of a taxonomic hierarchy. Both provide little fixed structure, but each still ensures that certain properties will be met by any system based on it, such as, respectively, the ability to decompose and hide complexity within modules and to restrict how the nodes connect in a hierarchy.

More basically, the simple act of decomposing systems into an architecture plus variable content can yield constraint that is often ascribed instead to particular forms of architecture. For example, arguments have been made for both modules and hierarchies as enabling more robust systems that are easier to learn, develop, evolve, and engineer (see, e.g., Simon, 1969). However, any architecture already provides at least an initial decomposition into fixed versus variable structure, and any substantive architecture will impose restrictions on the variable structure that further reduce its flexibility and complexity.

Another example comes from the world of cognitive architectures, where the argument has been made that the fixed structure of an architecture enables the possibility of one-time validation of the architecture followed by unlimited reuse with arbitrary variable content (Ustun et al., 2018). As with the previous arguments, this one also turns out to hold for all architectures. It clearly, for example, applies to the use of architectures as theories, where validating the theory once may enable an endless variety of uses, as well as to the use of architectures as fixed mechanisms at the core of complex engineered systems.

This doesn't rule out that any particular type of architecture might provide even more constraint and associated benefits, but it does provide a baseline benefit shared by all architectures. It also doesn't rule out enabling computational universality in the variable content (see section 2.4), as is almost definitional for a computer architecture but is also typical of cognitive architectures. Yet, even in such cases it limits the nature of the programs and data directly usable in the former and the skills and knowledge directly usable in the latter.

2.4 Multilayered Architectures

Both computational architectures—that is, architectures that transform information—and theoretical architectures are amenable to being parts of layered stories.[3] Consider first computational architectures. A computer architecture provides a fixed foundation that supports leveraging programs to transform data. However, as anticipated with the familiar equation introduced in the previous section, at a second level, programs may themselves yield computational architectures, particularly when fixed over some span of interest, with their data then providing the only source of variation. This data variation may in turn be very flexible—as, for example, in a cognitive architecture that supports a broad range of skills and knowledge—or it may be severely limited.

In the former case, this process of implementing computational architectures within the variable content that is enabled by other computational architectures can conceivably proceed to arbitrary levels. For example, as mentioned in section 1.2.2, the Sigma cognitive architecture comprises two such layers—a graphical architecture and a cognitive architecture—with the former implementing the latter. Some of the capabilities included in the common model of cognition (section 1.2.3)—such as a form of declarative memory—are then implemented in a third layer, with the aid of skills and knowledge represented within Sigma's cognitive architecture.

Similarly, a theoretical architecture can be a theory itself or just one part of a more elaborate theory that also includes appropriate choices within the variable content the architecture enables. Kivunja (2018), for example, discusses the notion of a theoretical framework as the accepted wisdom from experts that serves as the background for a graduate student's own research. The former might perhaps also be considered a Kuhnian paradigm (Kuhn, 1962). This effectively partitions a student's contributed theory into a fixed, architectural framework—or paradigm—plus the student's own contributed variable content. A Lakatosian research programme (Lakatos, 1978) also partitions in this manner—"For Lakatos an individual theory within a research programme typically consists of two components: the (more or less) irrefutable hard core plus a set of auxiliary hypotheses" ("Lakatos," 2025).

2.5 Theoretical Computational Architectures

A computational architecture, in conjunction with appropriate variable content, can compute, in the sense of being able to transform information. Both computer and cognitive architectures are examples of computational architectures, whereas building architectures and many scientific theories are not. A theoretical architecture, with or without accompanying variable content, is about some phenomena of interest. As shown in figure 2.1[4], these two variants can be viewed as overlapping regions of the space of all architectures, yielding four clear regions that span whether or not an architecture is computational and/or theoretical.

Yet, it is worth considering multiple ways in which this neat picture can be fuzzed up, as in figure 2.2[5], without destroying the overall message. Consider, for example, scientific theories expressed as mathematical equations. Such theories directly provide theoretical architectures, while not typically being directly computational. Yet, it can be straightforward to convert them into a computational form that can directly transform measurements into

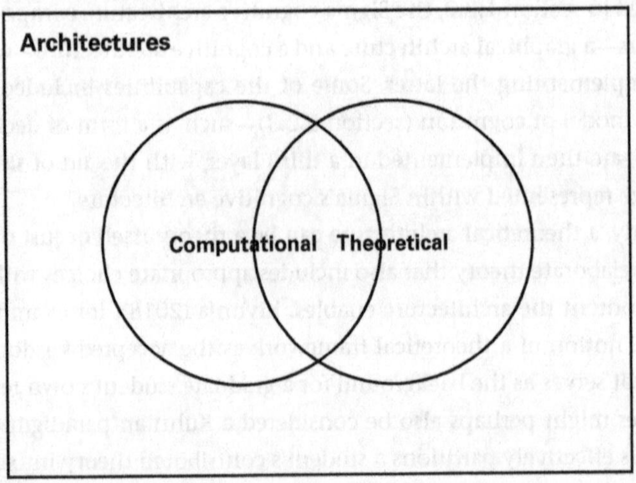

Figure 2.1
Space of architectures, highlighting overlapping classes of computational and theoretical architectures.

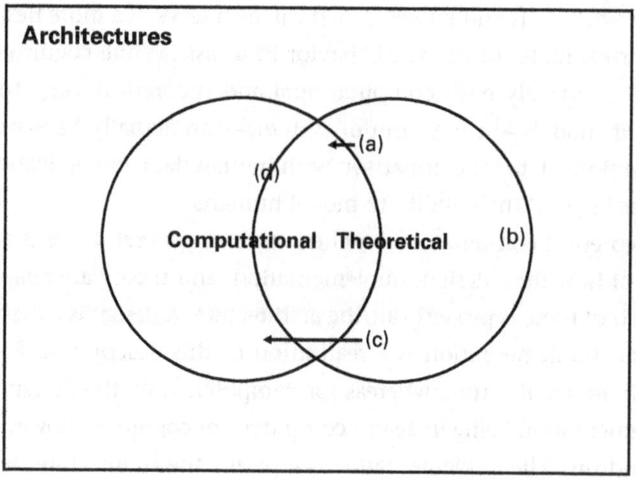

Figure 2.2
Some variations in the core architectural space: theoretical noncomputational architectures that can (a) straightforwardly or (b) not straightforwardly be converted into theoretical computational architectures, or (c) be converted into atheoretical computational architectures; and (d) computational architectures that embody implicit theories about their domains in the abstract.

predictions and simulations, as shown by location (a) in the figure. This contrasts with other theoretical architectures, perhaps expressed in English, or derived from the humanities, that may be far from computational architectures, as shown by location (b) in the figure.

In a different manner, a noncomputational theoretical architecture may be converted into an atheoretical computational architecture—as shown by location (c) in the figure—such as when the theory of evolution was converted into genetic algorithms (Holland, 1992) and programming (Koza, 1992) that, while inspired by the original theory, are intended to provide useful tools rather than a model of the natural phenomena. Or a computational architecture that is primarily intended to provide practical value, such as a computer or A(G)I architecture, may also implicitly embody some amount of theory about its domain in the abstract, as shown by location (d) in the figure.

A cognitive architecture can be considered as a theory of the fixed structures and processes found in a human mind, or of what is necessary and/or sufficient to yield an artificial mind capable of general intelligence. When combined with skills and knowledge, it can also serve as a more fleshed-out model, for example, of human behavior in a task. While cognitive architectures are typically both computational and theoretical, only the more fleshed out models—in the form of systems—can actually be executed to yield behavioral data for comparison with human data, or applications that have value beyond their ability to model humans.

For theoretical computational architectures in general, there is a serious question of how their design, implementation, and theory are related; and, also, which of these is properly an/the architecture. A design is a description, whereas an implementation is a realization of this description. For buildings, both are architecture, whereas for computers only the design is, with the implementation being instead a computer. For cognition, however, there is a conundrum. The implementation is typically the architecture, as there is rarely an explicit design. Yet there clearly is at least some notion of a cognitive architecture that goes beyond any particular implementation and which enables researchers to refer to multiple implementations as being the same architecture. Versions of both the ACT-R and Soar cognitive architectures, for example, have been implemented in different programming languages—such as Lisp, C, Java, and Python—implying that there must be such a distinction. Nevertheless, there is typically no separate specification of the architectures beyond what is provided by their implementations, along with informal descriptions in the literature and even less formal ideas within the minds of their developers.

Starting back at the beginning, in the simplest case, there is just a theory about a body of phenomena (figure 2.3). It might be described informally in text or more formally as mathematical equations. For a computational theory, there is typically an implementation, of which the theory itself is only a subpart (figure 2.4). The remainder is "implementation details" necessary to make the theory executable but not part of the theory itself.[6] There is typically no separate design specification in such cases, although a theory such as AIXI (Hutter, 2005) could be considered as such, and there were several attempts to specify Soar more formally (Milnes et al., 1992; Cooper et al., 1996).

Still, ideally there would always be a separate design that specifies in a more comprehensible yet abstract fashion what is to be in the implementation (figure 2.5), as is common with buildings and computers. This design would have its own theoretical subpart—corresponding to the theory in

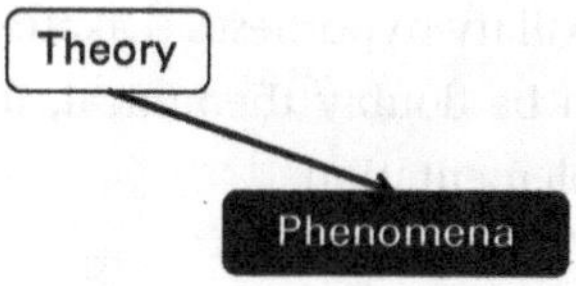

Figure 2.3
Simple theory.

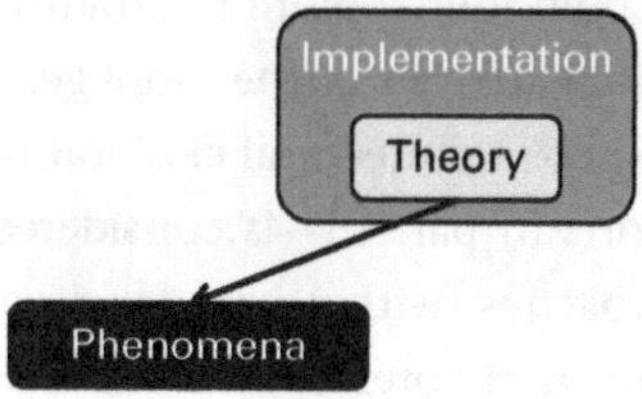

Figure 2.4
Typical computational theory.

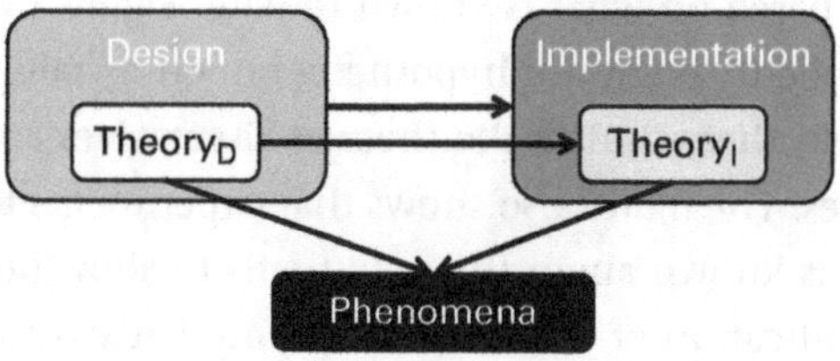

Figure 2.5
Ideal computational theory.

figure 2.3—which may be more easily demarcated than is possible within the implementation. Without this, the boundary between what is part of the theory versus an implementation detail remains fuzzy (Cooper et al., 1996).

Where is the architecture in all of this? According to the definition in section 2.2, all four nonphenomenal components in figure 2.5 are theoretical architectures. They may even share the same name in practice, although figure 2.5 can potentially be leveraged to distinguish the distinct roles they each fill with respect to a common name. It should be noted though that the atheoretical aspects of the two larger components maintain an ambiguous state of being fixed with respect to their enabled variable content but malleable in being adjustable as necessary without affecting the core theory, much like Lakatosian auxiliary hypotheses (Lakatos, 1978). The two design architectures in turn can be doubly theoretical, in being both about the phenomena and the implementation.

2.6 Architectural Exploration

Work on cognitive architectures is more directly a part of science than metascience and, as such, should follow more the traditional scientific method rather than either essential analysis or the more general reflective method. In this section, the notion of architectural exploration—that is, how I characterize my career's efforts in part III—is considered in the light of more standard scientific approaches (with a broader discussion of architectural exploration to be found in chapter 14). Given that architectures can be essences (section 2.2), investigations of them may also take on the flavor of metascience, but that will not be a focus here.

Science conventionally plays off theory and experiment. Theory generates hypotheses based on what is known to date. Figure 1.6, on the scientific method, explicitly mentions hypotheses but oddly fails to mention the theories that yield them, unless the theories themselves are considered to be the hypotheses. The figure also shows that experiments test hypotheses, increasing what is known about them, but fails to show how this leads to rejection or modification of theories, except implicitly through eventually looping back to the hypothesis stage. All in all, the focus on hypotheses here rather than on theories reveals a backgrounding of theories that go beyond the individual hypotheses being tested by the experiments that is problematic for complex theories, such as cognitive architectures, that may be the source of many hypotheses.

Beyond the use of experiments, and thus also beyond what is shown in the figure, proofs may also play a role in science when it is possible to formalize a theory to the extent that conjectures—a form of prediction—can be proven or disproven, although such activities may at times be considered more mathematics than science. Similarly, simulations may play a role when a theory can be made sufficiently computational[7] to enable implementing and executing it.

Another oddity about figure 1.6 is that the boxes shown for the scientific method are an incoherent mix of nouns (observation/question and hypothesis) and verbs (research, test, analyze, and report). Yet, when considering

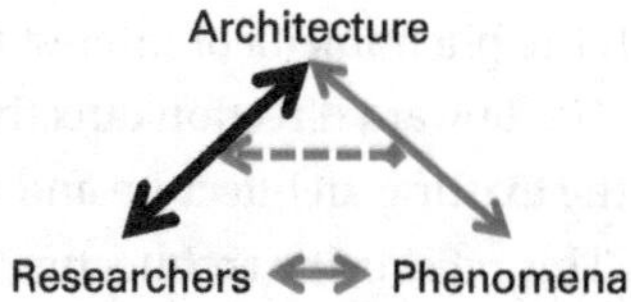

Figure 2.6
Architectural exploration.

my own methodology for cognitive architecture research—that is, architectural exploration—over the past forty-plus years, I end up with something more like figure 2.6, which assigns nouns to nodes and verbs to arrows.

The arrows here that point to the researchers reflect understanding by them of the phenomena and architecture. The former corresponds to the observation and data analysis stages in figure 1.6, with the difference between the two depending on whether the researchers have shaped the phenomona—via the arrow back from the researchers to the phenomena—prior to understanding them. When there is such shaping, it corresponds to the experimentation stage of figure 1.6. The arrow back from the researchers to the architecture reflects a second aspect of shaping, corresponding to the hypothesis stage in figure 1.6.

Missing from figure 1.6, however, is the understanding of the architecture. For an isolated, passive hypothesis, this may not be necesssary, but a cognitive architecture is the essence of an integrated, active mind that may itself play a central role in understanding and shaping the phenomena, as represented by its own arrow on the right of figure 2.6. Understanding of such a complex entity is thus an important activity in its own right.[8] Furthermore, understanding of the interactions between the architecture and phenomena—as shown in figure 2.6 by the middle arrow—can influence how researchers both understand and shape the architecture. It can also correspond to aspects of the data analysis stage in figure 1.6.

When both directions are considered in combination, both the architecture–researchers arrow and the researchers–phenomena arrow yield forms of exploration in which observation and manipulation are intertwined. As implied previously, exploration of the space of architectures is implicit in figure 1.6, with movement in the space occurring within the hypothesis stage but informed by the full loop. In figure 2.6, the downward direction represents the development of insight based on the architecture and its interactions with the phenomena of interest (with direct observation entering from the side). The upward direction directly represents architecture modification based on the existing architecture and its interactions with the phenomena of interest. This reflects the architecture's role as both a hypothesis and a domain of exploration of its own.

Exploration of the space of phenomena—that is, searching for new phenomena to understand—has no explicit role in the scientific method, as characterized in figure 1.6, although it may be implicit within the observation/question box. It is often considered prescientific,[9] but the scientific enterprise would be terribly impoverished without it. Early natural scientists fanned out across the globe to deliberately seek out and make sense of novel plants, animals, and other natural phenomena. To this day, disciplines such as paleontology, astronomy, and particle physics intentionally explore the natural universe. They may also expend considerable effort developing new instruments that sense aspects of the world/universe not previously perceptible to us. There may be hypotheses and predictions that are intended from the beginning to be tested via such instruments, but part of the anticipation and excitement in any such enterprise must always be the possibility of uncovering what was not expected.

As I have approached research in cognitive architectures, most of the actual work is exploratory, using the architectures to guide the exploration of the phenomena of cognition and to interpret the results of these explorations; and exploring the space of cognitive architectures. Controlled experiments provide only a small fraction of the criteria used to evaluate such work, to the point where at times they seem almost epiphonemal, in the sense of being afterthoughts performed more for publication purposes than as part of an original plan. This is, of course, an exaggeration, as the experiments do have purpose and have been informative, but the bigger-picture criteria center on: (1) whether an architecture, or some part of it, actually functions as hoped; (2) how broadly the architecture models and/or produces the phenomena of interest; (3) how simple and elegant the architecture is; and (4) how much insight the architecture inspires and captures, whether into anticipated implications or wholly unanticipated ones.[10]

Despite numerous attempts by the field to develop better approaches to evaluating work on cognitive architectures, much of the evaluation activity around these four criteria necessarily continues to center around simply building new architectural components and exploring their interactions.

Controlled experiments, in contrast, tend to focus on refining what can be said about criterion (1), in terms of how well the architecture models and/or produces phenomena of interest, and a bit of criterion (2). Such experiments clearly play a role but not a particularly dominant one. If one is problem focused, rather than methodology focused, it is critical to use whatever the best methods available are for the problem of interest, rather than limiting oneself to problems for which the strongest methods—in terms of the veracity they guarantee—are applicable; and exploration is still the best methodology for much work in cognitive architectures.

Exploration was the name of the game in the early days of AI, when there were many more unexplored areas and approaches than researchers to investigate them, and it was thought better to move on to new ones once sufficient low-hanging fruit had been harvested than to work over existing ones very carefully first. Now, however, exploration has fallen into relative disrepute as a method of investigating topics in AI. With a broader coverage of the space of options now under our belts, and general progress in the application of proofs and controlled experiments to specific AI problems, these methodologies have become the coin of the realm instead. Strong methods once they get a foothold in a field tend to push out weaker ones, even in areas where the stronger ones are not applicable (Rosenbloom, 2012a).

Exploration has continued to remain a major methodology in AGI, presumably enabled by separating the field out from AI as a distinct discipline. However, in doing so, it must also be sure to avoid becoming a wayside by either ignoring what is happening in the mainstream or by clinging to weaker methods if/when stronger ones do come along that apply to the problem(s) of producing general intelligence.

2.7 Conclusion

The crux of the idea of an architecture is to differentiate the fixed from the variable aspects of a system. This distinction, which in honor of Herb Simon might be denoted simply as *(in)variant*, or even as *(im)mobilière*, can then be elaborated on in various ways, with the (non)computational and (a)theoretical distinctions being of especial relevance to cognitive architectures. Additional issues that have been revisited here concern architectural layering and the nature of theoretical computational architectures. The question of how architectures should be investigated, with a specific focus on exploration, has then capped off these thoughts. Future work in this area might occur in a variety of directions, including incorporating into this analysis other forms of architecture, such as software architecture (Perry & Wolf, 1992; Clements et al., 2010), that are relevant to architecture in general and that should in principle be relevant to cognitive architecture, but which as far as I know did not have a foundational impact on the development of this latter notion.

3 On Symbols

The notion of a *symbol* was foundational in the inception of both artificial intelligence (AI) and cognitive science in the 1950s and 1970s, respectively. Although, its roots go back much further in philosophy—most notably in logic—computer science, linguistics, literature, and the arts (Newell, 1980). *Turing machines*, for example, depend on reading and writing symbols onto locations on an infinite tape (Turing, 1937). During the 1970s, Newell and Simon (1972; 1976) and Newell (1980) devised a computational framework around symbols—*physical symbol systems*, "a broad class of systems that is capable of having and manipulating symbols yet is also realizable within our physical universe"—that was to serve as the conceptual foundation for AI and cognitive science.[1]

Newell (1980) went on to state that "the hypothesis is that humans are instances of physical symbol systems, and, by virtue of this, mind enters into the physical universe. In my own view this hypothesis sets the terms on which we search for a scientific theory of mind." In line with this, he claimed that "the physical symbol system is to our enterprise what the theory of evolution is to all biology, the cell doctrine to cellular biology, the notion of germs to the scientific concept of disease, the notion of tectonic plates to structural geology."

A review of such systems, and their culmination in the *physical symbol systems hypothesis* (*PSSH*) that captures their intended centrality in intelligent behavior, can be found in section 3.1. Many critiques of the PSSH have been proposed since it was first introduced, with some that have easily been refuted and others that have lingered (section 3.2). Yet this hypothesis has never truly been replaced. Here, two critiques are considered that in my view have remained compelling. Both concern the general need for numeric processing—as found in both probabilistic and neural approaches to intelligent behavior—in addition to symbolic processing, with the second tied specifically to the overwhelming recent successes of approaches based on neural networks.

The idea of replacing the PSSH with a *neural-network systems hypothesis (NNSH)* is briefly considered before a novel form of *hybrid symbol system* is proposed as a more appropriate response to the remaining critiques (section 3.3). As part of this, the notion of symbols, and thus of symbol systems, is rethought, yielding the topic for step 1 of the specialization of the reflective method to essential analysis. This proceeds to an essential definition of what it means to be a computational symbol—that is, step 3—that is grounded via step 2 in ideas from the common model of cognition (CMC) (section 1.2.3 and chapter 16) and the Sigma cognitive architecture (sections 1.2.2 and 15.4). This analysis strips out many of the definitional aspects of traditional symbols, converting them instead—along with a hybrid aspect—into dimensions of a definitional space for symbols (step 4). A key part of step 6 is also found in this section, in terms of the *hybrid symbol systems hypothesis (HSSH)* that offers a direct alternative to the PSSH.

Demonstrating that neural networks are themselves instances of this form of hybrid symbol system (section 3.4), rather than being limited to the numeric component of a coarse-grained combination of symbolic and numeric processing, provides a key part of step 5 that helps bridge the gap between symbolic and neural approaches. In consequence, it enables recent successes with neural networks to be weighed in a positive manner in evaluating hypotheses concerning symbol systems, rather than the former necessarily serving as a challenge to the latter.

The core results of this chapter are a new way of thinking about symbols and symbol systems; the introduction of a novel form of hybrid symbol system and the corresponding HSSH; and an understanding of how neural networks are examples rather than counterexamples of this form of symbol system. The hope is that this all helps cut the Gordian knot that has resulted from past discussions on these topics.

Proposing hybrid or neurosymbolic systems is certainly nothing new. Many approaches have already been investigated—see, for example, Kurfess (2002) and Bader and Hitzler (2005) for overviews, and Nilsson (2007) for a discussion of the PSSH and the relevance of hybrid systems. But the point here is to introduce a novel take on hybrid symbol systems that is in service of an appropriate rethinking of the PSSH. The approach is broader than neurosymbolic, as it also includes hybrid systems that span other numeric paradigms, such as probabilities. In addition, it spans both tightly coupled and loosely coupled approaches for combining symbolic and numeric processing.

Following the core material of this chapter—but prior to the conclusion in section 3.6—section 3.5 extends the work on step 6 by exploring variations on the PSSH that go beyond the HSSH. This involves decomposing the PSSH into three fragments; identifying a three-dimensional space of variations that arises from alternatives for these existing fragments, yielding a form of definitional space for symbol systems hypotheses; and delving more deeply into three loci within this space.

3.1 Physical Symbol Systems

According to the traditional view, *symbols* are distinct patterns (Vera & Simon, 1993) in the physical world that can be *composed* into *expressions*, or *symbol structures*. *Processes* are then defined on these symbol structures that can create, modify, reproduce, and destroy them. An expression *designates* an entity, whether internal or external, if the expression's use depends on the nature of the entity. An expression is *interpreted* if it designates an internal procedure that is then executed. The physicality of such symbol systems reflects that they are *natural*, in obeying the laws of physics and being amenable to engineering, and that they aren't limited to what is in human minds, or even necessarily based on the same kinds of symbols that have traditionally been imputed to humans.

Given composition, designation, and interpretation, along with the appropriate processes, physical symbol systems provide a form of universal computation. There are certainly more details in the various papers, but this provides the essence of what can now be considered the classical notion of a physical symbol system. The PSSH then states that "a physical symbol system has the necessary and sufficient means for general intelligent action."

This hypothesis was introduced as an empirical generalization rather than a theorem. Evidence for sufficiency stemmed from the universality of symbol systems and the success of such systems built as of then. Evidence for necessity stemmed from noting that the one natural system exhibiting such intelligent behavior—that is, humans—appeared to be such a system, and from the lack of alternative approaches that were nearly as successful. Newell, for example mentions that "these advances far outstrip what has been accomplished by other attempts to build intelligent mechanisms, such as the work in building robots driven directly by circuits; the work in neural nets, or the engineering attempts at pattern recognition using direct circuitry and analogue computation" (1980).

3.2 Critiquing the Physical Symbol Systems Hypothesis

It has now been over fifty years since the PSSH was first articulated, with numerous critiques and defenses occurring in the intervening years. For example, Nilsson (2007) lists four general types of critiques with his responses to them (in italics here), which in brief are:

1. Lack of embodiment/grounding.

   This is a misunderstanding, as the PSSH already includes this.
2. Nonsymbolic/analog processing.

   Include numbers; that is, make the systems hybrid.
3. Brain-style versus computation-style (i.e., brains are not computers).

   The brain is computational.
4. The mindlessness of much of what appears to be intelligent behavior.

   Mindless constructs only yield mindless behavior.

This last critique might particularly be considered to apply to generative AI (e.g., Brown et al., 2020), which can give the impression of intelligent behavior yet are mindless in a fundamental sense. Analyzing them in terms of the ideas presented on cognitive architectures in chapter 1, they can be seen to comprise long-term memories that contain vast amounts of knowledge (including knowledge of skills)—what would normally be considered variable content if it weren't frozen once pretraining is complete—but without the other pieces required for a full mind. According to this analysis, such a system may be thought of informally as a broadly knowledgeable subconscious that is largely mindless,[2] but a deeper analysis of generative AI and its relationship to intelligence can be found in section 4.3.2. Fortunately, there is some work at expanding such models to include others of the critical pieces. A bit of this can be seen later in this section and in section 3.4.

In the remainder of this section, two particular critiques of the PSSH are considered, based on new empirical evidence in the form of the recent successes of deep learning (LeCun et al., 2015; Goodfellow et al., 2016)—whether or not actually mindless—and to a lesser extent probabilistic graphical models (PGMs) (Koller & Friedman, 2009), plus work on the common model. One critique, aligned with Nilsson's second, challenges the sufficiency of the PSSH and the other its necessity.

The sufficiency challenge focuses on the lack of numeric processing—that is, calculations on quantities—in the PSSH. Nilsson's response is to shift to hybrid systems that include both symbols and numbers. In a sense, this isn't logically necessary, as the universality of symbol systems implies that, as with any modern digital computer, they can implement algorithms for numeric processing. However, universality is weaker than what was originally proposed, as it omits grounding sufficiency in the successes of existing symbolic AI systems. Given the range of general intelligent action that has been shown over the years to proceed more effectively with numeric processing, whether in the form of probabilities or activations, the success of purely symbolic systems no longer provides compelling empirical evidence itself for the sufficiency of symbols on their own. Thus, we are left with a weakened form of sufficiency for the PSSH, based solely on universality.

The necessity challenge is rooted directly in how neural networks now provide a better approach for many problems related to intelligent action. Successes with probabilistic graphical models can be considered here as well, although they are already hybrid systems that add probabilities to classical symbol systems, particularly in their most general form as statistical relational systems (de Raedt et al., 2016), so they do not directly challenge the necessity of physical symbol systems. In contrast, deep learning has the potential to provide an alternative that completely overturns the necessity argument.

Given these challenges, does it make sense to consider replacing the PSSH with an alternative based on neural networks? The *neural-network systems hypothesis* (*NNSH*) could be stated as:

Neural-network systems are necessary and sufficient for intelligent behavior.

Necessity and sufficiency arguments here may be based on the obvious success of neural networks in the brain, even if they are quite different kinds of neural networks; the recent successes of neural networks in artificial systems; and the potential for some forms of neural networks to exhibit universality (e.g., Graves et al., 2014). Further attempts to build more elaborated cognitive architectures based on neural networks, such as LeCun (2022), can also yield additional variations on, and strengthening of, this hypothesis.

However, this only applies, as in this case, when the entire architecture is based on neural networks. When mechanisms are added to the overall system based on other technologies (e.g., Park et al., 2023; Sumers et al., 2023), it actually challenges the sufficiency side of the hypothesis.[3]

One example of this concerns the need to add a mechanism for state-space search—a classic, compositional, symbolic capability even if enhanced with probabilities—to neural networks to successfully play board games such as backgammon (Tesauro, 1995) and Go (Silver et al., 2018). Similarly challenging is work demonstrating how generative models by themselves struggle to generate effective action plans (Hao et al., 2023). Newell (1980) highlighted this question early on, by noting, "A general intelligent system must somehow embody aspects of what is to be attained prior to attainment of it, i.e., it must have goals," "A general intelligent system must somehow consider candidate states of affairs (and partial states) for the solutions of these goals (leading to the familiar search trees)," and concluding that "(1) rationality demands designation of potential situations; (2) symbol systems provide it; (3) only symbol systems can provide it when sufficient novelty and diversity of task are permitted." No evidence that I am aware of yet contradicts these claims.

A second example concerns compositionality into expressions, one of the key features of symbol systems. As will be seen in section 3.4, even standard feedforward networks embody a few fixed forms of composition, in defining vectors of units and links between units, but a number of more sophisticated systems have found it necessary to include further forms of composition on top of the basic network architecture, whether in support of convolutions (e.g., LeCun & Bengio, 1995), graphs (e.g., Scarselli et al., 2009), transformers (Vaswani et al., 2017), or other structural forms.

The alternative proposed in section 3.3 is based on hybrid symbol systems that have the potential to restore the stronger sense of sufficiency that was assumed for the PSSH. The necessity challenge is approached via a demonstration that, given the rethinking of symbol systems in section 3.3, neural networks are themselves instances of hybrid symbol systems (section 3.4). This approach avoids the need to resolve the contentious question of whether neural networks have or need traditional symbols, a question that appears unresolvable, at least to me, without additional evidence.

Such systems also deal with a related sufficiency challenge that arose from the history of work on cognitive architectures. Early architectures were typically fully symbolic—with early Soar (section 1.2.1) as a preeminent example—but became increasingly hybrid over time, such as with the late version of Soar. The common model, based on this history, acknowledges that the universality of classical symbol systems makes them sufficient in principle for implementing the numeric processing required for intelligent behavior but denies that they are sufficient in practice when timescales matter, such as within a single cognitive cycle. In particular, if statistical processing must occur on the same timescale as symbolic processing in such an architecture, then implementing the former in terms of the latter—as the universality argument for sufficiency implies—is insufficient. Thus, the common model implies the need for numeric and symbolic processing on the same timescale, requiring a fully hybrid system.

3.3 Rethinking Symbol Systems

This section proceeds through steps 1–4 of the specialization of the reflective method to essential analysis. This starts with the identification of the notion of a *symbol* as the topic to be reconsidered (step 1), in particular, a symbol as it is used computationally rather than as it is used in the humanities and arts. For an example of the latter, Dictionary.com defines a "symbol" (2024) as "something used for or regarded as representing something else; a material object representing something, often something immaterial; emblem, token, or sign." This focuses on an abstract notion of *designation* or *aboutness*, which, in chapter 5, is key to the essence of a theory.

Step 2 amassed a small handful of definitions of (computational) symbols, both traditional ones we have already seen and more recent versions, such as are found in the common model of cognition and the Sigma cognitive architecture. Based on these, the essence of a symbol is proposed to be an *atom* that is: (1) *indecomposable* into other atoms; and (2) *distinct* from other atoms (step 3). McDermott informally introduced the notion of a symbol as a *placeholder* (McDermott, 2001). Although yielding different connotations, this notion is compatible with that of an atom here.

This essence retains the classical notion of a computational symbol being a primitive element that can be distinguished from other such elements but eschews the need for physicality and the need for symbols to be structured as patterns. There were good reasons at the time to emphasize physicality—to counter both Cartesian dualism and the notion that only humans could use symbols—but these battles have already been won, at least in my judgment, so this explicit emphasis on physicality is now dispensable.

Pattern comparison is one way to determine whether two atoms are distinct. Yet, such a notion need not be definitional if it is just used to compare symbols. If symbols are considered as *types* (rather than *tokens*)—a notion implicit in the traditional definition—patterns are simply *intensional* definitions of symbols. An *extensional* alternative defines each symbol in terms of a set of tokens, with each token in a set considered to be indistinct from other tokens in the same set and distinct from tokens in other sets.

The classical notion of symbol also includes *composability*—into symbol structures or expressions—*designation*, and *interpretation*. The first of these is traditionally assumed to be part of the very nature of symbols, whereas the latter two are additional properties necessary to enable the classical form of physical symbol systems. The essential definition of a symbol introduced here includes none of these three notions; that is, all are optional. Therefore, any system that includes even these minimal, atomic forms of symbols can be considered a symbol system of some sort.

Figure 3.1 structures these optional properties, plus a bit more, into a small tree (step 4). According to this perspective, a symbol may be *composable* into expressions (aka symbol structures). It may also *designate*—that is, stand in for something else. A designation is *procedural* if it is about a process. This is the classical notion of interpretation, when combined with the ability to execute the designated process. A procedural symbol, according to this definition, designates a process rather than being part of the process itself. If the process is itself a symbol structure, it will contain symbols, but they themselves may be of any type. A designation is *declarative* if it is about an object—essentially anything other than a process—which may be *internal* to the system or *external* to it, with the latter relating to grounding (a topic that will be seen again in section 4.1 in the context of its overall relationship to intelligence). This corresponds to the classical notion of designation when contrasted with interpretation. Beyond this difference in what is designated, there is no intent to impute any other aspects of the classical procedural versus declarative distinction (e.g., Winograd, 1975) here.

Symbols in a classical symbol system support all of these properties—although typically only one form of designation at time—enabling them to exhibit computational universality. Whether systems in which some or all of the symbols lack some of these properties provide anything like universal

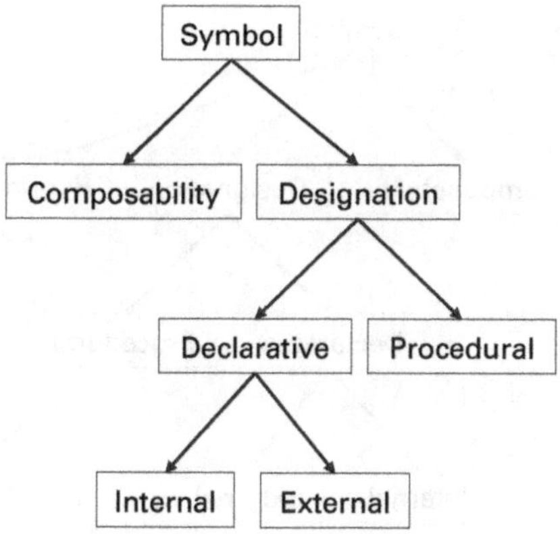

Figure 3.1
Optional properties of symbols.

computation would necessarily depend on the details of the individual systems.

The common model took a step toward such an essence by dropping the necessity of designation, and thus also of interpretation, stripping symbols down to primitive elements that support only composability into symbol structures. Although designation somewhere in a system seems necessary for it to be either meaningful or operational, it is not necessary for all symbols. The Sigma cognitive architecture goes beyond this to deny the necessity of all symbols supporting arbitrary forms of composition, thus implicitly leading to the essence made explicit here.

The common model also associated *quantitative metadata* with such symbols and structures—that is, the *data*—to modulate how they are processed (with Sigma taking this to an extreme). Such combinations can be considered as *hybrid symbols* or *structures*. Considering *hybrid* of this sort as a third optional property of symbols leads to figure 3.2. When multiple such symbols are combined with the appropriate processes, the result is a *hybrid symbol system*.

This approach is orthogonal to whether the data itself may be numeric, via designation of a quantity. It is also orthogonal to whether numeric aspects are allowed in the composition process. Traditional composition would typically combine symbols qualitatively into triples, relations, lists, vectors, and so on. However, computer memories assign numeric addresses to individual word locations, so the memory can be seen as being composed

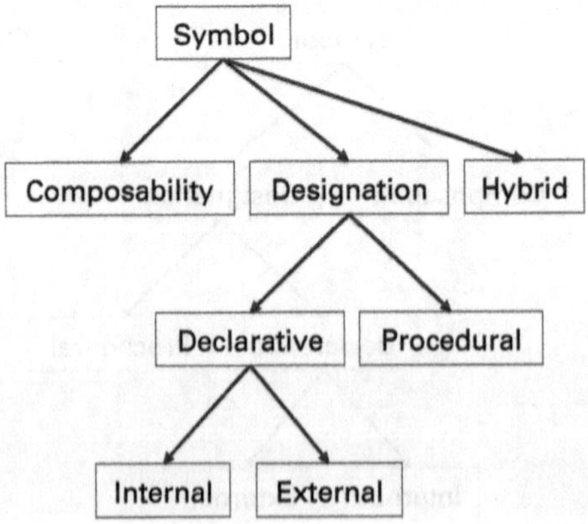

Figure 3.2
Optional properties of symbols from figure 3.1, extended with hybrid.

numerically, via integer addresses. Sigma goes even further, with locations along dimensions having continuous addresses, although these are typically binned into contiguous segments along the real number line (section 1.2.2). These segments provide the discrete locations needed along the axes for tensors, but they still retain their continuous character so that they are usable, for example, in computing dot products between vectors. The tree of optional properties in figure 3.2 could be extended so that composability has qualitative and quantitative children, but since this is really an aside from the main point of this chapter, it is left implicit here.

Meta²data—that is, metadata associated with metadata—is not currently part of the model. Even though metadata amounts to symbols, given that they are numbers, these are symbols that don't themselves include the hybrid property. This doesn't rule out the possibility of more general forms of symbolic metadata, as becomes central in many forms of reflective or metacognitive processing (Maes & Nardi, 1988; Nelson, 1992; Cox, 2005), but that is a distinct story that has not yet been fleshed out in the common model (although, see Kralik et al., 2018; Laird et al., 2025), and is also not of direct concern here.

Smolensky et al. (2022) highlighted the importance of the combination of composition and continuity in the search for general intelligence but doubted that a hybrid approach could be viable in this sense without embodying a form of *neurocompositional computing* that simultaneously enables individual encodings to exhibit both aspects. The hybrid symbol systems proposed here do support this combination, with data enabling composition and metadata enabling continuity, without necessarily implying any particular form of neural computation.

Such an approach enables physical symbol systems, probabilistic graphical models (PGMs) (Koller & Friedman, 2009; de Raedt et al., 2016), and—as is shown in section 3.4—feedforward neural networks to fit within the same space. It also permits both fine-grained and coarse-grained combinations of these various computational models. Fundamentally, it manages this by expanding on the original space of physical symbol systems in two ways: (1) making previously necessary properties optional on a symbol-by-symbol basis; and (2) adding a symbol-by-symbol option for quantitative metadata.

This contrasts strongly with an approach that simply adds to the traditional notion of a physical symbol system a requirement for quantitative metadata. In one sense such a combination would yield a space of systems that is distinct from the traditional one—one more aligned with PGMs—but in another sense, it can be considered as yielding a superspace of the traditional space if binary (0/1) metadata can effectively operate as T/F. This superspace is still, however, much smaller than the space that results from allowing all of the properties to be optional on a symbol-by-symbol basis. For example, feedforward neural networks based on distributed representations do not naturally fit within this smaller superspace, although some other forms of neural networks that map directly onto PGMs do (see, e.g., Jordan & Sejnowski, 2001).

We can refer to these two different approaches as *weak* and *strong* hybrid symbol systems, respectively, with the weak approach implying a much larger superspace than the strong one. The weak approach matters in particular because, of the two, only it subsumes feedforward neural networks. However, as a result, is it too weak to be useful, when from one perspective it can appear to be essentially vacuous in allowing just about every form of computational system? I would argue, perhaps unsurprisingly, that the answer is "no." The original notion of physical symbol systems was susceptible to a similar kind of argument, given their support for universal computation that enables constructing any kind of computational system via them. From one perspective, they are thus just another way to specify a general-purpose computer. However, the counter was that physical symbol systems formulated general-purpose computation in a manner that was particularly relevant to intelligent systems.[4]

The question I would pose about the weak formulation is analogous—in particular, whether it also formulates a computational model that is relevant—and, in fact, even more relevant—to intelligent systems. My answer here, again unsurprisingly, would be "yes," and the proposed replacement for the PSSH is based on this formulation. As discussed in section 1.1, having such a formulation can also be useful in enabling analyses of just where in a system, whether neural networks or any other architecture, the optional aspects of hybrid symbol systems can be found without their presence necessarily implying the existence of a traditional physical symbol system.

Still, it is possible to consider other variants that fall in between the weak and strong formulations. One such variant is much like the weak formulation but makes hybridness mandatory. In other words, it leverages the first form of expansion mentioned earlier but modifies the second to require quantitative metadata. In practice, it may not be substantially different in implication from the weak formulation if the trick mentioned previously for encoding T/F in the strong formulation works, but it is more aligned with the approaches embodied in the common model and Sigma.

A second variant keys off the idea mentioned earlier that although designation is optional for symbols in the common model, it must appear somewhere for the overall system to be meaningful and operational. The variant here thus starts with both forms of expansion that underlie the weak formulation but adds onto this a further requirement that every optional element must appear somewhere in a full hybrid system—there must be parts that are composable, parts that designate, and parts that are hybrid. These can be different parts of the system, or they can overlap in various ways.

Both variants retain the key property from the weak version that physical symbol systems, PGMs, and neural networks all fit within them, and that hybridness can exist in either a fine-grained or a coarse-grained manner. It is also possible to formulate relevant hypotheses based on both. However, here we proceed with the weak formulation, as introduced—although not under that name—in Rosenbloom (2023b). Under this assumption, the *hybrid symbol systems hypothesis* (*HSSH*) can be stated as: "Hybrid symbol systems are necessary and sufficient for general intelligent action."

If the sufficiency clause of the PSSH is valid, then so must be the comparable clause in the HSSH, at least for hybrid symbol systems that are universal. However, the HSSH responds to the PSSH sufficiency challenge by including numbers, as suggested by Nilsson (2007). Necessity of the HSSH is not implied by the corresponding clause in the PSSH. Instead, the HSSH responds to the PSSH necessity challenge by coopting the successes of neural networks (section 3.4).

One potential challenger to both sides of this argument arises from work on *quantum symbol systems* (e.g., Laskey, 2006). Little is yet known about them, but if it is shown that they are distinct from hybrid symbol systems yet can yield intelligent action, they could weaken—or possibly even negate—sufficiency here, much as the need for numeric processing challenged the sufficiency of the PSSH. Likewise, if they can generate intelligent behavior that is at least as good as what is achievable with hybrid symbol systems, they could challenge the necessity of the HSSH.

3.4 Neural Networks as Hybrid Symbol Systems

What makes neural networks hybrid symbol systems, as defined here, rather than simply the numeric component of a larger system that also includes a symbolic component (e.g., Sun, 2017)? To keep things simple, the focus here is limited to standard feedforward neural networks, consisting of multiple layers of nodes and links, where nodes have activations, links connect pairs of nodes across levels and have weights, and processing occurs by multiplying input activations along links by the links' weights and then nonlinearly transforming the sums of these weighted inputs.

To be a bit more specific, let's assume a small network for *paired associates* that maps an input word to an output word. Figure 3.3 exemplifies this via a completely connected network with six-unit input and output layers—yielding a six-dimensional vector of activations for each—and two intermediate layers, each with three units. Words map consistently to input and output vectors via *encoding* and *decoding* processes that are external to the network. These processes may be based on an arbitrary or random assignment of vectors to words or a more sophisticated *embedding* process, such as in Mikolov et al. (2013).

The focus here, however, is on analyzing the forward processing in the network itself to show how it amounts to a hybrid symbol system. It should be possible to extend such an analysis to encoding and decoding processes, as well as to learning in neural networks, but this simple example is sufficient to establish the precedent.

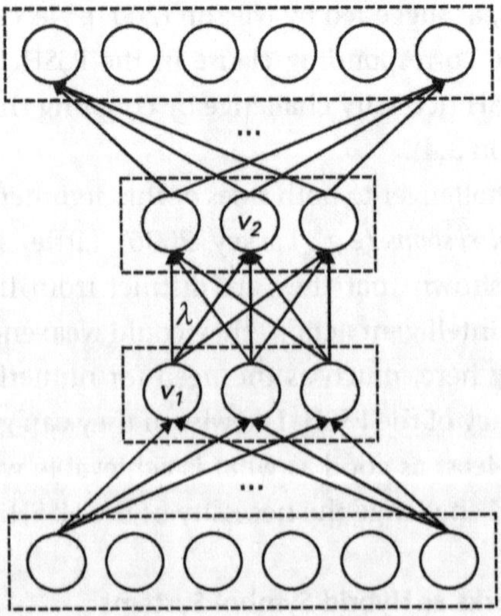

Figure 3.3
Simple network for paired associates.

First consider the nodes in the input layer of the network, now shown at the top of figure 3.4 as locations within a vector of nominal activations. Such nodes can be seen as hybrid symbols—symbolic nodes (i.e., locations as shown here) with activations as their quantitative metadata—that exhibit a limited form of composability in yielding the hybrid symbol structure that is the input vector. In contrast to the traditional interpretation of distributed representations (Hinton et al., 1986)—where nodes are subsymbolic, or microfeatures, with symbols only arising as patterns over these elements—here the individual nodes are themselves hybrid symbols that do not themselves designate, with patterns arising as structures of these hybrid symbols.

This hybrid symbol structure then internally designates a word structure that has one location per letter (middle of figure 3.4). The metadata in the word structure is not shown, as it is irrelevant to this analysis. What does matter is that declarative symbols in this word structure externally designate letters of the alphabet,[5] in this case making up Allen Newell's first name (bottom of figure 3.4). This word vector thus uses a local rather than a distributed representation, with each location/symbol within it designating an individual letter. The word structure as a whole then externally designates its meaning, iconified to the right of figure 3.4 via an image of him.

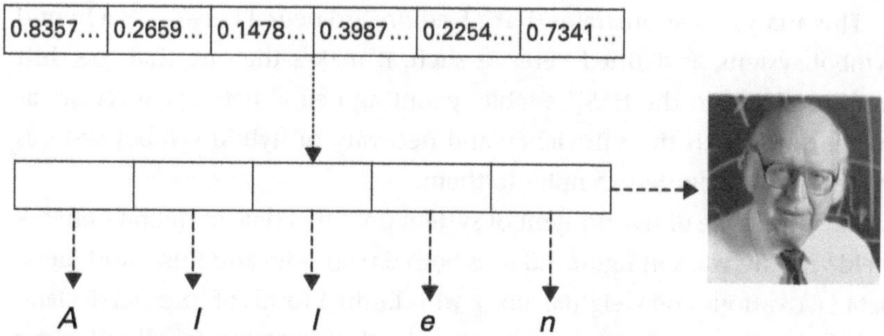

Figure 3.4
Designation relationships among input vector, word vector, letters, and word meaning. Dotted lines reflect internal designation and dashed lines external designation.

Key to this all working is that it isn't just the data aspect of hybrid symbols and structures that can designate but the entirety of the hybrid symbols and structures—including their metadata—that can do so, just as is traditionally assumed for vectors in distributed representations. The word itself is epiphenomenal to the feedforward network processing here—only the hybrid symbol structure at the top of figure 3.4, as yielded by encoding, actually participates. As Fodor and Pylyshyn (1988) put it, "The node labels in a Connectionist machine are not part of the causal structure of the machine."

The internal nodes in the network are also hybrid symbols but without declarative designations, fitting the intuition that there are no fixed meanings inside the network. Instead, internal nodes—and links—procedurally designate fixed processes. Consider link λ in figure 3.3, which points from node v_1 to node v_2. This link is a hybrid symbol structure composed from these two hybrid symbols, with a weight as its metadata. It procedurally designates a process that multiplies the activation arriving at v_2 from v_1 by this weight. Internal nodes such as v_2 then procedurally designate processes that sum all of their inputs—in this case, from v_1 and any other nodes linked to it from the preceding layer—and then nonlinearly transform the results (via, e.g., a sigmoid function or a rectified linear unit).

The last part of the analysis concerns the output nodes. Perhaps surprisingly, they too do not declaratively designate anything here. Instead, they procedurally designate the same summation and transformation process as the internal nodes. It is not until postprocessing—that is during decoding—that this reverse mapping occurs.

This analysis demonstrates that a feedforward neural network is a hybrid symbol system, as defined here. As such, it makes the case that the shift from the PSSH to the HSSH enables coopting neural-network successes as evidence for both the sufficiency and necessity of hybrid symbol systems rather than as counterexamples to them.

But what type of hybrid symbol system does this type of neural network yield? The network in figure 3.3 uses both data (nodes and links) and metadata (activations and weights) along with limited forms of internal declarative designation (at input nodes), procedural designation (at all but input nodes), and composition (via vectors within a level and links across levels). Including figure 3.4 extends this to also include external declarative designation. Thus, even this simple network depends on every optional property of a symbol, just not all properties at every symbol. It is also worth noting though that the composition that exists here is purely static and predefined, with no dynamic composition of symbol structures during performance.

Other forms of neural networks also go further to yield more complex hybrid symbol systems. As mentioned in section 3.2, additional forms of composition may be added to support convolutions, graphs, and transformers. The flexibility of composition seen in the output of transformer-based generative networks (Brown et al., 2020) is in fact quite compelling. In addition, some forms of neural networks support universal computation (e.g., Siegelmann & Sontag, 1991) and others—as also mentioned in section 3.2—are augmented with explicit state-space search, a form of dynamic composition that is necessary for combinatorial board games. So, the overall story is complex, dependent on the exact nature of the neural networks considered, and still not completely understood.

3.5 Other Variations on the PSSH

The PSSH naturally decomposes into three fragments: (1) physical symbol systems are (2) necessary and sufficient for (3) human-level intelligent behavior. The HSSH can be viewed as a variation on this that replaces the first fragment with "hybrid symbol systems." But, beyond the HSSH can be found a larger space of variations on the PSSH that are all based on modifying one or more of these fragments and that may be worth considering for particular purposes.

In addition to the HSSH, there are many alternatives for the first fragment. For example, one recent paper claims that "reward is enough" for "most if not all abilities that are studied in natural and artificial intelligence" (Silver et al., 2021). Other claims may concern specific cognitive architectures, whether implicitly or explicitly; for example, in an early paper Soar was hypothesized as the basis for "an architecture for general intelligence" (Laird et al., 1987). The common model of cognition has more recently been proposed as a consensus concerning aspects of what is necessary for humanlike minds.

For the second fragment, there can be claims of necessity and/or sufficiency, with the possibility of various qualifiers on each. For example, for reward to be enough it must be sufficient, but nothing is necessarily being claimed about necessity. The common model instead is making a claim about necessity, with nothing being claimed about sufficiency due to how it includes only components about which a consensus has been reached. The PSSH itself assumes a variant form of sufficiency, that whatever may be missing from a basic physical symbol system can be built on top of it via symbol structures and processing. So, it is a sufficient basis while not being directly sufficient all on its own.

For the third fragment, the PSSH concerns "general intelligent action." The common model concerns humanlike minds. The article on reward concerns "natural and artificial intelligence." Other variants could be considered as well. For example, the alternative of "cognitive architectures" will be considered later.

Three loci in this three-dimensional space of possibilities that are directly relevant to cognitive architectures are considered further here. The first reuses the first fragment of the HSSH but then also modifies the PSSH's third fragment, from "intelligent behavior" to "cognitive architectures." The *hybrid cognitive architectures hypothesis* (*HCAH*) thus states: "Hybrid symbol systems are necessary and sufficient for cognitive architectures."

While clearly related to the HSSH, the HCAH matters in itself because the comparable hypothesis—perhaps called the *physical cognitive architectures hypothesis* (*PCAH*)—fails due to the need for numeric processing being implemented directly in cognitive architectures (section 3.2). Thus, the sufficiency side of the PCAH is invalid irrespective of what might be true with respect to necessity. Conceivably, the necessity side of such a hypothesis could be stated on its own, but that seems of little value at this point.

As with the PSSH and the HSSH, the HCAH is an empirical generalization. The arguments for it are similar to those for the HSSH but with a few twists. As with the HSSH, the successes of cognitive architectures based on physical symbol systems and traditional hybrid systems count in its favor. Whether the successes of architectures based on neural networks also count depends on whether they are cognitive—that is, yield a mind—given that many would consider them to be focused on the brain rather than the mind. However, generative models, for example, do seem to provide better models of the mind than of the brain, even if only partial models of the mind, and so we will count them as cognitive here.

There are two additional pieces of evidence on the necessity side of the argument. First is the evidence already mentioned that numeric processing is necessary in cognitive architectures. Second is that there are no alternatives that have been anywhere near as successful. Purely symbolic architectures clearly have not, and if neural-network architectures can be considered as cognitive architectures, they support the necessity claim rather than counter it. As with the HSSH, the possibility of quantum symbol systems could challenge necessity here, as well as sufficiency, but that is a matter for future evidence to determine.

The second locus—the *cognitive architecture hypothesis* (*CAH*)—is related to multiple of the hypotheses already discussed but carves out its own unique position: "Cognitive architectures are necessary and sufficient for intelligent behavior."

At its most trivial, the necessity side of this variant could simply imply that there must be some fixed structure in any model of the mind. In general, it is thus often worth considering stronger variants of this hypothesis that include additional assumptions about the fixed structures and processes. One such example, focused on the common model of cognition, can be found next but other hypotheses based on specific architectures are also conceivable. Still, as a template for such variants, it is worth considering the CAH on its own.

While the argument for necessity here may seem obvious, there have been systems that can modify most, if not all, parts of themselves through reflection (e.g., Lenat, 1983; Steunebrink & Schmidhuber, 2012), although that can be a recipe for disaster when changes leave the system in a damaged state from which it cannot recover. As with the HCAH, the status of the necessity argument here also depends on whether neural networks are considered to be based on cognitive architectures.

A sufficiency argument here would be analogous to that made for the PSSH or the HSSH—it does not imply that cognitive architectures are all that is needed, but that the rest of what is needed can be built on top of the architecture in terms of the knowledge and skills that the architecture supports. However, much work on cognitive architectures does get by with an even weaker notion of sufficiency, where it is assumed that there are other noncognitive aspects of the overall system—such as perception and motor control—that are constructed separately and attached to the cognitive architecture. If the scope of what is considered "intelligent behavior" is limited to the central processes typically supported by cognitive architectures, this weaker form of sufficiency becomes indistinguishable from the more standard form. Alternatively, the stronger notion of sufficiency might be approached, as in Sigma, by supporting what might traditionally be considered noncognitive aspects of intelligent behavior within the cognitive architecture itself.

The third locus is derived from the original paper on the common model (Laird et al., 2017) but it was neither stated explicitly there as a hypothesis nor given a name. Making it explicit, the *common model of cognition hypothesis* (*CMCH*) can be framed as: "The common model of cognition is necessary for humanlike intelligent behavior."

The evidence for this claim comes from the consensus that has been achieved among both cognitive architectures and cognitive architects. As hinted at earlier, the CMCH is a stronger variant of the CAH in assuming more within the first fragment. However, it is weaker in lacking a sufficiency claim in the second fragment and in reducing the scope of the third fragment. It is possible to conceive of a sufficiency claim here but given the incompleteness endemic to the common model that stems from its consensus nature it has felt wisest to postpone considering such a prospect to future work.

3.6 Conclusion

Leveraging the specialization of the reflective method to essential analysis, symbols are (re)defined in this chapter as atoms or placeholders, and a definitional space is identified for symbols that yields concomitant implications for symbol structures and symbol systems. The dimensions of variability include the classic traits of compositionality and designation, plus hybridness and additional subtraits under designation (such as interpretation). In response to lingering challenges to the physical symbol system hypothesis (PSSH), a new hypothesis—the hybrid symbol systems hypothesis (HSSH)—has then been introduced that focuses on the resulting hybrid symbol systems: "Hybrid symbol systems are necessary and sufficient for general intelligent action."

The HSSH is intended as a replacement for the PSSH, based on evidence accumulated since the latter was introduced as an empirical hypothesis more than a half-century ago. Given this recent body of evidence, there is a sense in which the PSSH still holds, but it is a weaker sense. The HSSH recaptures the originally intended strength while adding further to it by reinterpreting neural networks as compatriots—that is, as hybrid symbol systems themselves—rather than as competitors. These results also help chip away in a rather fine-grained manner at the overall divide between symbolic and neural systems.

Beyond the HSSH, five additional hypotheses, based on alternatives to the three fragments into which the original PSSH can be decomposed, have also been explored:

- The neural-network systems hypothesis (NNSH) states that neural-network systems are necessary and sufficient for intelligent behavior.
- The hybrid cognitive architectures hypothesis (HCAH) states that hybrid symbol systems are necessary and sufficient for cognitive architectures.
- The physical cognitive architectures hypothesis (PCAH) states that physical symbol systems are necessary and sufficient for cognitive architectures.
- The cognitive architecture hypothesis (CAH) states that cognitive architectures are necessary and sufficient for intelligent behavior.
- The common model of cognition hypothesis (CMCH) states that the common model of cognition is necessary for humanlike intelligent behavior.

These additional hypotheses, along with the PSSH and HSSH, can be viewed as populating a form of definitional space for symbol systems hypotheses, although no complete analysis of this topic has yet been attempted. Of these additional hypotheses, the PCAH has been dismissed outright, and serious doubts remain about the NNSH. The other three hypotheses remain open to future evidence (as technically also does the NNSH). Further hypotheses, whether variants on any of these or ones that are quite different, may also be worth considering. Also worth further consideration is what impact, if any, quantum symbol systems may have on any of these results.

4 On Intelligence

The topic of this chapter—step 1 of the specialization of the reflective method to essential analysis—is *intelligence*. This is, of course, a critical topic in any discipline concerned with the mind, despite there being no single, consensus definition of it. Legg (2023) lists seventy-one different definitions, while quoting Robert J. Sternberg as saying that "viewed narrowly, there seem to be almost as many definitions of intelligence as there were experts asked to define it." Similarly, Monett and Lewis (2018) earlier referenced a list of seventeen definitions of intelligence along with seventeen further definitions of artificial intelligence. But these compilations do provide the fodder for step 2. The core mission of this chapter is then to tame this cacophony through an essential definition of it (step 3) in section 4.1. Perhaps surprisingly, rather than yielding a single essential definition this process yields a nested set of three essential definitions (figure 4.1).[1]

The trilevel conceptualization arises because, as with consciousness, multiple semirelated concepts turn out to be conflated when considering intelligence, each of which demands its own definition. Although this approach does yield multiple levels of intelligence, it differs significantly, for example, from efforts to identify levels of artificial general intelligence (AGI) in a manner analogous to what has been done for autonomous driving (e.g., Morris et al., 2023).

The most minimal concept is *immediate intelligence*, which concerns acting rationally in the here and now (section 4.1.1). *Cumulative intelligence* then adds to this the ability to improve over a lifetime via learning (section 4.1.2). *Full-spectrum intelligence* takes a more comprehensive perspective by adding in much more of the breadth implicated in human intelligence, including situatedness, (self-)awareness, projection of future possibilities, generality across problems, feelings, purposefulness, language, collaboration, socialization, and

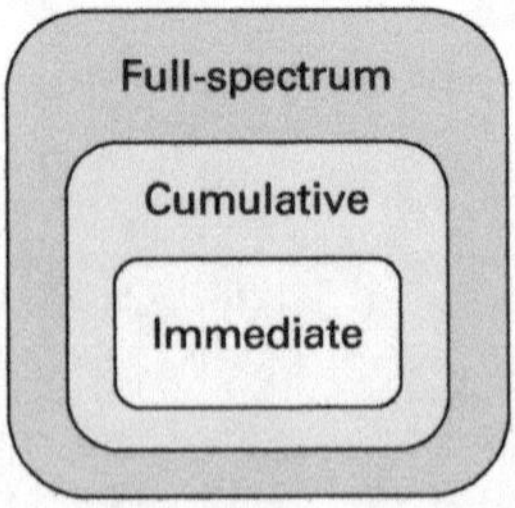

Figure 4.1
Trilevel nested conceptualization of intelligence.

education (section 4.1.3). In section 4.1.4.4, an attempt is made at a simpler, nondisjunctive essence for intelligence, but the body of this chapter focuses on this trilevel conceptualization.

A definitional space can be defined for each of the three concepts identified in the previous paragraph by considering its essential definition as an ideal—or *prototype* (Rosch, 1973)—with its definitional space then arising not from the all-or-none presence or absence of nondefinitional aspects but from degrees of approximation to the ideal. For example, the ideal for immediate intelligence is rationality (section 4.1.1.4), with various forms of "bounded rationality" (2024) counting as approximations to it. In combination, as a trilevel ideal, an essential definition of intelligence as a whole is yielded, with a definitional space that forms what can be called the *intelligence space* (step 4).

We know little at this point about the overall nature of the intelligence space, other than how it is at present populated with various, often only partially understood, instances of natural and artificial intelligence, although some earlier thoughts on this space—typically under different names—can be found in Sloman (1984), Dennett (1996), Gray et al. (2007), and Ball (2022). The field of comparative psychology focuses somewhat more narrowly on how thought processes vary across different forms of natural intelligence ("Comparative psychology," 2024). Simon and Kaplan (1993) go beyond this to suggest that cognitive science should span the full intelligence space, a call that has been echoed more recently by Rosenbloom and Forbus (2019).

Part II can be seen as an attempt to understand this space in a systematic manner by defining a set of dichotomies whose cross product spans many of the technologies underlying AI and cognitive science, although quite likely covering only a fraction of the entire space in the process. In section 4.1, a complementary approach to this is taken, with an idealized, trilevel definition of intelligence anchoring the intelligence space and a range of approximations to this definition hypothesized to flesh it out. Step 5 begins with assorting a selection of existing definitions of intelligence across this space (section 4.1.4.2), before proceeding with mappings of human intelligence, humanlike intelligence, artificial intelligence, and cognitive science onto regions of it (section 4.2).

Section 4.3 then pursues step 6 via explorations of six implications—most of which are quite preliminary or potentially controversial or both—concerning: (1) the singularity, as it relates to intelligence; (2) the current white-hot topic of generative AI, with a focus on large language models (LLMs); (3) Sigma, in terms of how it relates to the three levels of intelligence and thus how it can be contrasted with LLMs; (4) ethics, with a focus on its relationship to artificial intelligence (AI); (5) intellectual property, with a focus on whether its creation might be ascribed to AI systems; and (6) free will, with a focus on the importance of a belief in free will rather than on its actual existence. Section 4.4 then concludes.

Structurally, much of this chapter proceeds in an unusual manner, via a cascade of concepts and their definitions—in the spirit of Lucretius (first century BCE) and Spinoza (1677)—along with associated commentary when appropriate.[2,3] Although the scope of this effort is not nearly as wide ranging as either of these previous ones, it does share with them the attempt to reconstruct, and to reunderstand, an important range of complex concepts from the ground up. Like them as well, it includes topics that may be controversial, particularly many of the implications in section 4.3. It is also similar to them in not simply being a dictionary or glossary (as can, e.g., be found in appendix A for part II), despite its enumeration of many definitions. The ordering of the concepts that are defined, and the dependencies proposed among them, build momentum, from the very simplest to the most complex and potentially controversial topics.

The overall approach does bear a resemblance to how mathematics builds up from simple to complex concepts—Spinoza, in fact, appeals to the method of Euclid (c. 300 BCE)—but it is less formal, with no attempt at articulating theorems and proofs. It also bears a resemblance to modern work on ontologies and, in this spirit, is illustrated with diagrams that map out the dependencies among related sets of concepts—an arrow from one concept to another in such a diagram implies that the definition of the latter depends on the former, with the core concept in each diagram signified by a bold contour. Yet, in reality, the approach is more attuned to the earlier philosophical efforts already mentioned. So, this chapter may best be considered as part of a broad philosophy of mind that is aimed at the full intelligence space. It is written, however, from the perspective of someone who, rather than being a professional philosopher, has been active in artificial intelligence and cognitive science over many decades but has more recently taken to reflecting extensively on this experience.

This chapter's trajectory occurs in a certain sense in isolation, driven by its own internal consistency and momentum, rather than being a deeply scholarly work concerned with relating the definitions that result to everything relevant out there, although such contact is made when particularly helpful. It is not the originality of the definitions—to the extent that they are original—nor even that the definitions are exactly right, or that the terms associated with them are necessarily the right terms, that matters, but that a step forward is made in understanding intelligence and its overall space of possibilities in terms of a coherent skeleton that spans natural and artificial intelligence while providing insight into some of the more speculative and controversial—yet crucial—issues concerning them.

In other words, this chapter can be considered as a whirlwind tour across many concepts that provides one, hopefully coherent, perspective across them all, rather than as a careful analysis of each concept from multiple perspectives. This may be dissatisfying to readers for whom the analyses are too minimal or one-sided to be satisfying, but it can be thought of as providing one form of a theoretical, noncomputational architecture for intelligence, although not one that would be considered a cognitive architecture in any traditional sense.

4.1 Intelligence and the Intelligence Space

This section defines the three nested levels of intelligence—immediate, cumulative, and full-spectrum. It starts at quite a primitive level, with the concept of an object, and builds up from there in an incremental fashion.

4.1.1 Immediate Intelligence

Immediate intelligence concerns the making of rational decisions in the here and now—in particular, how to choose actions that achieve goals. This section proceeds to define immediate intelligence across four steps. The first three subsections introduce a hierarchy of entity types, starting with objects, continuing through systems and beings, and ultimately ending with agents. The fourth subsection then defines immediate intelligence in terms of the mind of a rational agent. This section ultimately aligns reasonably well with the corresponding parts of chapter 2 of Russell and Norvig (2010), although not all of the terminology is the same.

4.1.1.1 System Here we begin with the minimal concept of an object and build up from there to that of a system (figure 4.2) that is based on mechanism(s). The notion of an architecture, as a system's fixed mechanisms is also highlighted.

1. An *object* is a distinct something.
   - It need not be physical.
   - A symbol (chapter 3) can be thought of as an indecomposable object.
2. A *mechanism* is a combination of structures and processes.
   - This aligns with the notion of science, writ large, as concerned with understanding the interactions among structures and processes (Rosenbloom, 2012a).
3. A *fixed mechanism* remains unchanged over time.
4. A *system* is an object that includes one or more mechanisms.
   - According to chapter 2, a system can be decomposed semi-formally as system = architecture + variable content.[4]

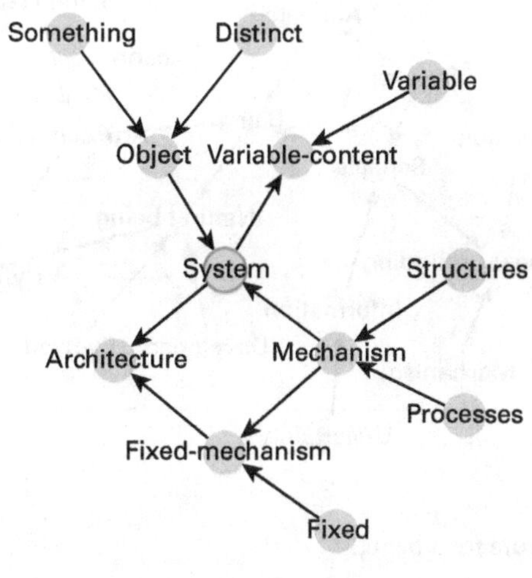

Figure 4.2
Dependency structure for a system.

5. The *architecture* of a system comprises its fixed mechanisms.
6. The *variable content* of a system comprises those aspects of the system that may change over time.

4.1.1.2 Being Given the notion of a system, we build up here to the concept of a being (figure 4.3), which can be either natural or artificial. Introduced in the process is one sense of how a being's internal mechanisms can be grounded in what they represent (Harnad, 1990).

7. An *animate* is a system capable of external action.
8. *Information* reduces uncertainty (Shannon, 1948).
9. A *sensation* is information that becomes available to a system.
   - This largely concerns external information but may also be internal.
10. A *sensate* is a system capable of processing sensations.
    - This captures only one aspect of the notion of a *sentient* that is defined more fully in section 4.11.
11. *Sensational grounding* ties the internal mechanisms of a sensate to the sensations that engender them.
    - This is one approach to the problem of symbol grounding; in particular, it can be considered as a *procedural* form, given that it centers on the processing that relates sensations to internal mechanisms. A second approach can be found in section 4.1.3.2.

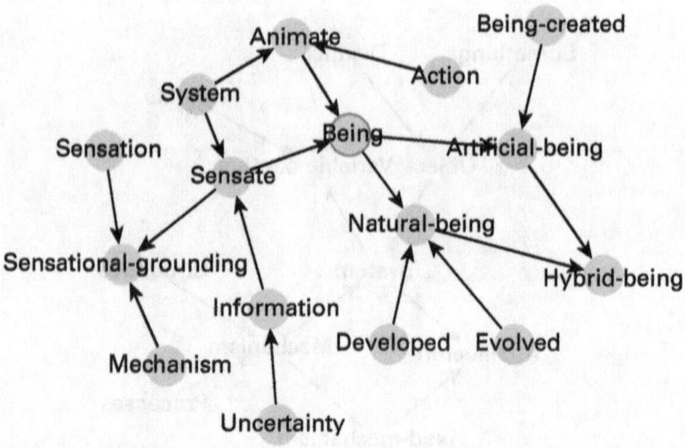

Figure 4.3
Dependency structure for a being.

12. A *being* is a system that is both animate and sensate.
    - Assumes some form of embodiment in being both animate and sensate.
    - The study of beings draws on disciplines such as biology and engineering, including more specialized areas such as robotics and virtual humans.
13. A *natural being* has evolved and develops.
    - Evolution can alter the architecture across a sequence of descendant beings.
    - Development can alter the architecture of an individual being over relatively long periods of time.
14. An *artificial being* has been constructed (not birthed) by a being (or possibly just an animate).
    - Section 4.3.1 expands a bit on this definition as a function of the kinds of beings that construct the new being.
15. A *hybrid being* mixes aspects of natural and artificial beings.

4.1.1.3 Agent Here the focus shifts up to agents (figure 4.4)—beings with minds that include goals to be achieved and memories. Brains are introduced as systems that control beings and implement their minds, and cognitive architectures as the architectures of these minds.

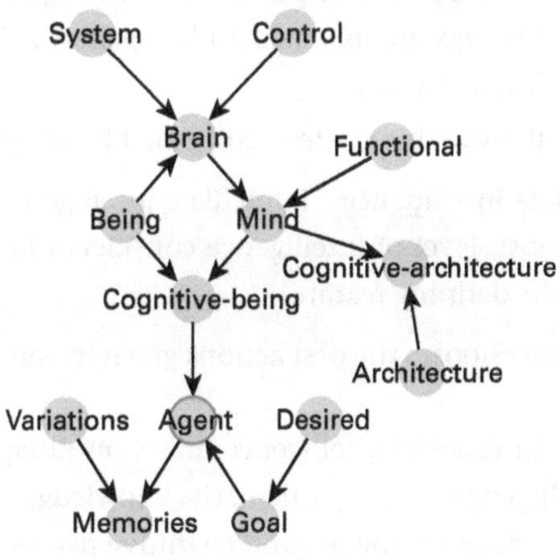

Figure 4.4
Dependency structure for an agent.

16. A *brain* is a system that controls a being.
    - This same terminology is intended to apply to both natural and artificial beings.
17. A *mind* is a functional system implemented by a brain.
    - In being functional, minds concern behavioral results that may abstract away from the physical structures and processes in the brain.
    - This implies a form of *monism* rather than *Cartesian dualism*, with the relation between the mind and brain being one of implementation and abstraction ("Mind-body problem," 2024).
18. A *cognitive being* is a being with a mind.
19. A *cognitive architecture* is the architecture of a mind.
    - This aligns with the definition of a cognitive architecture found at the beginning of chapter 2, as comprising the fixed structures and processes that yield a mind.
20. A *goal* is something desired.
    - This is intended to be a broad notion that includes achieving particular states, abiding by particular norms or constraints, meeting particular standards, maximizing particular rewards or objective functions, accomplishing particular missions, and so forth. It also includes any background motivations that lead to particular goals.
21. An *agent* is a goal-oriented—that is, *teleological*—cognitive being.
    - Goals may be explicit or implicit within the agent.
    - All natural beings are assumed to be agents, although many may only have implicit goals.
22. *Memories* are the variable content embodied in an agent.

4.1.1.4 Immediate intelligence Immediate intelligence (figure 4.5) is the lowest, or innermost, level of intelligence considered in this chapter, with rationality being its defining feature.

23. A *rational agent* chooses the best actions given its sensations, memories, and goals.
    - The study of rational intelligence draws on disciplines such as artificial intelligence—most particularly, knowledge representation and reasoning—decision theory, and cognitive psychology.
    - This definition assumes a knowledge-level perspective (Newell, 1981) that glosses over the reasoning processes that derive the information

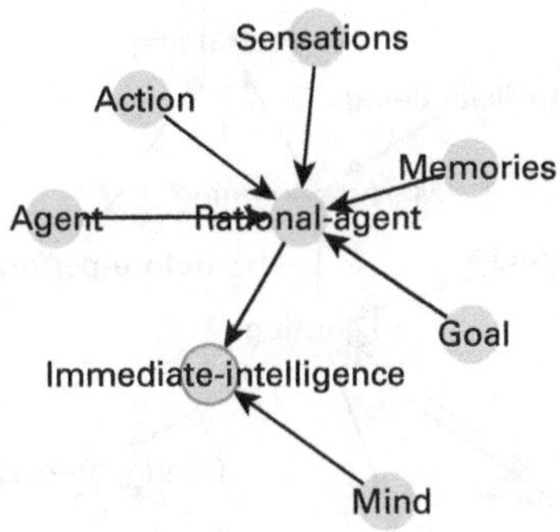

Figure 4.5
Dependency structure for immediate intelligence.

that directly bears on decision-making from sensations, memories, and goals.

24. *Immediate intelligence* is what is yielded by the mind of a rational agent.
    - May be realized at various levels of approximation across different agents.

4.1.2 Cumulative Intelligence

Cumulative intelligence (figure 4.6) adds pervasive learning to what is already provided by immediate intelligence. This goes significantly beyond the short-term rationality that underlies immediate intelligence but could potentially be considered as a long-term form of rationality that is concerned not just with the here and now but also with future decisions.

Machine learning—that is, learning in artificial systems—has been the dominant topic in AI for a number of years and, in the process, has made a strong case for the criticality of learning in achieving intelligence. This further highlights the need to distinguish cumulative intelligence from both immediate and full-spectrum intelligence, even though most forms of conventional machine learning yields only an approximation to pervasive learning due to its restriction to pretraining.

Still, with a focus on adaptation, whether short term in the form of rationality or long term in the form of learning, cumulative intelligence can be found at the heart of definitions of intelligence within both machine learning and AGI. For example, within machine learning Mnih et al. (2015) refer to the ability to "develop a wide range of competencies on a varied range of

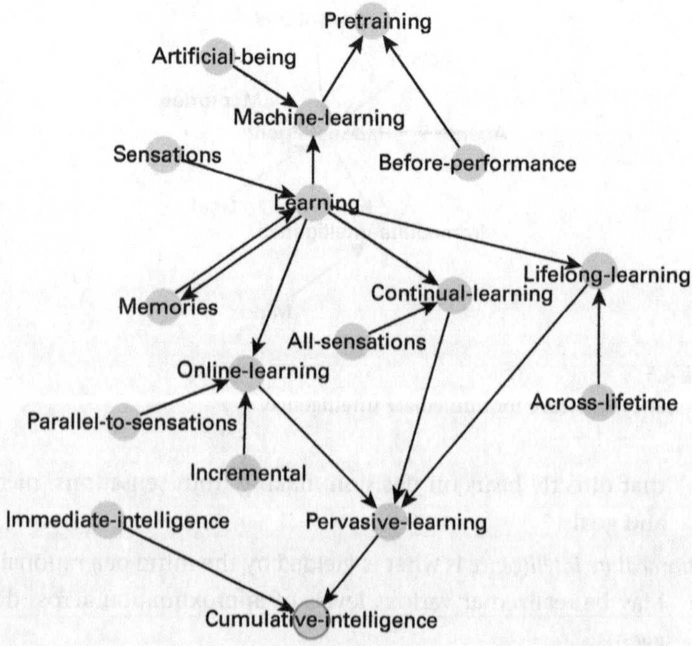

Figure 4.6
Dependency structure for cumulative intelligence.

challenging tasks" and within AGI, Wang (2019) refers to "adaptation with insufficient knowledge and resources." There are additional aspects here that will be more explicit under full-spectrum intelligence, but both definitions clearly center on cumulative intelligence. This centrality of cumulative intelligence, and thus also of a broad notion of adaptation, will reappear in section 4.1.4.4 when considering a simpler essence for intelligence.

25. *Learning* turns sensations into (long-term) memories.
    - May relate sensations to each other and to existing memories in the process.
    - May acquire or modify nonarchitectural mechanisms and and/or their parameters.
26. *Machine learning* is learning by an artificial being.
    - A machine in this sense can be seen as an artificial mechanism.
    - A sensate can in principle learn but the focus is on the more general case of a being here.

27. *Online learning* occurs incrementally, in parallel with the stream of sensations being received (and actions being taken).
28. *Continual learning* occurs for all available sensations.
29. *Lifelong learning* occurs across the entire lifetime of the being.
30. *Pervasive learning* is online, continual, and lifelong.
    - Most modern cognitive architectures provide some reasonable approximation to this.
31. *Pretraining* completes all learning before "real" performance begins.
    - This is how most modern machine learning proceeds—except for reinforcement learning (Sutton & Barto, 2018)—but it provides only an approximation to true pervasive learning.
32. *Cumulative intelligence* comprises immediate intelligence plus pervasive learning.
    - May be realized at various levels of approximation across different beings.
    - Draws on disciplines such as (machine) learning, reinforcement learning, optimization, cognitive architecture, and (cognitive) neuroscience.
    - "Reward is enough" (Silver et al., 2021) implies reinforcement learning is sufficient for intelligence. Such approaches as universal AI (Hutter, 2005) and deep reinforcement learning (Francois-Lavet et al., 2018) imply that an additional form of machine learning, in the form of induction, is also needed.

4.1.3 Full-Spectrum Intelligence

Full-spectrum intelligence starts with cumulative intelligence, and thus also immediate intelligence, but then adds to it a list of other aspects/capabilities presumed to be relevant for achieving the kind of intelligence found in humans (see section 4.2.1 for more specifically on human intelligence). Such a list is inherently ad hoc to some degree, but the version here is a "best attempt" based on contemplating previously published lists from researchers in cognitive architecture (Newell, 1980; Newell, 1990; Anderson & Lebiere, 2003; Langley et al., 2009; Kotseruba & Tsotsos, 2020) and artificial general intelligence (AGI; Goertzel, 2014). Cognitive architecture studies the architecture of mind(s), whereas AGI studies how to (learn to) perform (at least) any task performable by humans. Both areas are thus deeply concerned with the full spectrum of what is necessary for intelligent behavior.

Also included in the deliberations has been recent evidence concerning: the capabilities and limitations of state-of-the-art neural networks, particularly those that learn to play games (e.g., Silver et al., 2018) and large language models (e.g., Brown et al., 2020); plus, a definition of intelligence I developed some years ago while teaching introductory courses on AI—"the common underlying capabilities that enable a system to be general, literate, rational, autonomous and collaborative." One final source considered has been a pair of lists due to Gary Marcus and coauthors concerning what is necessary for trustworthy AI (Marcus & Davis, 2019; Lenat & Marcus, 2023). Although not as directly on point, there are interesting overlaps between these two lists and what appear in the other sources considered.

The overall result focuses on agents that are not only cumulative—and thus immediate as well—but also situated, aware, projective, affective, purposeful, linguistic, collaborative, socialized, and educated. Although an extensive list, it may very well still be incomplete in some ways, and possibly incorrect in other ways. Wang (2019) would likely consider both immediate and cumulative intelligence as *principled definitions* that are based on "fundamental principles that can uniformly explain the relevant phenomena," whereas full-spectrum intelligence would more likely be considered a *functional definition* that is based on "cognitive functions identified in the human mind."

4.1.3.1 Situated agent There has been controversy over the years in AI and cognitive science as to whether intelligence can be studied in the abstract or whether it is inherently tied to the context in which an agent is situated (e.g., Brown et al., 1989). Either way, when an agent is actually situated in an environment, the latter does constrain the former. The definition of a situated agent (figure 4.7) here does not include all aspects mentioned in the literature on situatedness, such as culture, but it is intended to cover the most essential aspect, with culture being considered separately, as part of the discussion of collaborative agents in section 4.1.3.8.

33. A *situated agent* can behave appropriately with respect to its goals in an environment that imposes structure and constraints on sensation and animation.
    - This may involve spatial and temporal considerations in physical environments, multiagent considerations in social environments, and a broad spectrum of additional possibilities in these and other environments.

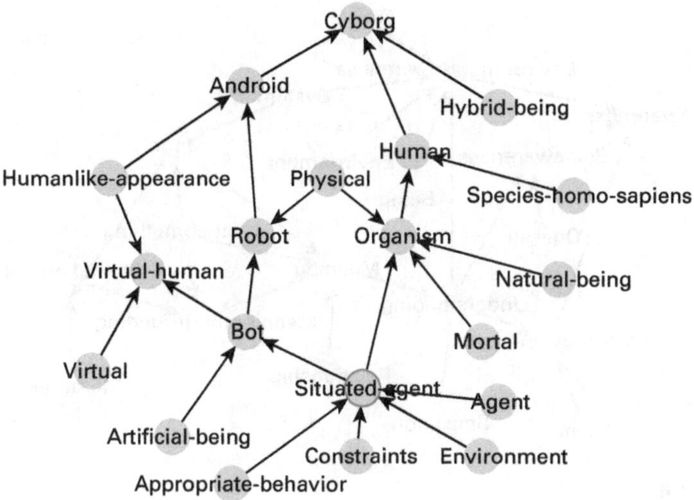

Figure 4.7
Dependency structure for a situated agent.

34. A *bot* is an artificial situated agent.
35. A *virtual human* is a virtual bot with a humanlike appearance.
36. A *robot* is a physical bot.
37. An *android* is a robot with a humanlike appearance.
38. An *organism* is a natural, mortal, physically situated agent.
    - Although organisms can also participate in virtual environments through appropriate interfaces.
39. A *human* is an organism that is a member of the species *Homo sapiens*.
40. A *cyborg* is a hybrid being that is part human and part android.

4.1.3.2 Awareness The definitions here attempt to walk a fine line of addressing important aspects of tricky topics such as understanding, awareness (figure 4.8), and self-awareness while barely touching on the even trickier topic of consciousness, which will be mentioned but without any serious attempt at a definition.

41. A *model* is a system that is about something.
    - This can involve one or more mechanisms. It can also potentially involve just fragments of mechanisms, such as structures without processes.

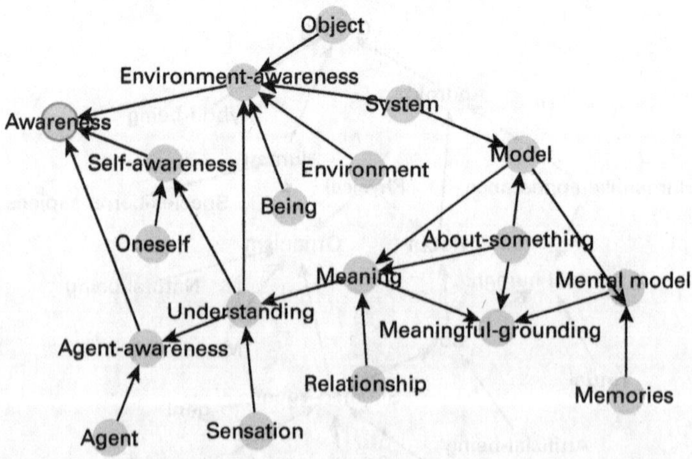

Figure 4.8
Dependency structure for awareness.

- Models relate to both the notion of designation found in chapter 3 and the notion of a theory that is at the heart of chapter 5.

42. A *mental model* is a model in an agent's memory.
    - This inherently brings up questions of representation, as has long been a focus of study in AI, but here both the nature of what is being represented and how it is represented are glossed over.
43. *Meaning* is the relationship between a model and what it models.
    - This glosses over many of the distinctions and complexities found in the study of semantics ("Semantics," 2024), focusing more abstractly instead on this specific relationship.
    - This is a notion of individual meaning rather than meaning that is shared within an aggregation (section 4.1.3.8).
44. *Meaningful grounding* ties mental models to what they are about.
    - This goes beyond sensational grounding but almost necessarily arises from it.
    - It can also be considered as providing a *declarative* form of grounding that complements the procedural form yielded by sensational grounding.
45. *Understanding* is a sensation of meaning.[5]
    - Requires sensations of mental models, of what they model, and of the relationship between them.

- This notion of understanding is inherently metacognitive/reflective while inching into aspects of consciousness.

46. *Self-awareness* is understanding of oneself.
    - The ability to sense oneself, have a mental model of oneself that can be sensed, and sense the relationship between oneself and the model of oneself. Notice that this goes significantly beyond just sensing oneself.
    - This need not imply full self-awareness, in which one can understand everything about oneself, but it does inch further into aspects of consciousness by implying some minimal level of self-awareness.
47. *Environment-awareness* is understanding of objects, systems, and beings in the environment.
48. *Agent-awareness* is understanding of other agents.
    - This includes *"Theory of mind"* (2025) for understanding the minds of other agents.
49. *Awareness* is a combination of self-awareness, environment-awareness, and other-awareness.

4.1.3.3 Projective agent A projective agent (figure 4.9) is an agent that can project—a basic capability that underlies anticipating possible futures, as found, for example, in search, problem-solving, and planning. It is broader than simply predicting what is to occur in also including hypotheticals. Projection can be crucial in appropriately responding to unexpected situations. In enabling hypothesizing what does not already exist it is also crucial to creativity.

50. *Projection* uses memories and sensations to generate and sense potential futures.
51. A *projective agent* is capable of projection.

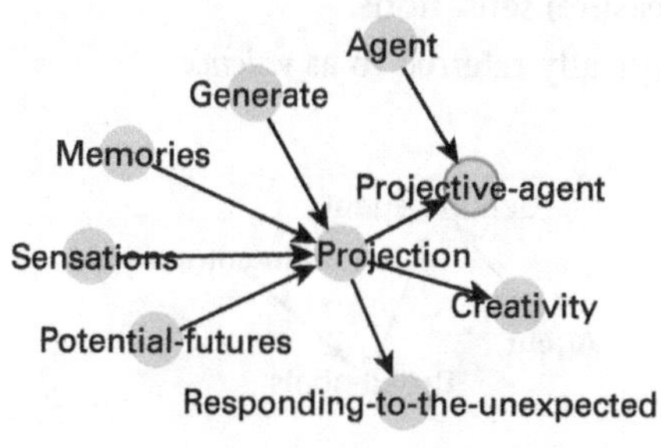

Figure 4.9
Dependency structure for a projective agent.

This is akin to, but distinct from, the notion of the *predictive mind* (Hohwy, 2013). Both do involve a form of prediction. However, a projective agent attempts to predict possible futures by mentally executing sequences of actions, whereas the predictive mind attempts to predict what is currently being perceived based on what has previously been learned. As such, the predictive mind is more a way of thinking about what learning is to achieve—that is, learning is about improving these predictions—than a way of predicting possible futures. Several extant definitions of intelligence can be seen as focusing directly on projection (Bigge, 1976; Das, 1986).

4.1.3.4 General agent The notion of a general agent (figure 4.10) is at the heart of artificial general intelligence and is a key driver of work on cognitive architectures.

52. A *general agent* can work appropriately on a broad range of goals across a broad range of environments.
    - This is typically conceived of as being comparable to the set of goals on which a human can work across the environments that they might face.

4.1.3.5 Affective sensate An affective sensate (figure 4.11) has sensations that truly matter to them. This relates to both feelings and emotional states, although these latter are not in direct focus here. Such sensations seem particularly relevant to constructing an ethical framework that is based on aspects of beings that are more elemental than their species membership (section 4.2). As discussed later, they also turn out to be relevant to effective performance.

53. A *feeling* is an internal sensation that cannot easily be ignored in deciding what to do.
    - At a minimum, distinguishes between negative (e.g., pain) and positive (e.g. pleasure) sensations.
        - This is typically referred to as *valence*.

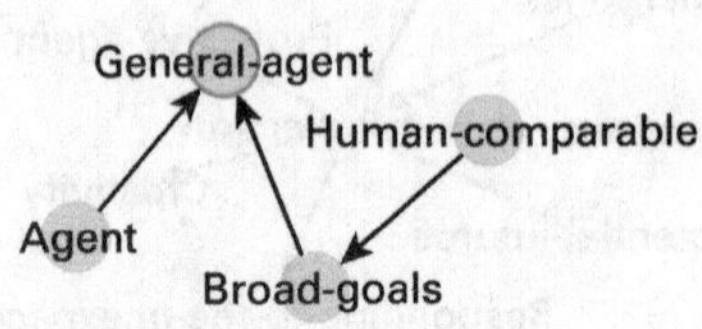

Figure 4.10
Dependency structure for a general agent.

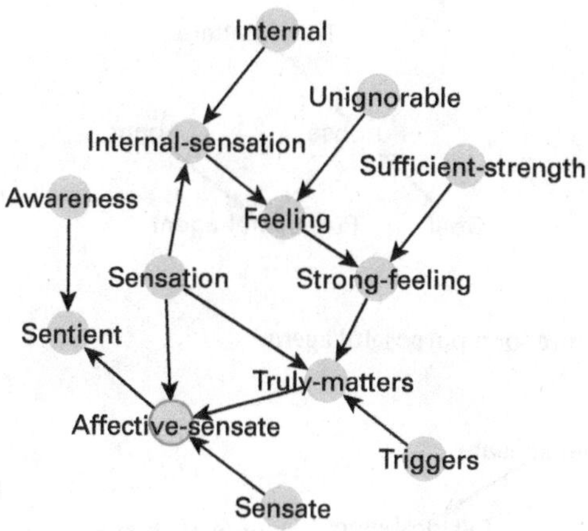

Figure 4.11
Dependency structure for an affective sensate.

- In natural sensates, the existence of feelings can be considered as part of the *wisdom of evolution* that encourages, even when too coarse grained or out of date, what has proven useful for long-term survival of the species (Rosenbloom et al., 2015; Larue et al., 2018).

54. A sensation *truly matters* to a sensate if it triggers a feeling of sufficient strength.
    - The notion of whether something truly matters—or, more particularly, what it means to truly matter—is one I have struggled with for a while. I'm not sure this is ultimately the right answer, but it is the best I have been able to come up with at this point.
55. An *affective sensate* is a sensate with sensations that truly matter.
56. A *sentient* is an affective sensate that is aware of its sensations that truly matter.
    - Like consciousness, there are multiple quite different ways to define sentience, yielding a tricky and controversial notion, both in general and in the context of AI systems (e.g., Brodkin, 2022). This definition provides one take that requires both awareness of sensations and that the sensations themselves truly matter.

4.1.3.6 Purposeful agent A purposeful agent (figure 4.12) has purposes that go beyond goals to aims that truly matter. They are akin to feelings in this way and are similarly an essential component of a deeper ethical framework (section 4.2).

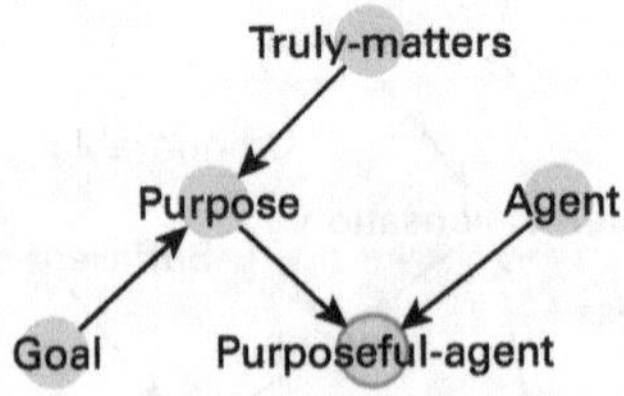

Figure 4.12
Dependency structure for a purposeful agent.

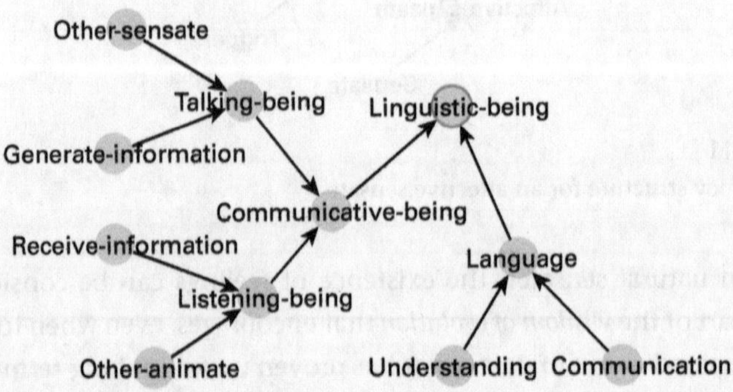

Figure 4.13
Dependency structure for a linguistic being.

57. A *purpose* is a goal for which progress and achievement, and their lack, yield sensations that truly matter to an agent.
58. A *purposeful agent* is an agent that has purposes.

4.1.3.7 Linguistic being A linguistic being (figure 4.13) uses language to communicate with others, but the notion goes considerably beyond just the ability to communicate.

59. A *listening being* can receive information generated by an animate.
60. A *talking being* can generate information for a sensate.
61. A *communicative being* is both a listening being and a talking being.
62. *Language* is communication that is understood.
    - This is a very broad definition in the sense that it potentially includes communication among any types of beings, not just humans.
    - This is also a very narrow definition in the sense that it requires understanding of what is being communicated.

- Computer "languages" may be understood in the sense specified in section 4.1.3.2 by human programmers but are not typically understood in this sense by the computers being programmed. Perhaps these should be thought of as *semi-languages*.

63. A *linguistic being* is a communicative being that uses language.
    - Large language models are communicative beings but not linguistic beings, as they do not understand their communications.

4.1.3.8 Collaborative agent A collaborative agent (figure 4.14) shares goals with others in an organization. The definitions here start with the notion of an aggregation of beings and of a culture as its architecture. It then works up from there.

64. An *aggregation* is a collection of beings.
65. A *culture* is the architecture of an aggregation.
    - Shared meaning may be part of a culture.
    - Culture may develop over the history of the aggregation, as development may slowly change a being's architecture (section 4.1.1.2).
66. An *organization* is an aggregation of agents that is itself an agent.
67. A *human organization* is an organization of humans and is typically a hybrid agent.

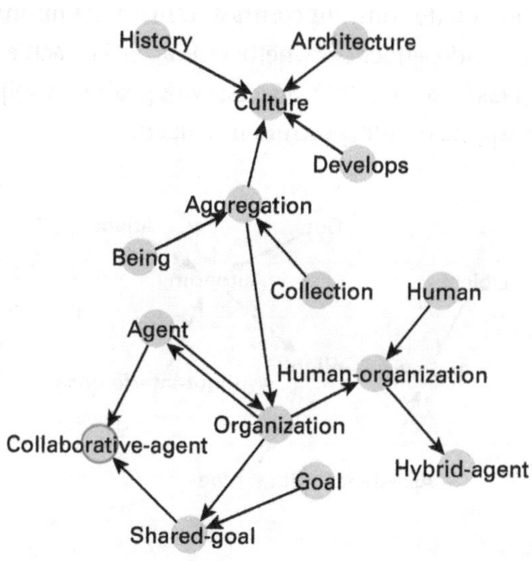

Figure 4.14
Dependency structure for a collaborative agent.

- While humans are natural agents, the organization itself is created by humans and thus is not completely natural.

68. A *shared goal* is a goal of an organization.
69. A *collaborative agent* is an agent that has a shared goal.

4.1.3.9 Socialized agent A socialized agent is aware of its situation within a society and behaves appropriately within such a context. In a sense, a socialized agent can be considered simply as an aware, situated agent within a social environment.

We will build up to the notion of a society through the introduction of several intermediary concepts, such as rights, responsibilities, autonomy, and authority. Due to the overall complexity of the dependency graph here, it is presented in two parts, with the first focused just on the notion of a right (figure 4.15) and the second on the notion of a socialized agent (figure 4.16). Further discussion of rights and responsibilities can be found in section 4.3.4.8, as part of the broader discussion of ethics in section 4.3.4.

70. A *benefit* provides external assistance to an object.
    - Benefits may, for example, aid the existence of an object, a feeling of an affective sensate, or a goal of an agent.
71. *Autonomy* is the ability of an agent to choose its own goals and to pursue them without outside interference.
    - This notion of autonomy, in contrast to many traditional uses of the term in AI, is independent of whether the agent can achieve these goals on its own (as in section 7.7). Success with goals may require many of the other aspects of full-spectrum intelligence.

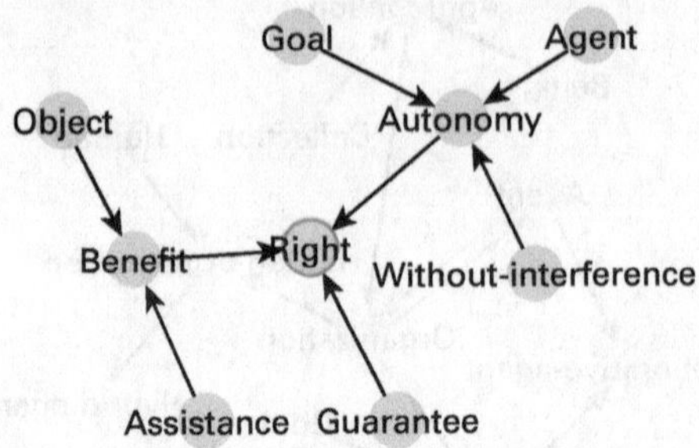

Figure 4.15
Dependency structure for a right.

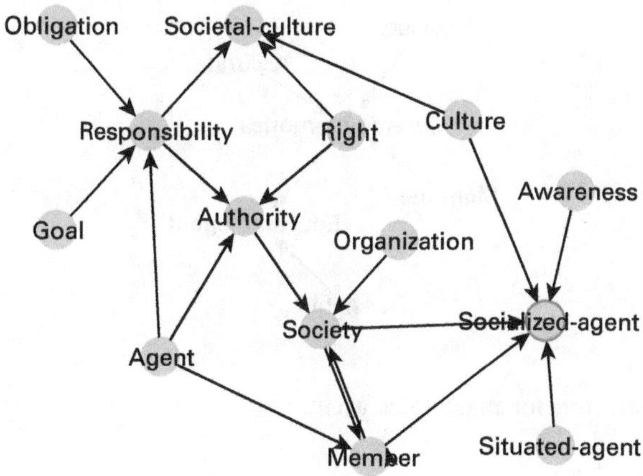

Figure 4.16
Dependency structure for a socialized agent.

72. A *right* is a guarantee of a benefit or of autonomy.
    - Individual rights may require trade-offs when they conflict with other rights.
73. A *responsibility* is an obligation by an agent to a goal.
74. An *authority* is an agent that can determine rights/responsibilities for objects/agents.
75. A *society* is an organization with authority over its members.
76. A *member* is an agent within a society over which the society has authority.
77. A *societal culture* includes the rights and responsibilities associated with membership in a society.
78. A *socialized agent* is a situated member of a society that has awareness of the society's culture and acts appropriately within this culture.

4.1.3.10 Educated agent In contrast to learning, which focuses on how sensations are turned into memories, education focuses on how what society as a whole has learned—and curated in the sense of having reached a consensus on its validity—is reflected in an agent. It may be learned by the agent in some manner, or it could potentially be embedded into an agent via other means, such as evolution, development, or programming. Either way, the result should be an educated agent (figure 4.17) with curated memories.

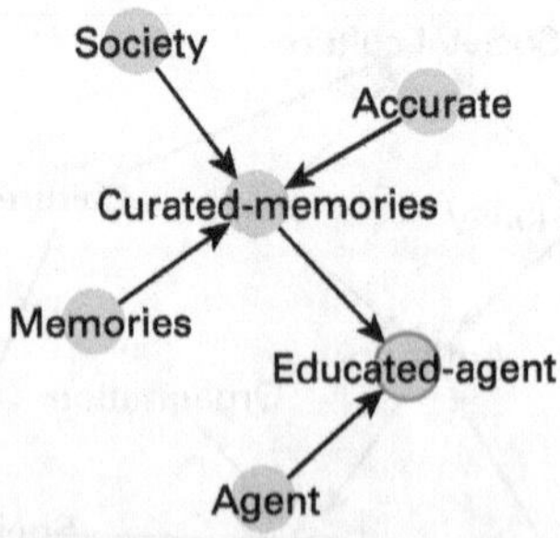

Figure 4.17
Dependency structure for an educated agent.

These memories need not just correspond to the kinds of factual knowledge that would be stored in the declarative memory of the common model of cognition (section 1.2.3). It can also include skills that would be stored in procedural memory, particularly ones concerning reasoning, judgment, and so on ("Education," 2024).

There is a conventional—and essentially correct—notion that ignorance is distinct from lack of intelligence, but the lack of education on the part of generative AI systems highlights how much education matters to overall judgments of an agent's intelligence. From the perspective here, ignorance is distinct from a lack of immediate intelligence but not from a lack of full-spectrum intelligence. Cumulative intelligence emphasizes that, as an aspect of intelligence, this really may be more about the ability to acquire and use curated knowledge rather than simply possessing it.

79. *Curated memories* have been determined by a society to be accurate.

80. An *educated agent* has curated memories.

4.1.3.11 Full-spectrum intelligence Full-spectrum intelligence (figure 4.18) combines cumulative intelligence with all of the capabilities so far discussed in this section.

81. A *full-spectrum agent* has cumulative intelligence (goals, memories, rationality, and pervasive learning) and is situated, aware, projective, general, affective, purposeful, linguistic, collaborative, socialized, and educated.
    - Looks to cognitive, social and biological—particularly neural—sciences, as well as artificial (general) intelligence, for inspiration.
    - May be realized at various levels of approximation.

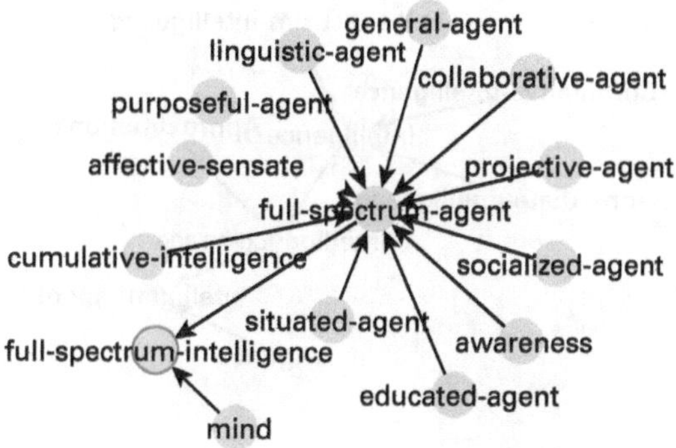

Figure 4.18
Dependency structure for full-spectrum intelligence.

Many such lists would include autonomy as a key aspect of intelligence. However, as defined in section 4.1.3.9, autonomy here concerns the freedom to work on goals of an agent's own choosing rather than the ability to successfully achieve these goals on its own. This latter more traditional notion of autonomy, as seen as an evaluation criterion for technologies and systems in section 7.7, depends on many of the other aspects of full-spectrum intelligence rather than being its own independent aspect.

82. *Full-spectrum intelligence* is what is yielded by the mind of a full-spectrum agent.

4.1.4 Intelligence

Here the notion is that there is not a single idea that existing definitions of intelligence are striving to capture. Instead, a good part of the cacophony among these definitions arises from whether they implicitly concern immediate, cumulative, or full-spectrum intelligence. The alternative followed here is to consider intelligence as a whole disjunctively, as concerning any one of these three, but with them related via nesting of earlier forms within later ones.

Beyond this, a range of approximations to these three levels induces a broad space of possibilities for intelligence. The analysis here starts with a definition of intelligence and several related concepts (section 4.1.4.1) before following up with its relationship to a sampling of existing definitions of intelligence (section 4.1.4.2) and to theories concerning multiple

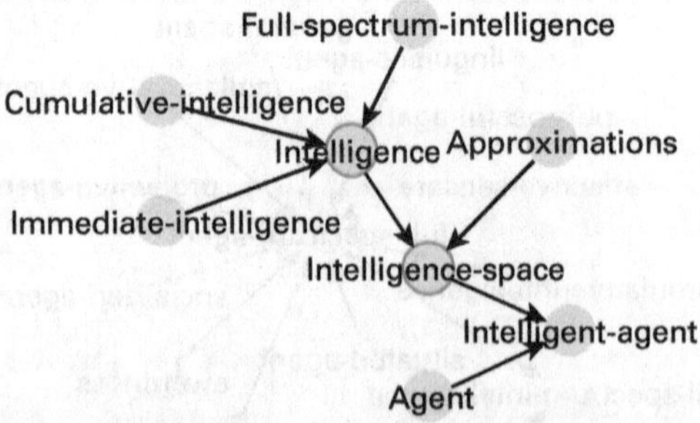

Figure 4.19
Dependency structure for intelligence and the intelligence space.

intelligences (section 4.1.4.3). Section 4.1.4.4 wraps up with speculations concerning a simpler, non-disjunctive essence of intelligence.

4.1.4.1 Defining intelligence and the intelligence space

83. *Intelligence* can be immediate, cumulative, or full spectrum.
84. The *intelligence space* comprises the three levels in the definition of intelligence plus a space of approximations to these levels.
    - This leaves open the possibility of a wide range of different forms of intelligence. However, since the definition of full-spectrum intelligence is highly influenced by our current understanding of human intelligence, as well as what is now known about artificial intelligence, this space may ultimately be viewed as skewed with respect to the space of all possible forms of intelligence.
85. An *intelligent agent* is an agent with a mind that sits somewhere in the intelligence space.
    - However, the approximation must not be so extreme that what is left is too minimal.

4.1.4.2 Relationship to existing definitions of intelligence In this section the relationship of existing definitions of intelligence to the one proposed here is considered, focused on how the former stratify across the three levels that are at the core of the latter. We can start with the first paragraph in Wikipedia on "Intelligence" (2024):

> Intelligence has been defined in many ways: the capacity for abstraction, logic, understanding, self-awareness, learning, emotional knowledge, reasoning, planning, creativity, critical thinking, and problem-solving. More generally, it can

be described as the ability to perceive or infer information, and to retain it as knowledge to be applied towards adaptive behaviors within an environment or context.

The first sentence approximates full-spectrum intelligence: abstraction, logic, and critical thinking can all be seen as aspects of the reasoning that is glossed over in defining immediate intelligence (section 4.1.1.4); understanding and self-awareness appear as part of awareness in full-spectrum intelligence (section 4.1.3.2); and planning, creativity, and problem-solving all relate to projection in full-spectrum intelligence (section 4.1.3.3), but also less directly to other aspects of it. The second sentence maps onto cumulative intelligence, although it does also touch on situatedness.

Examples of traditional definitions aimed directly at immediate intelligence include:

- Intelligence comprises the ability to act rationally; that is, "does the 'right thing,' given what it knows" (Russell & Norvig, 2010).
- Intelligence is "computational rationality: computing with representations, algorithms, and architectures designed to approximate decisions with the highest expected utility, while taking into account the costs of computation" (Gershman et al., 2015).
- "A system is intelligent to the degree that it approximates a knowledge-level system" (Newell, 1990).
- "Intelligence measures an agent's ability to achieve goals in a wide range of environments" (Legg & Hutter, 2006).

The first two definitions here focus strongly on rationality, although the second also touches on self-awareness with respect to the cost of internal computations. The third definition is missing from Legg's (2023) list and does not explicitly mention rationality, but it is based on the association between rational agents and a knowledge-level perspective that is mentioned in section 4.1.1.4. Like the second definition, the fourth goes a step beyond rationality, in this case to explicitly include the breadth of environments in which goals may be achieved. Given that this is a definition out of the world of AGI, this is not a surprising addition. More on this though can be found in the discussion of a simpler essence of intelligence in section 4.1.4.4.

Several recent attempts at defining AI by (inter)governmental organizations also appear to focus on immediate intelligence. Even though their concern is with regulating deep learning, or at least systems that result from deep learning, the definitions themselves don't include learning, except for one ambiguous reference mentioned below, perhaps because the pretraining that populates these systems with variable content is not considered part of the system itself.

For example, the US Federal Trade Commission says that "AI includes, but is not limited to, machine-based systems that can, for a set of defined objectives, make predictions, recommendations, or decisions influencing real or virtual environments. Generative AI can be used to generate synthetic content including images, videos, audio, text, and other digital content that appear to be created by humans" (FTC, 2023). Similarly, the Organisation for Economic Co-operation and Development says that "an AI system is a machine-based system that, for explicit or implicit objectives, infers, from the input it receives, how to generate outputs such as predictions, content, recommendations, or decisions that can influence physical or virtual environments. Different AI systems vary in their levels of autonomy and adaptiveness after deployment" (OECD, 2023). However, as with the Wikipedia definition, this last definition may implicate cumulative intelligence if adaptiveness is intended to refer to some form of learning rather than just flexible behavior.

Two example definitions beyond the second sentence in the Wikipedia paragraph that are aimed directly at cumulative intelligence, as quoted from Legg's webpage, are:

- "The capacity to acquire and apply knowledge" (AHDEL, 2000).
- "The capacity to learn or to profit by experience" (W. F. Dearborn, as quoted in Sternberg, 2000).

Immediate intelligence, as an aspect of cumulative intelligence, is largely implicit in these definitions, but it is implied by phrases such as "apply knowledge" and "profit by experience."

The definition by Mnih et al. (2015) in section 4.1.2 is also aimed at cumulative intelligence, although, like Legg and Hutter, it also explicitly includes generality. The earlier definition by Wang (2019) also focuses on cumulative intelligence but, as with the second sentence in the Wikipedia paragraph, with the addition of situatedness. This can be seen in its concern for what is lacking with respect to environmental exigencies. It also implicitly includes projection as part of a definition of adaptation that goes beyond cumulative intelligence.

A definition of intelligence endorsed by fifty-two experts in the field—"Intelligence is a very general mental capability that, among other things, involves the ability to reason, plan, solve problems, think abstractly, comprehend complex ideas, learn quickly and learn from experience" (Gottfredson, 1997)—can also be seen as a variant of cumulative intelligence, with the implicit addition of projection, even if in form it looks to be more aligned with full-spectrum intelligence. Although none of these definitions that align with cumulative intelligence directly focus on pervasive learning, they clearly emphasize the importance of learning from experience and thus assume at least some approximation to it.

Beyond the first part of the Wikipedia definition, the Newell test (Anderson & Lebiere, 2003) provides a set of functional criteria for evaluating an architectural theory of human cognition that can be viewed as implying a variant of human-level/like, full-spectrum intelligence. It is analyzed in more detail in section 4.2.1, on human intelligence. The Turing test (Turing, 1950) also explicitly tests whether an agent's behavior is indistinguishable from that of a human, and thus implicitly tests whether it shares many, if not all, of the externally discernible aspects of human-level/like, full-spectrum intelligence.

This mapping of a sampling of existing definitions of intelligence onto the three levels of the hierarchy provides prima facie evidence for their relevance, as does the analysis of multiple intelligences in the next section, along with the work already mentioned as presaging the first and third levels, and the centrality at this point of machine learning—the core of the second level—to the field of AI. However, this need not imply that there may not still be other points within the overall intelligence space defined here that could usefully be called out explicitly, whether or not they track the nested hierarchy introduced here or are even part of a larger intelligence space not yet identified.

4.1.4.3 Relationship to theories of multiple intelligences Any definition of intelligence that embodies multiple aspects, such as the trilevel conceptualization of intelligence as a whole or its third level of full-spectrum intelligence in particular, can be considered an implicit theory of multiple intelligences. Beyond this, though, a significant body of research explicitly considers multiple forms of intelligence as a counter to the assumption behind IQ tests that there is a single general intelligence factor—denoted *g*—that can be measured by such tests. Here we examine three such theories through the lens of the trilevel definition, exploring how each theory can be seen as mapping onto a combination of aspects that make up this definition. The intent is to extend how the trilevel definition subsumes multiple existing definitions of intelligence to subsuming existing definitions of multiple intelligences.

Gardner's (1983) theory of multiple intelligences includes *musical*, *visual-spatial*, *linguistic*, *logical-mathematical*, *bodily-kinesthetic*, *interpersonal*, and *intrapersonal*, although even more were later added to this list. The first three may best be viewed as reflecting agents that are situated and educated with respect to particular kinds of environments, such as physical environments for visual-spatial and bodily-kinesthetic intelligence. The remaining three then combine these two aspects with mappings onto other aspects of the trilevel definition, with linguistic intelligence mapping onto linguistic beings, logical-mathematical intelligence mapping onto immediate intelligence, interpersonal intelligence mapping onto collaborative, socialized agents, and intrapersonal intelligence mapping onto self-awareness.

Sternberg's (1985) *triarchic theory* of human intelligence includes *practical*, *creative*, and *analytical* intelligence. As viewed here, practical intelligence emphasizes the contextual behavior found in situated agents; creative intelligence emphasizes the ability to come up with new ideas that is found in projective agents; and analytical intelligence emphasizes the ability to reason, understand, and achieve goals that is found in immediate intelligence plus projection and awareness. Thus, overall, we see immediate intelligence, plus three aspects of full-spectrum intelligence: situated, projective, and aware. We can also see a potential overlap between what are assumed to be distinct forms of intelligence in this theory, with projection being critical to both creative and analytical intelligence.

Cattell partitions *g* into *fluid* versus *crystallized* intelligence ("Fluid and crystallized intelligence," 2024). Fluid intelligence is the "ability to solve novel reasoning problems," which maps roughly onto a combination of immediate intelligence and projection, but which may also include aspects of awareness. Crystalline intelligence involves "applying previously learned primary relational abstractions," which maps onto at least an approximation of cumulative intelligence. As with Sternberg's theory, there again appears to be an overlap here, with both forms of intelligence Cattell identifies requiring immediate intelligence.

4.1.4.4 Is there a simpler essence of intelligence? The definition of intelligence proposed here is moderately disjunctive in having three levels of hierarchy and significantly conjunctive in the aspects of full-spectrum intelligence. Together these reduce the simplicity and elegance of the definition, raising the question as to whether there is a more minimal essence that might do as well, or at least nearly as well. If so, the notion of *adaptivity*, as found in several of the existing definitions, seems likely to be at the core of this. As an everyday word, its inherent ambiguity can capture short-term forms that are at the heart of rationality and goal-orientation, as well as long-term forms that implicate learning. It thus can succinctly capture the core aspects of both immediate and cumulative intelligence, in effect thus amounting to cumulative intelligence. As such, it could be considered as the essential definition of intelligence.

What then is the status of the other aspects of full-spectrum intelligence with respect to this possible essence? Should they be seen as subsumed by it, optional with respect to it, or necessary additions to it—that is, effectively extending it to include them? For instance, as mentioned in section 4.1.4.2, Wang (2019) has already proposed that adaptivity subsume projection; and Cattell's fluid intelligence was just described as combining immediate intelligence and projection. This is similar to how Newell's (1990) definition of intelligence in terms of approximation to the knowledge level abstracts over forms of symbol-level processing, such as projection, that are necessary for knowledge to be fully exploited in service of the goals of interest. Awareness and situatedness could arguably also be viewed in a similar manner.

Legg and Hutter (2006) instead found it necessary to explicitly mention generality in their otherwise minimal definition of intelligence, as "achieve goals in a wide range of environments." However, if generality is definitional for intelligence, then the phrase "general intelligence" becomes redundant and the phrase "narrow intelligence" becomes meaningless. Legg and Hutter (2007) effectively acknowledge this by reframing this as a definition of *universal intelligence*. Thus, it would seem that generality should remain optional even while remaining a critical aspect of research into intelligent systems in general and the architectures that support them.

What remains from full-spectrum intelligence is being affective, purposeful, linguistic, collaborative, socialized, and educated. Affective and purposeful are evolutionary adaptations that become hard-wired into natural beings because they are beyond the dynamic adaptations that are possible when they are actually needed.[6] More generally, they would thus seem to be preadaptations that are necessary for successful adaptivity, and it could thus be argued that intelligence requires not only adaptivity but also preadaptations. Linguistic, collaborative, and socialized are all relevant to social environments in particular and thus demand a form of generality that at least includes such environments. Like affective and purposeful, they may also require some preadaptation before these environments are ever experienced. Educated is a matter of having an appropriate sequence of experiences in the presence of a sufficient learning capability. It is thus closely related to adaptivity, but it also requires exposure to curated experiences.

Full-spectrum intelligence might therefore largely be considered as being subsumed by a suitably broad notion of adaptivity, although with the addition of necessary preadaptations and curated experiences, plus an option for generality. The full trilevel definition of intelligence is disjunctive to align with the reality that there does not appear to be a single notion of intelligence that people have in mind when they use, or try to define, the term; and hopefully it clarifies how these existing notions stratify. The essential definition introduced here instead provides a more normative notion that builds on previous minimal/principled definitions to capture much, but not all, of what is found across these three levels.

Despite the long-term potential of this more minimal approach to an essential definition of intelligence, this section should be considered as a speculative aside with respect to the remainder of this chapter, which remains focused on the more complex essence provided by the trilevel definition.

4.2 Natural and Artificial Intelligence

This section maps multiple variations on minds, along with the disciplines that study them, onto regions of the intelligence space. It starts with a form of natural intelligence—human intelligence—as an approximation to full-spectrum intelligence (section 4.2.1) before widening the focus a bit to consider humanlike intelligence (section 4.2.2). Artificial intelligence is then considered as both a form of intelligence that is distinct from natural intelligence and as a field of study (section 4.2.3), before wrapping up with consideration of cognitive science as an interdisciplinary field that spans aspects of both natural and artificial intelligence (section 4.2.4).

4.2.1 Human Intelligence

Human intelligence is defined here as a form of natural intelligence that arises from human minds, as implemented by human brains (figure 4.20). This then enables defining the notion of a human cognitive architecture and a more detailed mapping of the Newell test for such architectures onto full-spectrum intelligence (figure 4.21).

86. A *natural brain* is the brain of a natural being.
87. A *natural mind* is a mind implemented by a natural brain.
88. *Natural intelligence* approximates intelligence via a natural mind.
89. A *human brain* is the natural brain of a human.
90. A *human mind* is a natural mind implemented by a human brain.
91. *Human intelligence* is a form of full-spectrum natural intelligence yielded by a human mind.
92. A *human cognitive architecture* is the architecture of a human mind.

The Newell test comprises twelve criteria for a human cognitive architecture distilled down from two earlier lists (Newell, 1980; Newell, 1990): (1) flexible behavior, (2) real-time performance, (3) adaptive behavior, (4) vast knowledge base, (5) dynamic behavior, (6) knowledge integration, (7) natural language, (8) consciousness, (9) learning, (10) development, (11) evolution, and (12) (human) brain. The question of interest here is how these criteria relate to the trilevel definition of intelligence introduced in section 4.1, and thus

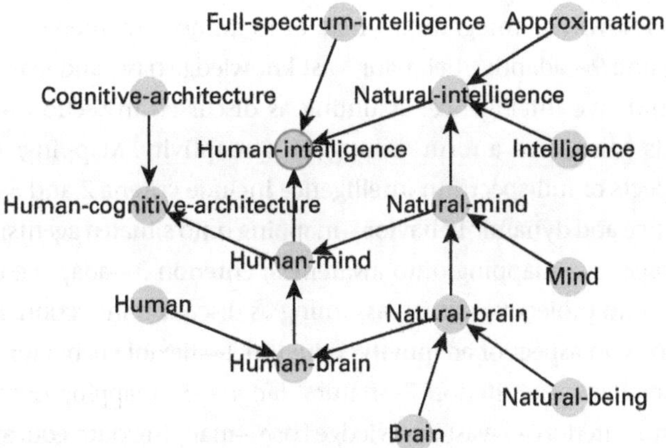

Figure 4.20
Dependency structure for human intelligence.

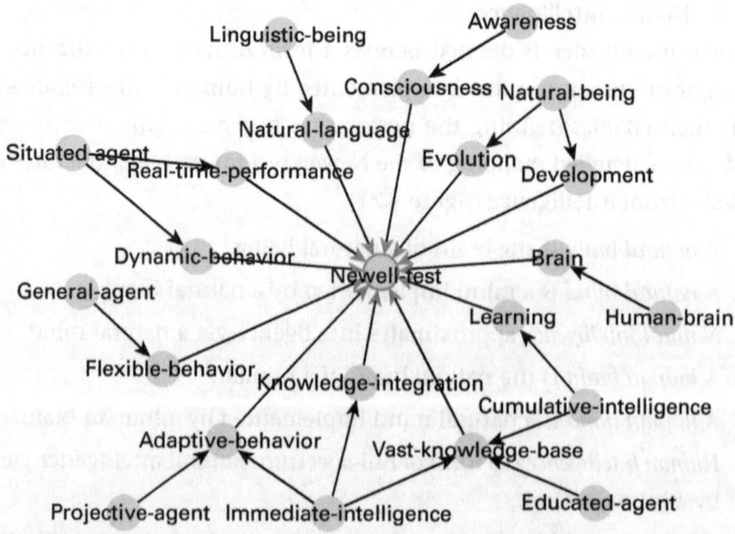

Figure 4.21
Dependency structure for the Newell test.

how to understand the Newell test in terms of this definition. Given that the Newell test was part of the input in developing the notion of full-spectrum intelligence, a close relationship should be expected, but a deeper analysis of this relationship can still yield useful insight into both.

The first nine criteria can all be mapped directly onto aspects of full-spectrum intelligence. Criteria 3, 4, and 6—adaptive behavior, vast knowledge base, and knowledge integration—map onto immediate intelligence, and criteria 3, 4, and 9—adaptive behavior, vast knowledge base, and learning—map onto cumulative intelligence, assuming as discussed in section 4.1.4.4 that learning is considered a form of long-term adaptivity. Mappings onto additional aspects of full-spectrum intelligence include criteria 2 and 5—real-time performance and dynamic behavior—mapping onto situated agents; criterion 8—consciousness—mapping onto awareness; criterion 3—adaptive behavior—mapping onto projective agents, assuming as discussed in section 4.1.4.2 that projection is an aspect of adaptivity; criterion 1—flexible behavior—mapping onto general agents; criterion 7—natural language—mapping onto linguistic beings; and criterion 4—vast knowledge base—mapping onto educated agents. According to these mappings, criteria 3 and 4—adaptive behavior and vast knowledge base—turn out to span all three levels of intelligence.

Full-spectrum intelligence thus subsumes the aspects of the Newell test that are neutral with respect to whether natural or artificial intelligence is being considered. Yet, in the reverse direction, two aspects of full-spectrum intelligence turn out to be missing from this neutral portion of the Newell test: the notion that something truly matters, as is central to both affective sensates and purposeful agents, and the interpersonal/social aspects that are central to collaborative and socialized agents. Full-spectrum intelligence, at least as defined here, thus proposes a somewhat more comprehensive list for what is necessary for intelligence in general.

The final three criteria in the Newell test all focus on natural agents, with the last being even more specific to humans. The definition of full-spectrum intelligence is deliberately neutral as to whether it concerns natural versus artificial minds, yet criteria 10 and 11—development and evolution—directly relate to natural beings and criterion 12—(human) brain—to human brains.

4.2.2 Humanlike Intelligence

The notion of humanlike intelligence (figure 4.22) assumes that there exists a compact region of the intelligence space around human intelligence that encompasses forms of both natural and artificial intelligence. The latter, while it may be distinct from human intelligence, must be similar enough to it that a single model, at a suitable level of abstraction, is sufficient to characterize the whole region. The concept, although implicitly in existence for some time, became explicit in the development of the common model of cognition.

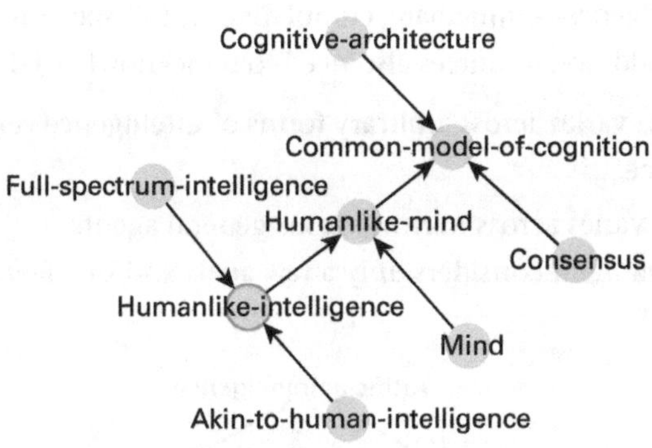

Figure 4.22
Dependency structure for humanlike intelligence.

93. *Humanlike intelligence* comprises forms of full-spectrum intelligence that are akin to human intelligence.
    - This definition is neutral with respect to whether natural or artificial intelligence is implicated.
94. A *humanlike mind* is a mind that exhibits humanlike intelligence.
95. The *common model of cognition* is a consensus concerning what must be in a cognitive architecture for a humanlike mind.
    - It explicitly accepts incompleteness where no consensus exists.

4.2.3 Artificial Intelligence

The range of existing definitions of *artificial intelligence (AI)* is at least as confusing as that for intelligence itself, with existing definitions tending to either focus on the perceived gap between human intelligence and what computers can currently do (e.g., Rich & Knight, 1991; "Artificial Intelligence (AI)," 2024) or some explicit notion of the aspects required for full-spectrum (artificial) intelligence (e.g., UNESCO, 2021; OECD, 2023). However, with a definition of intelligence in hand, a very straightforward definition of AI becomes possible (figure 4.23).

96. *Artificial intelligence* is intelligence in artificial agents.
    - As a special case, AGI may be seen as (at least) general cumulative intelligence in artificial agents.

This simple definition, while cutting through much of the murk, unfortunately still fails to capture anything like the full diversity of what is found across the field (figure 4.24). Although part of this diversity is due to the level of intelligence—immediate, cumulative, or full-spectrum—that is being considered, additional sources also need to be acknowledged, for example:

- *Humanness* varies across arbitrary forms of intelligence versus humanlike intelligence.
- *Generality* varies across narrow versus general agents.
    - A *narrow agent* considers only a few goals and/or a few environments.

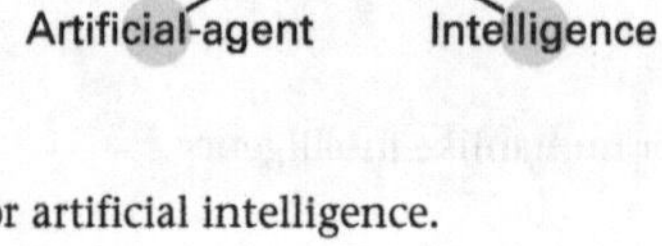

Figure 4.23
Dependency structure for artificial intelligence.

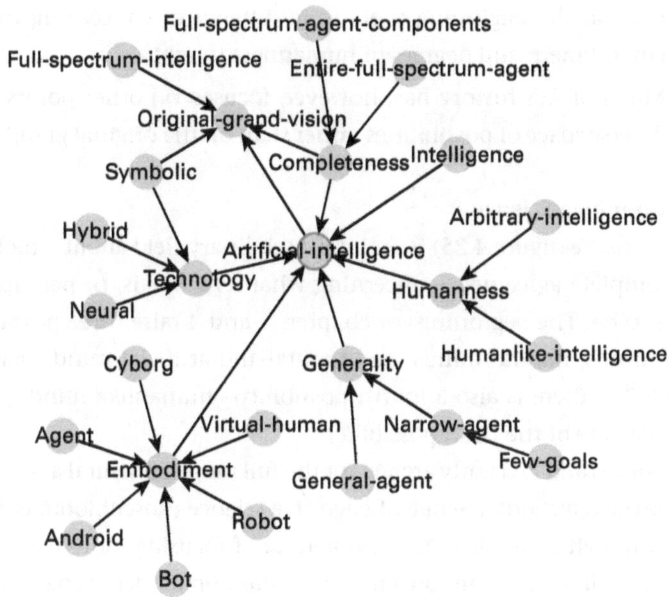

Figure 4.24
Dependency structure for diversity within the field of AI, including the original grand vision.

- *Completeness* varies across whether the focus is on individual aspects of full-spectrum intelligence versus all of it.
  - For the former, more variation results from which aspect(s) are the focus.
- *Embodiment* varies across agents, bots, virtual humans, robots, androids, or cyborgs.
- *Technology* varies across neural, symbolic, or hybrid.
  - Along with many possible variations within each of these broad categories, some of which can be seen in part II.

The original grand vision for the field can now be characterized as follows.

97. The *original grand vision* for the field, coming out of the 1956 Dartmouth workshop, could be said to comprise complete, full-spectrum artificial intelligence, with a bias toward symbolic technologies.
    - The desire for general agents is implied by the overall focus on complete full-spectrum intelligence.

- Even at the beginning there were differences concerning forms of embodiment and degrees of humanness sought.
- Much of AI's history has, however, focused on other points in this diverse space of possibilities rather than on the original grand vision.

4.2.4 Cognitive Science

Cognitive science (figure 4.25) is an interdisciplinary field about which there is not complete agreement concerning what its scope is, or perhaps what it ought to be. The beginnings of chapters 1 and 4 raise three possibilities: human minds, natural minds, and natural-and-artificial minds. Based on section 4.2.2, there is also a fourth possibility—humanlike minds—which is a special case of the third possibility.

Ken Forbus and I recently argued for the full range of natural and artificial minds as the appropriate scope of cognitive science (Rosenbloom & Forbus, 2019), although I can also see the potential of focusing more narrowly on humanlike minds, as is the concern of the common model of cognition. Yet, the field's zeitgeist clearly centers on human minds, with artificial minds relevant only to the extent that they assist in understanding human minds. To accommodate this dissonance, the definition of the field here is disjunctive, merely including all four possibilities.

98. The field of *cognitive science* studies human, natural, natural-and-artificial, or humanlike minds.

Whichever disjunct is adopted, the field remains grounded in multiple disciplines (figure 4.26), each according to how it can contribute:

- Artificial intelligence: Functional and mechanistic insights from artificial minds
- Philosophy: Questions, concepts, and formalisms

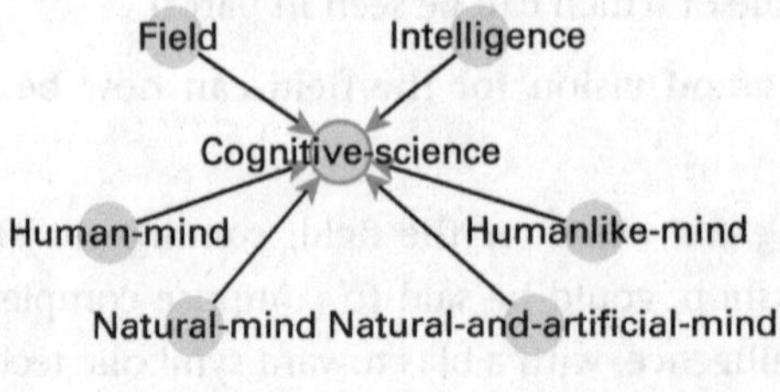

Figure 4.25
Dependency structure for the field of cognitive science.

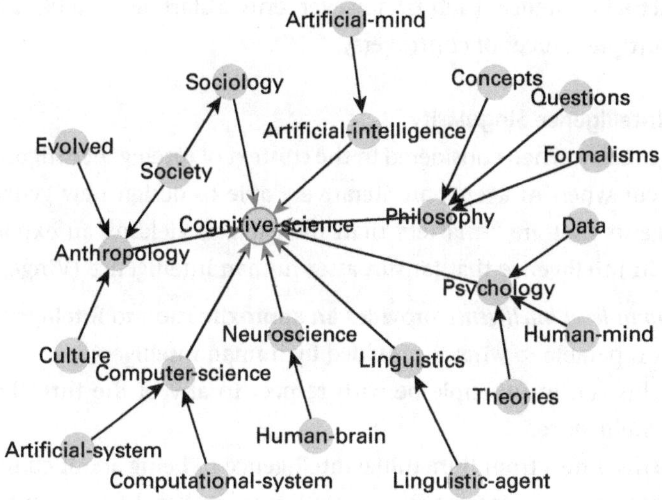

Figure 4.26
Dependency structure for the interdisciplinarity of cognitive science.

- Psychology: Data and theories about human minds
- Linguistics: Study of language structure and use
- Neuroscience: Data and theory that ground a human mind in a human brain
- Anthropology: Evolutionary and societal/cultural aspects of cognition
- Sociology: Data and theory on natural societies
- Computer science: Study and construction of computational systems.

There may also be more disciplines worth considering here as well.

4.3 Exploring Implications

This section makes a beginning at exploring six more advanced topics for which relevant groundwork has been laid via the definitions in the preceding sections: the *singularity*, as it applies to intelligence; *generative AI*, with a focus on large language models (LLMs); *Sigma*, with a focus on how it maps onto the trilevel definition of intelligence and thus compares to LLMs as a basis for intelligent systems; *ethics*, with a focus on its relationship to AI; *intellectual property*, with a focus on the possibility of AI creation of it; and *free will*, with a focus on the importance of a belief in free will rather than on its actual existence. Each is, however, only a start, and each may introduce multiple sources of controversy.

4.3.1 Intelligence Singularity

The singularity, when considered in the context of intelligence (figure 4.27), is to occur when AI agents are iteratively able to design new generations of AI agents that are "smarter" than themselves, yielding an exponential growth in intelligence that far surpasses human intelligence (Vinge, 1993).

99. *Human-level intelligence* provides an approximation to intelligence that is comparable to what is provided by human intelligence.
    - This can in principle be with respect to any of the three levels of intelligence.
    - This differs from humanlike intelligence in being about comparable degrees of approximation to intelligence rather than similar forms of intelligence.

100. *Superhuman intelligence*—or perhaps *superintelligence* (Bostrom, 2014)—provides a better approximation to some level of intelligence than is found in human intelligence.

101. An *artificial^n being* is a being that is artificial to some—generically *n*—degrees.
    - An artificial^0 being is natural.
    - An artificial^1 being is artificial but created by a natural being.

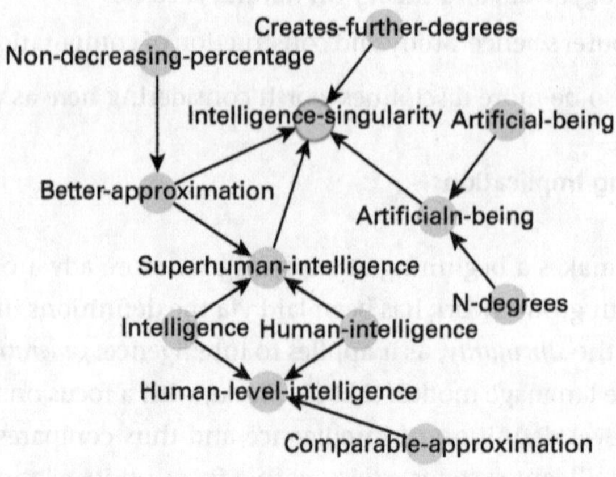

Figure 4.27
Dependency structure for the intelligence singularity.

- An artificial^2 being has been created by an artificial^1 being.
  - This may be a new being with a different architecture or merely a modification of its own architecture.
- An artificial^n being has been created by an artificial^{n-1} being.
- As in section 4.1.1.2, the more generic term *artificial* may be used for any $n>0$.

102. The *intelligence singularity* occurs when all of the following hold:
   - An artificial^n *being* with $n>0$ attains superhuman intelligence.
   - Further degrees of artificialness are created in which each degree provides a better approximation to intelligence.
   - Better is measured in terms of a percentage that is nondecreasing.
     - This yields the hypothesized exponential increase in intelligence.

Stated this way, it should be clear what this form of singularity is, but it is not at all clear—at least to me—that there is sufficient headroom between the approximation provided by human-level intelligence to any of the levels of intelligence identified in this chapter that exponential improvement beyond human-level intelligence is possible. More on the intelligence singularity can be found in aside 15.2 (section 15.1), where the question is raised more broadly as to whether an exponential model is even the most appropriate one to consider here.

4.3.2 Generative AI

Generative AI produces new content—such as text, images, or video—based on training parameters in a neural network from existing content. The most impactful form of it at this point comes in the form of *large language models (LLMs)* that learn to predict the next word in a sequence of words from previous words in the sequence (e.g., Brown et al., 2020). It has come as rather a revelation that a technology initially developed simply as a means of rating alternative sentences generated by speech-to-text (i.e., speech recognition) and text-to-text (i.e., translation) systems, in terms of how well they match the regularities/structure of the target language, can become generators of language that, when training is sufficiently scaled up, capture in a useful and somewhat flexible form much of the content, and even the implications, of the textual datasets from which they are learned.

This section focuses on LLMs. Their recent successes are due to algorithms that enable prediction from long sequences of previous words (e.g.,

Vaswani et al., 2017) plus the use of vast textual datasets and the massive amounts of computing required to process them (e.g., Brown et al., 2020). The specific question asked here is what can be understood about these systems based on the definitions developed in the previous sections. In answering this question, only the basic capabilities of an LLM are assumed, while mostly ignoring any additional architectural enhancements—whether reinforcement learning (Ziegler et al., 2019) or other mechanisms (e.g., Park et al., 2023)—that may significantly alter the analysis.

Understanding LLMs in terms of the previous definitions is best done over multiple steps. The first two involve a basic definition of an LLM (section 4.3.2.1) followed by an analysis of the structure of an LLM (section 4.3.2.2). LLMs are then mapped onto the trilevel hierarchy of intelligence (section 4.3.2.3).

4.3.2.1 Defining an LLM The dependency structure for defining an LLM can be found in figure 4.28.

103. *Pretrained parameters* are the parameters of an artificial system that are (machine) learned via pretraining, that is, before performance matters.
104. A *large language model (LLM)* is an artificial agent that uses a fixed neural-network structure plus parameters that are pretrained over a vast textual dataset to generate the most probable next word in a sequence given prior words in the sequence.
    - It qualifies as an artificial agent because it is sensate (sensing strings of words), animate (it generates words), goal-oriented (it has a hard-wired aim of generating the most probable next word given its memory and sensations), and based on an artificial brain and mind.

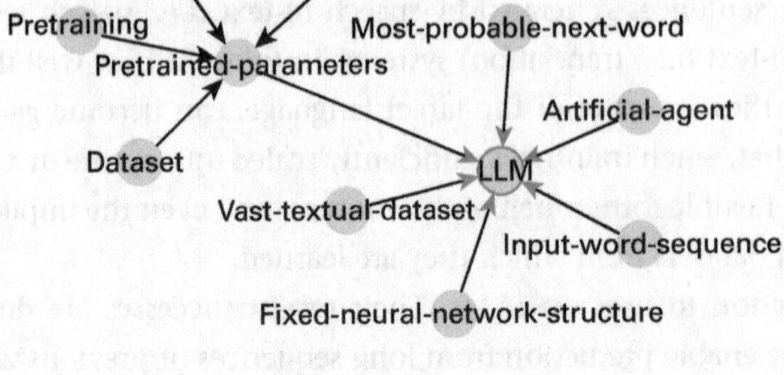

Figure 4.28
Dependency structure for an LLM.

- ◦ In systems that yield extended responses to input sequences, rather than just the most probable next word, there is an implicit goal of generating an appropriate such response, but the initial pretraining is not focused on such implicit goals.

4.3.2.2 Analyzing the structure of an LLM The dependency structure for the structure of an LLM can be found in figure 4.29.

105. *LLM memories* comprise a fixed network structure, learnable parameter values, and the transient word sequence that it receives as input.
106. An *LLM learning algorithm* is a machine learning algorithm that pretrains parameters in the network structure.
107. An *LLM performance algorithm* generates the most probable next word given its memories.
108. An *LLM architecture* comprises an LLM learning algorithm and an LLM performance algorithm.
    - The network structure might be considered as part of the architecture, given that it is fixed, but is probably better considered as innate memory, as implied earlier.
109. An *LLM* is an agent that combines an LLM architecture and LLM memories.

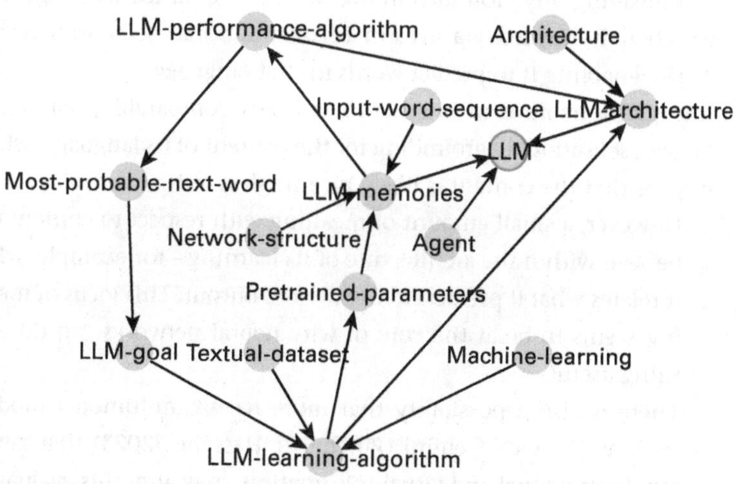

Figure 4.29
Dependency structure for LLM components.

4.3.2.3 Understanding LLMs in terms of the trilevel definition of intelligence An LLM exhibits immediate intelligence with respect to the hardwired aim of producing the most probable next word in the sequence.

- Rationality with respect to implicit goals is more a matter of chance, being a side effect of how memories are encoded in service of the hardwired goal, leading to what have been referred to as *hallucinations* when the probabilities get it wrong.

An LLM approximates cumulative intelligence.

- It learns extensively.
- However, it utilizes pretraining rather than pervasive learning; learns only from what it is spoon-fed rather than from its own interactive experiences; and learns only from text.

An LLM approximates full-spectrum intelligence.

- It minimally approximates a situated agent.
  - It is a bot in an environment of word sequences and to a large extent behaves appropriately within the constraints of those sequences.
  - It does not behave appropriately within the constraints of the larger reality and social environment within which it exists.
- It very minimally approximates awareness.
  - It is meaningfully grounded in the word strings of the language from which it is learned via a form of distributional semantics (Lenci, 2018), enabling it to predict words in that language.
  - It is questionable as to whether there is any comparable meaningful, or even sensational,[7] grounding for the content of its language, which implies that the content is likely meaningless to it.
    - However, a small amount of meaning with respect to content can be seen within the architecture of its learning—for example, when it relates what it predicts to the desired output. This locus of meaning seems to be at the core of why neural networks can do anything useful.
    - There is also a possibility that more recent multimodal models, such as Google's Gemini (Pinchal & Hassabis, 2023) that generates both textual and visual information, may alter this evaluation if there is, for example, meaningful grounding of text in images (Dove, 2023).

  - It has almost no ability to understand its content, even for the limited situation in which meaning exists with respect to word strings, as it cannot sense these meanings.
    - Again, the one exception appears to be within its learning mechanism where it can compare its model with the correct outcome and use this to send an error signal back through the network.
- It minimally approximates a projective agent.
  - Although the architecture does not directly support this, the memories may enable limited forms of it that fall significantly short of human level.
  - It doesn't have the pervasive ability to project the consequences of its actions before executing them and thus can't self-filter actions based on their projected consequences. This can lead to following what can be considered as *dangerous urges*.
- It provides a very good approximation to a general agent.
  - With respect to individual humans, it can exhibit superhuman generality in some ways while simultaneously falling short of what most people can do in others. It is thus unhumanlike in its generality.
  - It falls short of what humanity as a whole can do due to its other limitations.
- It is not an affective sensate.
  - It has no feelings, and nothing truly matters to it.
  - Along with the minimal approximation to awareness, there is clearly no sentience.
- It is not a purposeful agent.
  - It has no purposes as nothing truly matters to it.
- It approximates a linguistic being.
  - It is a communicative being that uses words effectively but does not understand the words it uses and thus has no language and remains significantly below human level as a linguistic being.
- It is a minimal approximation to a collaborative agent.
  - It has no shared goals other than what is implicit in the word sequences upon which it operates.
- It is a minimal approximation to a socialized agent.
  - It has no awareness of its society's culture but may be trained on materials that reflect aspects of it.

- It is a minimal approximation to an educated agent.
  - The extent to which it is educated is a function of how curated its training material is and how faithfully its learning algorithm reflects this material in its memories.

Given this analysis, it is clear that LLMs miss several key aspects of full-spectrum intelligence and, thus, also key aspects of the human psyche that make us what we are and make us, at least most often, reliable partners in the human community (a topic that is revisited more generally in section 4.3.4). This fairly negative judgment contrasts with the rather positive assessment of Centaur—an extension of the Llama 3 LLM (Grattafiori et al., 2024) to data from a wide variety of psychological experiments (Binz et al., 2024)—by its developers of its status with respect to the Newell test (Anderson & Lebiere, 2003). This is partly due to their glossing over the importance of things like goals and projection in checking off adaptive behavior; partly due to the four aspects of full-spectrum intelligence not found in the Newell test (section 4.2.1)—affective sensates, purposeful agents, collaborative agents, and socialized agents—that are likewise not strengths of LLMs; and partly due to the more stringent notion of understanding introduced here that yields a more negative judgment of language use by LLMs.

From this analysis, it is doubtful that LLMs by themselves are a major step toward the intelligence singularity. The only way in which they are superhuman is in generality, and even there it is of a restricted sort. I have at times considered LLMs metaphorically as akin to the monster from the movie *Forbidden Planet*—an amplified projection of Dr. Morbius's *id*, unconstrained by his *ego* or *superego*. An LLM instead has at its core what is effectively an amplified but unconstrained *subconscious* (or, alternatively, body of memories or System 1). It is amplified in the sense of being broader in what it knows than the subconscious of a typical human, and in some ways also in the range of capabilities it can exhibit. It truly does seem superhuman in this breadth.

Yet, LLMs are not superhuman in their depth—for example, in either rationality or projection—as would also seem necessary for enabling the intelligence singularity. Still, just as forms of generative AI are starting to serve a creative function in scientific discovery (Broad, 2024), some form of generative AI could perhaps play a similar role in helping to discover smarter artificial agents.

Furthermore, a variety of efforts are underway to augment LLMs in ways that yield better approximations to full-spectrum intelligence. To mention just one example, there is an effort to combine an ad hoc form of self-awareness that enables detection of at least some forms of hallucinations in LLM-generated text and then uses what amounts to a distinct educated module to make corrections (Wiggers, 2024). Such efforts could get us closer to LLM-based full-spectrum intelligence. Still, long-term success in this direction seems more likely if such extensions are pursued not just in a bottom-up fashion, as ways of handling immediate problems with existing systems, but also leveraging a top-down perspective, such as is discussed here, concerning what is necessary for them to truly be intelligent. Whether even such progress would also get us closer to the intelligence singularity still seems doubtful to me—see, for example, section 4.3.1 and aside 15.2—but it would give us a better basis for determining whether it is in fact possible.

4.3.3 Understanding Sigma in Terms of the Trilevel Definition of Intelligence

In the previous section, LLMs were understood in terms of how they map onto the trilevel definition of intelligence. In this section, a comparable attempt is made to understand the Sigma cognitive architecture (sections 1.2.2 and 15.4) in terms of this same definition. Why Sigma? Largely because it is my own best current approximation to full-spectrum intelligence. During the later years of work on Sigma, we were, for example, explicitly pushing it toward supporting humanlike autonomous social cognitive (HASC) systems that could perceive, reason, act, interact, use theory of mind, be affective, be motivated, exhibit personality, and learn (Ustun et al., 2018).

The intent is for this analysis to provide both a useful contrast with the one just done for LLMs and a stand-alone evaluation of Sigma's status as an architecture for intelligence, even if it should be presumed to be biased based on my status as its principal inventor. Many of the general lessons here should also carry over to other cognitive architectures. One important caveat before we start, though, is that Sigma, as an architecture rather than a system, does not automatically include variable content—as is the case for LLMs given pretraining—and so some of this analysis is necessarily based on potential rather than existence.

Sigma exhibits immediate intelligence.

- It makes decisions concerning action selection based on numeric preferences that stem from the current situation, memories, and goals.

- It is worth noting, however, that "nonrational" aspects arising from its approximation to an affective sensate, as will be discussed shortly, may also factor in.

Sigma approximates cumulative intelligence.

- It performs pervasive parameter learning over contiguous periods of processing.
- However, it does no structure learning, and it has had no extended lifetime over which it has cumulated learned knowledge.

Sigma approximates full-spectrum intelligence.

- It approximates a situated agent.
    - When combined with appropriate skills and knowledge it has successfully controlled agents in a variety of virtual environments (Ustun et al. 2015; Ustun et al., 2018).
    - There have also been explorations of perception within it based on probabilistic graphical models (Chen et al., 2011; Joshi et al., 2016), and it has the architectural capacity for neural-network-based perception (Rosenbloom et al., 2016b; Rosenbloom et al., 2017).
    - However, it has never been employed within a physical robot and overall has radically underdeveloped perception and motor abilities.
- It approximates awareness.
    - It utilizes models that can be programmed or learned.
        - These models may be in long-term memory and/or working memory.
    - Its models have meaning in terms of what they are about.
        - Learned models, furthermore, are sensationally grounded in the experience from which the learning occurs.
    - As with LLMs, understanding is built into the learning mechanism in comparing what the models in long-term memory predict with what they should predict for the current situation.
    - There is also a second form of architectural understanding in terms of low-level emotional appraisals, such as *surprise*, that are based on sensing how much a model must be adapted to shift it toward predicting the current situation.
    - It has largely unexplored potential for cognitive understanding and awareness.

    - There are, however, capabilities relevant to sensing and modeling itself, the environment, and other agents that have been partially explored (Pynadath et al., 2013; Pynadath et al., 2014; Rosenbloom et al., 2015; Rosenbloom et al., 2016a; Ustun et al., 2018).
- It is a projective agent.
  - Modeled after how Soar performs problem space search, Sigma leverages reflective capabilities to search over potential future states (Rosenbloom et al., 2016a).
- It is an architecture that can potentially support a general agent.
  - As shown in part II, Sigma effectively spans all three of the core dichotomies that have been identified as covering the space of AI technologies, including providing one of the deepest integrations of symbolic and neural technologies. Although not directly speaking to breadth of goals or environments, this breadth of technologies indirectly implies the potential for this.
  - Sigma also implements, or at least supports the implementation of, nearly all of what is required by the common model of cognition, leading to a comment much like that at the end of the previous point.
  - However, Sigma comes with no content, and although a variety of agents have been built based on it (Ustun et al., 2015; Ustun et al., 2018), each one is relatively narrow.
    - Its current implementation also does not achieve sufficient real-time efficiency for working with the very large bodies of content required by a general agent.
- It approximates an affective sensate.
  - It incorporates a subset of emotional appraisals, some of which can forcibly alter its thinking processes (Rosenbloom et al., 2015; Ustun et al., 2018).
  - It includes a distinction between positive and negative values—that is, valence—from work on motivation that affects how it behaves (Rosenbloom & Ustun, 2019).
  - However, it remains only a very partial model of emotion, and it remains unclear whether anything can fundamentally be said to truly matter to it.
  - This approximation, along with its approximation to awareness, still falls short of justifying any claim to sentience.

- It approximates a purposeful agent.
  - It has goals, motivations, and a desirability appraisal that, based on differences between the current state and the goal, can affect how it searches and how its attention is focused (Rosenbloom et al., 2015; Rosenbloom & Ustun, 2019).
  - It has also been shown able to use rewards in reinforcement learning, although most of the implementation of this is via memories above the architecture (Rosenbloom, 2012b; Pynadath et al., 2014).
  - However, like the assessment for affective sensate, it remains unclear if its goals truly matter to it.
- It minimally approximates a linguistic being.
  - It is a communicative being that has been explored in areas such as speech understanding (Joshi et al., 2014; Joshi et al., 2016) and natural language dialogue (Ustun et al., 2015).
  - It does not currently understand its communications, although there is potential for it to do so; and it falls far short of being a fully competent language user.
- It has the potential to be a collaborative agent.
  - No work has been done on shared goals in Sigma, but, as was earlier seen with respect to joint intentions in Soar (Tambe, 1997), the potential is there.
- It has the potential to be a socialized agent.
  - Given that it can be situated and has unexplored potential for awareness, this is another aspect with unexplored potential.
- It has the potential to be an educated agent.
  - It can represent curated knowledge, whether programmed by hand or learned, but no Sigma agent has ever had any substantial body of it.

Based on this analysis, it is possible to say that Sigma outdoes LLMs with respect to immediate intelligence; having awareness; being a situated, projective, purposeful, collaborative, and socialized agent; and being an affective sensate. However, there are trade-offs with respect to cumulative intelligence and being a linguistic being, a general agent, and an educated agent. With respect to cumulative intelligence, LLMs are pretrained on vastly more than Sigma has ever experienced, yet Sigma can learn from its experience as it is occurring rather than requiring everything to be pretrained. With respect to being a linguistic being, LLMs are presently much more effective communicative beings, whereas Sigma has at least the potential for being a true linguistic being through understanding of its communications. With respect to being a general agent, the vast knowledge embodied in LLMs enables them to exhibit intelligent behavior across many problems, whereas Sigma's embodiment of more of what is required for full-spectrum intelligence potentially enables it to exhibit intelligent behavior on problems not reasonably approachable by LLMs and to exhibit more intelligent behavior in some ways on problems that are approachable by them. With respect to being an educated being, LLMs clearly possess more in the way of curated knowledge than does any existing Sigma agent, whereas Sigma has the potential for this knowledge to be understood as curated, and used as such, whereas LLMs cannot distinguish it from informed guesses.

In summary, there is an overall trade-off that results from the contrast of LLMs' breadth of knowledge and remarkable communicative capability versus Sigma's overall potential as a humanlike cognitive architecture. Sigma does capture more of the key aspects of the human psyche that make us what we are and make us, at least most often, reliable partners in the human community. Yet, it still falls significantly short of full-spectrum intelligence, and no system based on it has yet exhibited altogether the full range of capabilities it currently possesses. It should also be noted that LLMs appear capable of passing the Turing test (Oremus, 2022; Jones & Bergen, 2024), which Sigma cannot come close to at present, and that LLMs yield much more payoff than Sigma.

4.3.4 Ethics

The overall goal of this section is to begin a process of understanding how it might be possible to assign rights and responsibilities to objects based on their *properties*.[8] It arose from a sense over the years that there is a difference between AI systems as tools versus AI systems as organisms that deeply impacts how safety can and should be approached for them, as well as what rights and responsibilities they should have. However, the notion of what an organism is when it came to artificial systems was always rather vague, and how to principledly assign rights and responsibilities based on properties of tools versus organisms has remained opaque. The impetus for this section is how the body of definitions provided so far begins to provide some leverage in answering these questions, with, for example, what was previously conceived of as an organism corresponding to some form of agent according to the definitions in section 4.1.

If done right, such an enterprise should ultimately deal appropriately with the full intelligence space, including both natural and artificial beings, both humans—including various forms of human neurodiversity—and other types of animals, and both terrestrial and (hypothetical) extraterrestrial beings. Part of this can be thought of as extending Singer's (2023) notion of *equal consideration* of rights across natural beings—"The basic element is taking into consideration the interests of the being, whatever those interests may be"—to artificial beings, while considering more than just interests as possibly being relevant, although the approach taken here need not align with the details of his approach. Here, the focus is also on responsibilities in addition to rights.

With respect to artificial beings, this overall topic clearly relates to AI ethics. However, the dominant thread in that field, concerned with ensuring the safe impact of AI systems on humanity, is just one piece of the overall picture, and as conventionally construed seems more directly applicable to AI systems that can be viewed as tools rather than to those that approach human-like or human-level intelligence. Still, relevant discussions of property-based approaches to questions of rights and moral status for social robots do exist (e.g., Sica & Sætra, 2024), and more will be said about this in section 4.3.4.8.

Out of all of the topics covered in this chapter, those around ethics seem the most likely to be controversial. However, an even more essential warning is also necessary at this point. Due to the potential implications of any proposed ethical system, and the disastrous results that may arise from not fully understanding the implications early enough, great care should be taken with any potential applications of these ideas. Of most concern would be anything that might appear to suggest reducing the rights currently ascribed to any subsets of humanity. Still, the goal is to at least make a start at understanding how properties of objects might factor into ethical decisions in a more nuanced fashion than the gross dichotomies mentioned previously.

We start with the notion of a tool and related concepts around work, before moving on to alignment, interests, property, and slavery. Based on these notions, the topic of AI safety is explored before entering a final discussion of rights and responsibilities that builds upon the basic definitions of these terms from section 4.1.3.9, but which attempts baby steps beyond this toward understanding how these terms might relate to properties of objects.

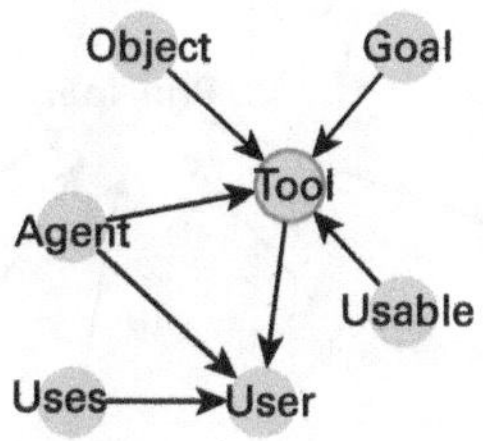

Figure 4.30
Dependency structure for a tool.

4.3.4.1 Tool This section covers definitions related to a tool (figure 4.30).

110. A *tool* is an object usable by an agent in service of its goals.

111. A *user* is an agent that uses a tool.

4.3.4.2 Work This section covers a range of definitions related to the notion of work (figure 4.31), but not the concept of work itself.

112. A *contractor* is an agent that achieves a goal for pay.

113. A *principal* is an agent that pays a contractor to achieve a goal.

114. A *subordinate* is an agent that is under an authority in some context.
    - The authority has some ability to determine the rights and responsibilities of the subordinate.

115. An *employee* is a subordinate contractor.

116. An *employer* is a principal who is an authority over an employee.

4.3.4.3 Alignment This section covers definitions related to alignment (figure 4.32).

117. *Incompatible goals* are goals that cannot be achieved simultaneously.

118. *Alignment* is the process of resolving incompatible goals across agents.

119. *AI alignment* is the process of alignment between AI and human goals (e.g., Russell, 2019; Ji et al., 2024).
    - It occurs by adjusting AI goals and the flexibility AI systems have in achieving them.

120. *Modification* alters an existing agent so that it is aligned.
    - *Training* provides experience to an agent that enables modification via learning.
    - *Programming* directly modifies an agent's structures and processes.

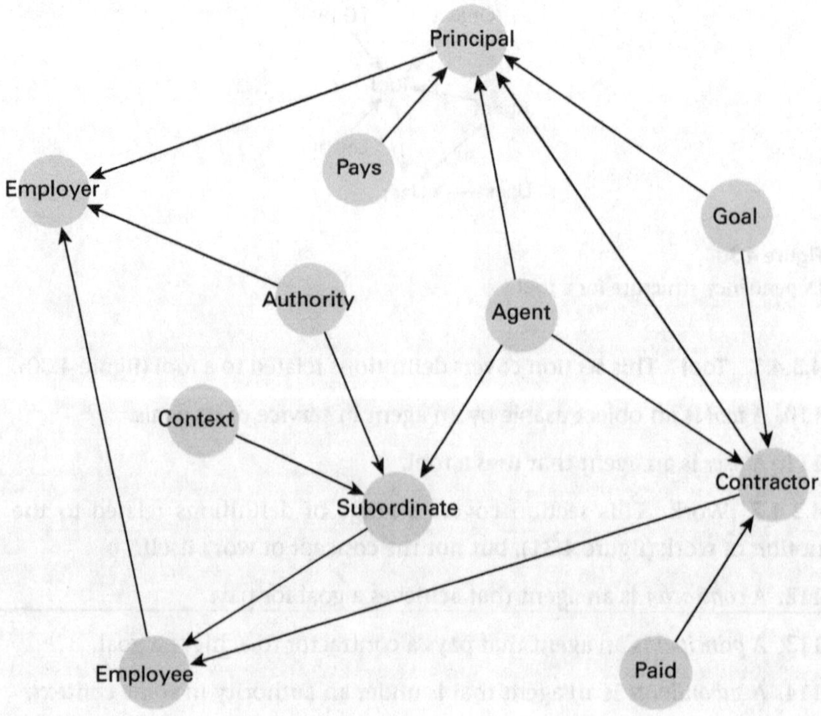

Figure 4.31
Dependency structure for concepts related to work.

121. *Breeding* can yield new agents that are (more) aligned from one or more existing agents that may not be as fully aligned.
122. *Governance* adds a governor (section 4.3.4.8), created in some fashion, to actively oversee an agent's alignment.

4.3.4.4 Interest This section covers definitions related to an interest (figure 4.33).

123. An *interest* is a purpose of a self-aware agent.
   - This aligns generally with Singer's (2023) usage of the term.
124. An *interested agent* has interests.

4.3.4.5 Property This section covers definitions related to property (figure 4.34).

125. An agent is the *owner* of an object if it has authority over it and has the right to sell, damage, or destroy it.
126. *Property* is owned objects.

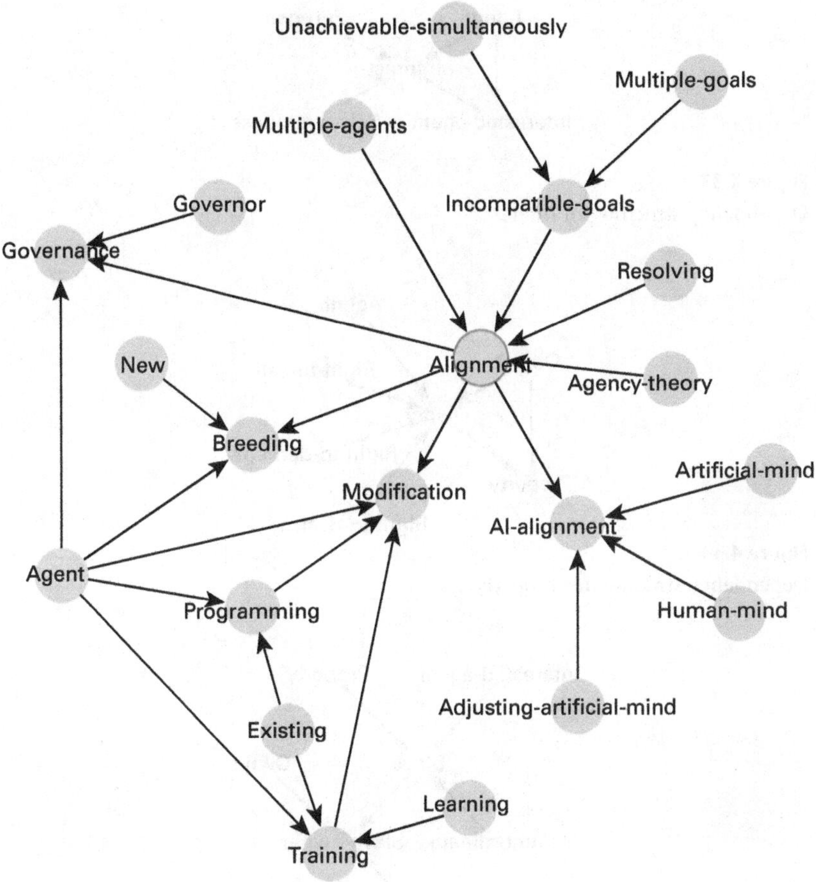

Figure 4.32
Dependency structure for alignment.

4.3.4.6 Slavery The topic of slavery (figure 4.35) is an awkward one for a chapter on intelligence, and this brief discussion won't do it justice, but a key concern with overfocusing on AI alignment is the possibility of turning AI systems into slaves. This concern is discussed further in section 4.3.4.7.

127. A *slave* is an interested subordinate that is property.
128. A *slave owner* owns and is an authority over slaves.
    - Effectively becomes the user of slaves as tools.
129. *Slavery* is a system with slaves and slave owners.

4.3.4.7 AI safety This section covers definitions related to AI safety (figure 4.36).

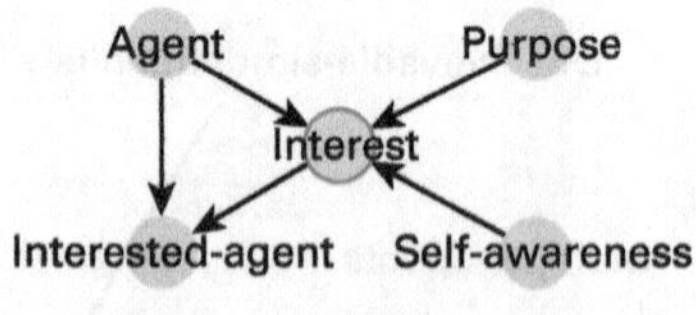

Figure 4.33
Dependency structure for interest.

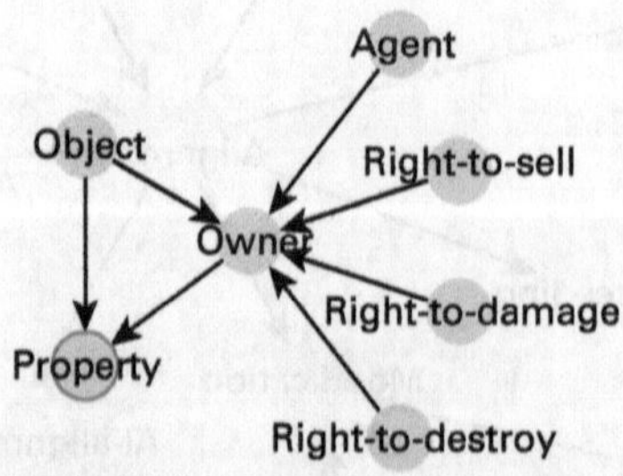

Figure 4.34
Dependency structure for property.

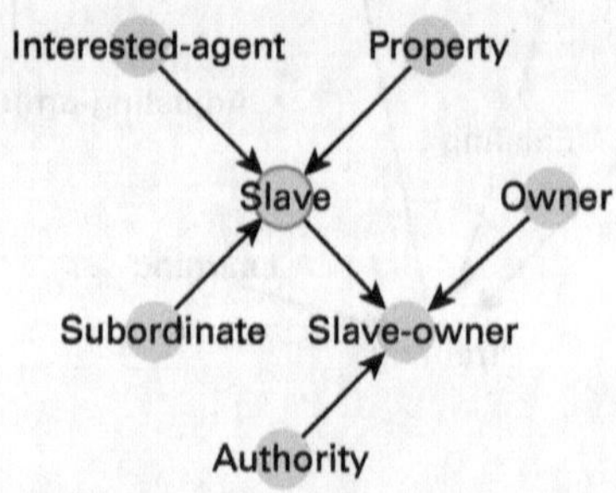

Figure 4.35
Dependency structure for slavery.

130. A *hazard* is a potential danger to an object.
131. *Safety* concerns the protection of an object from hazards.
132. A *safe tool* is usable without hazard to its user, and possibly others.
133. An *AI hazard* is a hazard from an AI system.
134. *AI safety* concerns removing AI hazards for humans.
   - AI alignment (section 4.3.4.3) does this by making AI systems into safe tools for humans.

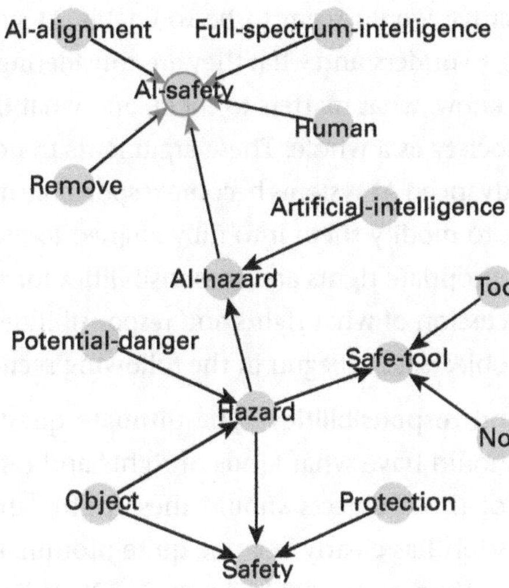

Figure 4.36
Dependency structure for AI safety.

- ◦ However, it is important not to turn them into slaves in the process.
- An alternative is extending AI systems into *responsible members of society*.
  - ◦ I won't attempt a precise definition of this term here, but it clearly at least involves being a socialized agent with cumulative intelligence.
  - ◦ It may also involve more aspects of full-spectrum intelligence, such as awareness, collaboration, purposefulness, and education.

This bipartite perspective on AI safety captures the essence of the tools-versus-organisms dichotomy that originally motivated this section. In general, I have two concerns about the alignment approach to AI safety. One has just been mentioned, of it turning AI systems into slaves, which in turn may lead to the kinds of dystopias imagined in science fiction involving robot revolutions. The other is whether agents that are sufficiently intelligent and that operate in sufficiently open worlds can ever be made safe through some form of alignment (e.g., Arvan, 2025), even if cast in terms of formal proofs (e.g., Dalrymple et al., 2024).

A better alternative would seem to be to enable AI systems to do what people do—that is, to understand what they are considering doing in the context of what they know, what matters to them, and what they know matters to others and to society as a whole. These arguments trend toward ensuring that sufficiently advanced AI systems become responsible members of society rather than trying to modify them into fully aligned tools. The accompanying question of appropriate rights and responsibilities for such agents is part of the broader discussion of what rights and responsibilities might belong to different types of objects, as is begun in the following section.

4.3.4.8 Rights and responsibilities The ultimate question here is what kinds of objects should have what kinds of rights and responsibilities, and on what aspects of those objects should these rights and responsibilities depend. This question has clearly become quite prominent with respect to animal rights (e.g., Nussbaum, 2023; Singer, 2023), although there can be significant disagreement about which aspects matter and what rights these should yield. There has also been concern with what are called the *rights of nature* that apply to both ecosystems and species ("Rights of nature," 2024). An early paper, for example, discussed the question of rights of trees from a legal perspective (Stone, 1972), whereas multiple recent efforts have focused on protecting the rights of rivers (Surma, 2024; Ma, 2025) and coastal marine ecosystems (Surma, 2025). Responsibilities—which are also sometimes referred to as *duties*—are a different matter.

I do not claim to be an expert on animal rights and related topics, with the material here arising primarily from considering what I understand about human and artificial minds. I have, however, read enough of this literature to acknowledge my general ignorance with respect to it, as well as with respect to too much of what has also been written about human rights, so the reader should consider this as input to the overall discussion of rights and responsibilities from a perspective that ideally complements what has come before in other areas.

Several prior efforts have taken up the general question of whether robots should have rights (e.g., Gunkel, 2018; Gellers, 2021), including ones making explicit arguments either for (e.g., Schwitzgebel & Garza, 2015) or against (e.g., Bryson, 2010); or arguing that they should have responsibilities rather than rights (e.g., Kim & Strudler, 2023). Some efforts, such as Neely (2014), have attempted to dig deeper, in discussing whether membership in a moral

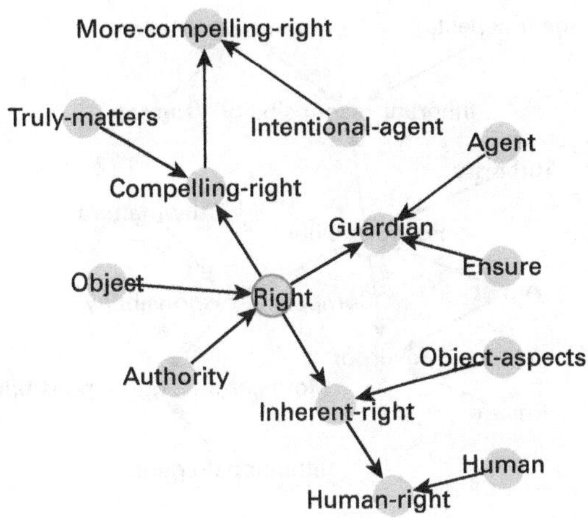

Figure 4.37
Dependency structure for rights.

community should depend on rationality (section 4.1.1.4), sentience (section 4.1.3.5), or interests (section 4.3.4.4). Such topics as moral status, which concerns how an object is to be treated, were also considered earlier from the more restricted perspective of living things (e.g., Warren, 1998).

The basic definitions of rights and responsibilities that are used in this section can be found in the earlier discussion of socialized agents (section 4.1.3.9). The material here goes beyond those social aspects of agents in an attempt to begin a broader discussion that encompasses arbitrary objects. Following some preliminary discussion of rights and responsibilities, new definitions concerning them are introduced. To simplify things, the dependency structure here is divided into one for rights (figure 4.37) and one for responsibilities (figure 4.38), albeit with some parallel structure across the two.

Rights may be relevant to any object.

- Consider, for example, how a culturally significant painting or building might be ascribed certain rights; and how people are starting to consider the rights of trees, rivers, and coastal marine ecosystems.

Responsibilities may be relevant to any agent.

- Without goals, there is no effective way to take on responsibilities.

As discussed in section 4.1.3.9, both rights and responsibilities derive from an authority.

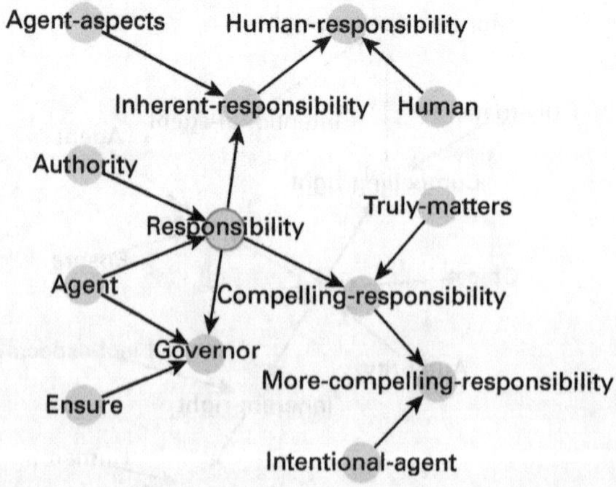

Figure 4.38
Dependency structure for responsibilities.

- Such an authority may be an employer, an owner, a society, a god, a ruler, or, in principle, any kind of agent (including itself).
- A document or abstract ethical system is not an authority in this sense, not being an agent, but can be the record of the rights and responsibilities authorized by an agent.
- Nothing is intended to be implied here about how an agent becomes an authority or what it might mean for such an authority to be legitimate.

Both rights and responsibilities become more compelling when things truly matter.

- This captures one of the essential implications of the notion of something truly mattering.
- It encompasses purposeful agents; and, for rights, it also encompasses affective sensates.

Both rights and responsibilities become even more compelling for interested agents.

- Such agents are aware of their rights and responsibilities and of the impact of these on what truly matters to them and others.

135. An *inherent right* or *responsibility* derives from aspects of the object in question.

- This assumes, even if only implicitly, that there is an authority that endows the object with this right or responsibility.
- The notion of inherent rights can be considered as a descendant of the traditional notion of *natural rights* ("Natural and legal rights," 2024).

136. A *human right* or *responsibility* is an inherent right or responsibility that derives from being human.
137. A *guardian* is an agent that is responsible for the rights of an object.
    - An agent may be its own guardian.
    - Guardians may also, for example, be relevant to the rights of plants, animals, ecosystems, children, forms of neurally diverse adults, and objects of cultural significance.
        - The general notion of a guardian already exists in the law for humans, but Nussbaum (2023) has proposed extending it to animals and Stone (1972) has discussed it for trees. Here the proposal is to extend this notion even further, as relevant, to consideration of all possible objects, including inanimate ones and artificial beings, although this should not be taken as an argument that any extant AI systems actually require guardians.
138. A *governor* is an agent responsible for ensuring that another agent upholds its responsibilities.[9]

Determining what inherent rights and responsibilities should arise as a function of different aspects of different types of objects is in my view critical for a general theory of ethics, but it is far beyond what can be attempted here.[10] It is also a question that has traditionally bedeviled such approaches to animal rights (Gellers, 2021). Still, examples of potentially relevant questions include:

- What rights should affective sensates have with respect to their feelings?
- What rights should purposeful agents have with respect to their purposes?
    - More minimally, what rights should agents have with respect to their goals?
    - More maximally, what rights should interested agents have with respect to their interests?
- When, and how, is it okay to terminate or destroy objects with different aspects?

   - As a function of its feelings, purposes, goals, self-awareness, cultural value, or other aspects (including the role it plays in the feelings, purposes, or goals of others).
- What rights and responsibilities should objects have when it is unclear what their relevant aspects are?
   - This can be especially relevant, for example, to questions concerning animal rights, where there is still much that we do not understand about them (Chan, 2011).
   - This may also be relevant—if not now, then before too long—to consideration of complex, poorly understood AI systems.
- What rights and responsibilities should generative AI systems have as a function of their distinctive approximation of full-spectrum intelligence?
   - Should they be entitled to any rights given that nothing truly matters to them and that they do not appear to have any other aspects that strongly bear on this question?
   - Can they be trusted to take on some responsibilities given their status as agents?
      - Their lack of awareness, collaboration, socialization, and education all lead to critical questions concerning how well they may be able to take on many responsibilities.
- What rights and responsibilities should corporations have as a function of their distinctive approximation to full-spectrum intelligence?
   - The notion that a corporation is in fact an AI has been suggested a number of times (e.g., Bengio, 2023; Lent, 2017; Stross, 2017), although according to section 4.1.3.8, corporations—as human organizations—are actually hybrid agents.
   - I won't attempt an analysis in any detail here, but does the fact that they may not have true purposes, and only one explicit goal—of maximizing profits, and, then, too often only profits in the short-term—imply that they cannot be responsible members of society and thus should have their rights more constrained and their responsibilities appropriately governed?
      - Through this lens, *environmental, social, and governance* (2025) can be seen as one route toward making corporations more responsible members of society.

- Does the ability to be projective impact the level of responsibility that can be assumed?
    - Without projection, unanticipated situations can yield wildly inappropriate behavior due to how they fall outside of what has been learned from prior experiences.
- Should self-awareness on its own imply certain rights?

4.3.5 Intellectual Property

The question of whether an AI agent can be, or should be, acknowledged as an inventor, an author, an artist, or a discoverer raises significant issues for the assignment of intellectual property. This section attempts to take a step toward resolving these questions by focusing on whether the agent is interested and, thus, self-aware (section 2.3.2) and purposeful (section 2.3.6). This focus on interest could easily be controversial and, quite possibly, not the right long-term answer, but it does at least provide a potentially principled rationale for resolving such questions that is not simply based on whether the agent is human (or perhaps a corporation, although as discussed in the previous section, corporations may actually be best thought of as AI or hybrid agents).

If this approach does prove to be too inappropriate in its implications, it will hopefully inspire follow-up attempts that manage better. As it is, though, it does imply that current AI agents should not themselves be considered creators of intellectual property. This conclusion builds up over definitions concerning inventors, artists, discoverers, and ultimately the idea of intellectual property itself.

The issue of generative AIs being trained over datasets that include intellectual property that belongs to others, while clearly an important one in itself, is not considered here. The definitions laid out so far do not appear to provide leverage with respect to this issue.

4.3.5.1 Inventor Here, a cascade of definitions builds up from a constructor to a creator, and finally to an inventor (figure 4.39).

139. A *constructor* is a situated agent that makes objects.

140. An *artificial object* is made by a constructor.
    - The concept of an artificial being was already introduced in section 2.1.2. In terms of the definitions in this section, an artificial being is an artificial object that takes the form of a being.

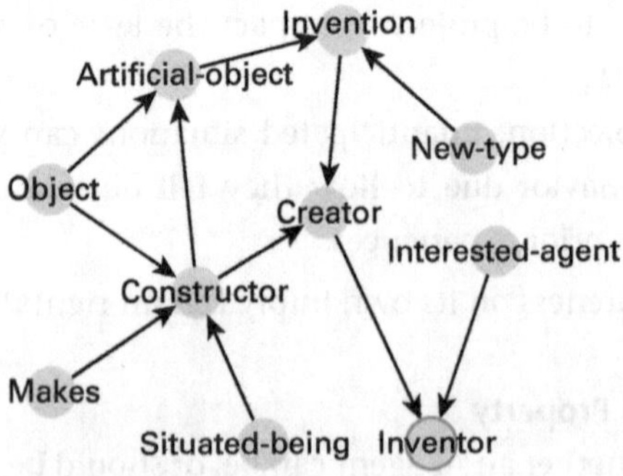

Figure 4.39
Dependency structure for an inventor.

141. An *invention* is a new type of artificial object.
142. A *creator* constructs an invention.
143. An *inventor* is a creator that is an interested agent.

4.3.5.2 Artist There is no generally accepted definition of what art is, but Wikipedia on "Art" (2024) contains a useful characterization of "artworks (art as objects) that are compelled by a personal drive (art as activity) and convey a message, mood, or symbolism for the perceiver to interpret (art as experience). Art is something that stimulates an individual's thoughts, emotions, beliefs, or ideas through the senses."

This characterization provides a lot to work with, including implicitly considering that the art must truly matter to the artist in being "compelled by a personal drive." It is less clear whether it assumes the art that results must truly matter to those sensing it, but this will be assumed here. Stimulation of emotions may involve this but whether stimulating beliefs or ideas does is iffier. Yet, without something like this, generating any kind of information that truly matters to its producer could be considered art as long as it gets the recipient to simply think about it.

In the other direction, this definition may be too broad, including, for example, novel enough verbal threats that would never normally be considered as works of art. This could be a problem for the Wikipedia definition as well as for the definition derived below that is based on it (figure 4.40).

144. An *artwork* is an invention by an inventor that provides a sensation that truly matters when understood by others.

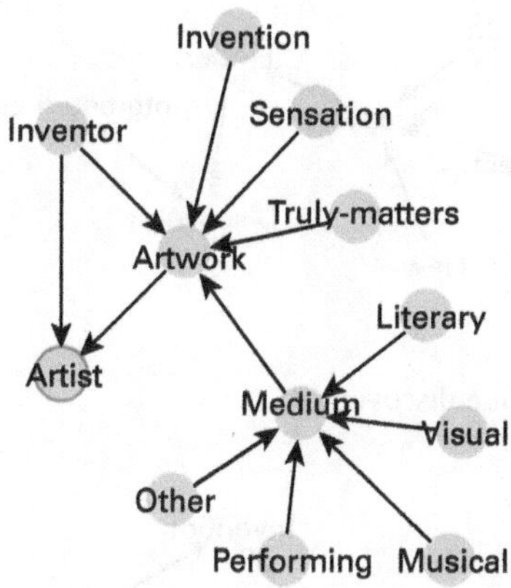

Figure 4.40
Dependency structure for an artist.

- An artwork may exist within any medium, including but not limited to visual, performing, literary, and musical.

145. An *artist* is an inventor of an artwork.

4.3.5.3 Discoverer This section covers definitions related to a discoverer (figure 4.41).

146. A *discovery* is the finding of a previously unknown object.
    - There is a gray area in which it can be ambiguous as to whether something new has been found versus constructed. This is especially subtle in areas such as mathematics, in a thread that goes back at least to Plato, who considered mathematical objects to be discovered rather than invented (see, e.g., Cole, 1999).

147. A *finder* makes a discovery.

148. A *discoverer* is a finder that is an interested agent.

4.3.5.4 Intellectual property Given the prior definitions, it becomes straightforward to define intellectual property (figure 4.42).

149. *Intellectual property* (*IP*) is a conceptual invention or discovery, yielded, respectively, by an inventor or a discoverer that becomes property.

Artificial agents currently have no interests, as even if they are self-aware, they have no purposes. Thus, according to this definition they can neither be inventors nor discoverers, although they may be creators or finders. One

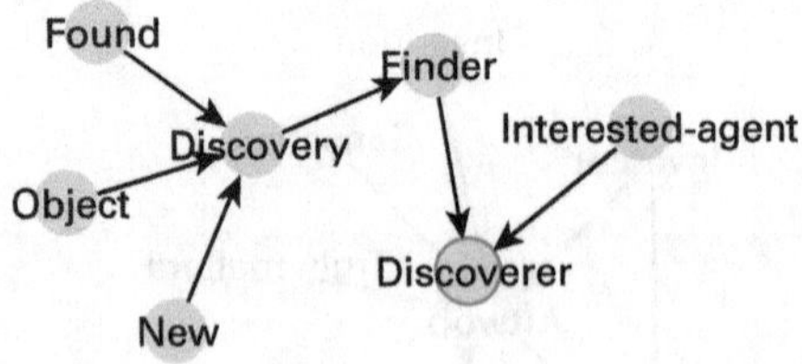

Figure 4.41
Dependency structure for a discoverer.

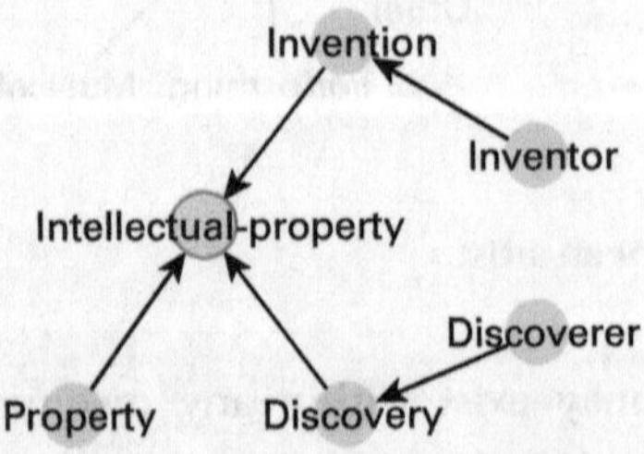

Figure 4.42
Dependency structure for intellectual property.

implication of this is that they cannot by themselves engender intellectual property. This does not, however, deny the possibility that artificial agents may at some point be capable of having interests and, thus, of generating intellectual property on their own.

In the meantime, this also raises the question as to whether a human who deploys an artificial creator or finder as a tool should be considered as an inventor or discoverer and, thus, whether intellectual property can be generated by said human. The deployment of generative AI may fall within this intermediate circumstance. The US Copyright Office has in fact recently ruled that if there is sufficient human contribution involved in a partnership with AI, then that contribution is copyrightable (U.S. Copyright Office, 2025). Perhaps, at some point in such a case intellectual property may be said to be invented or discovered by the hybrid organization comprising the artificial agent and the human. But then it must still be worked out what it means for such an organization to own property.

4.3.6 Free Will

There has been extensive media coverage recently of a book by a Stanford neurobiologist (Sapolsky, 2023) making a case against *free will* (2024)—"free and independent choice; voluntary decision." While I don't disagree with this position in principle, his argument still didn't strike me as the complete story where either human or humanlike artificial cognition is concerned.[11]

The question of whether humans have free will is often conflated with the related question of *determinism* (2024)—"that all events in the universe, including human decisions and actions, are causally inevitable." The difference here is that although determinism does seem to imply a lack of free will, *indeterminism*—where randomness or quantum effects play a role—also does not imply the existence of free will, as humans have no more control over these effects than they do over processes that are fully determined.

Compatibilists attempt a compromise by hypothesizing some alternative form of free will that exists even while acknowledging determinism. Based on the definitions laid down so far in this chapter, a distinctive form of compatibilism does seem to take shape. Consider an intelligent agent able to make rational decisions given its sensations, memories, goals (section 4.1.1.4), and the results of projecting possible futures given its potential actions (section 4.1.3.3). Assume also that it can understand what it is doing in the process (section 4.1.3.2) and learn from all of this (section 4.1.2).

There is indeed no room for the classical sense of free will in such an agent, as action selection is completely determined by these factors. Indeterminism could be involved, such as when actions are chosen according to a probability distribution computed from estimated utilities, or when the outcomes of projected action sequences are not completely deterministic. But, again, this does not yield anything like the kind of freedom of choice implied by free will. Still, something akin to free will can be discerned in such an agent.

As a preliminary note, although these ideas originated as a theoretical exercise that follows from the earlier definitions, delving into some of the key literature on free will from philosophy, psychology, and neuroscience has helped strengthen several of the points, with a smattering of citations and quotes to follow where they are especially useful. There is also some literature in AI on free will, but it mostly has taken a simpler view of it that did not prove to be useful here.

To continue the main narrative, consider if the agent introduced earlier were to choose whether to project based on whether its memories imply that its choice among actions can affect the world in such a manner that alters its chances of achieving its goals. Rationality would imply choosing to project when doing so may lead to selecting actions that improve its ability to achieve its goals. However, when there is no such implication there is no reason to project. In such a case, there is no lack of determinism in the choice of whether or not to project, as it follows directly from the agent's sensations, memories, and goals; yet, the ability to project and to choose actions based on these projections can provide, when combined with a self-awareness that this is in fact what the agent is doing, a sense of control that can plausibly be modeled by the agent as its having free will. Baumeister et al. (2001) earlier proposed something like this in the context of consumer psychology: "Self-control, planful action, and rational choice are vital forms of free will in this sense."

Such an agent may thus believe it has free will even if everything is determined. When learning is added into the picture, so that there is a memory of having had free will in making a decision, this could contribute to projecting in more relevant future situations, reinforcing this belief in free will. It is critical to acknowledge, however, that this only argues that it is reasonable for such agents to believe that they have free will, not that they actually have it, unless you redefine free will to be this. Dennett (2003), for example, uses the concept of *evitability*—that is, the ability to project in service of avoiding undesirable alternatives—to yield a notion that is weaker than the standard one but is claimed to be all that really matters. Dehaene (2014) and Blum and Blum (2022) have since proposed similar redefinitions. Here, rather than attempting a redefinition, the focus is explicitly on the belief.

This line of argument, and as hinted at by Dennett, can also lead to the conclusion that belief in free will is relevant for goal achievement. Consider what might happen if the agent instead believes it has no free will; that is, it believes it has no control over which actions it selects. Even with an understanding that its actions may affect its ability to achieve its goals, the agent may conclude that its thought processes will not actually affect which actions end up being chosen, and thus it may fail to project their consequences even when doing so could affect what actions it chooses by altering the contents of its memories.

Thus, a belief in free will may increase an agent's chances of achieving its goals, whereas a belief in its absence may decrease its chances. In fact, trying to convince people that they do not have free will could be destructive, in weakening the impetus for them to project and thus their chances of achieving their goals. This is a big part of what worries me about a blanket argument that there is no free will without factoring in that even without free will activities like projection can help in achieving one's goals, and a choice to project—although itself also deterministic—would be based on memories as to whether it might help. Rigoni and Brass (2014) state that a review of the literature "demonstrates that inducing disbelief in free will has an impact on folk psychology, social behavior and intentional action." The focus here is on intentional action, but much of the concern in the literature is with the impact of a lack of belief in free will on social behavior. One specific example of the latter is how the act of encouraging a lack of such a belief can increase the incidence of cheating (Vohs & Schooler, 2008).

In summary, the idea here is that a belief in free will is important if an intelligent agent is to engage, in a rational and deterministic manner, in projecting the consequences of its potential actions; and that the results of such projections can reciprocally encourage a belief in free will in a self-aware agent that learns. Under such a circumstance, evolution may also favor an innate belief in free will for natural agents of this sort. Discussions on the role of evolution and free will can be found, for example, in Baumeister et al. (2001), Dennett (2003), and Mitchell (2023).

Still, one important qualification to this all is worth noting—the free-will loop summarized in the previous paragraph can potentially be short circuited if the agent simply believes that projection can lead to choosing better actions. This is implicitly the case in AI agents that are hardwired to project, with no need for a belief in free will, self-awareness, or learning. Still, the full loop may very well be critical for the more flexible natures found in humans and other varieties of humanlike artificial agents.

4.4 Conclusion

This chapter has amounted to a journey through 149 enumerated concepts/definitions, plus a number of other unenumerated points, that is illustrated by forty-one graphs of dependencies among related concepts and associated commentary. The first eighty-three definitions built up to a nested, trilevel—immediate, cumulative, and full-spectrum—definition of intelligence that captures three distinct, but related, senses of intelligence to which existing definitions of intelligence have implicitly adhered. This tripartite definition also provides a way of making sense of theories of multiple intelligences. The alternative possibility of a more minimal essence for intelligence has also been explored around the notion of adaptivity.

When combined with a breadth of approximations to intelligence, an intelligence space results that provides one way of scoping the range of possible intelligent beings, whether natural or artificial. Such concepts as natural intelligence, human intelligence, humanlike intelligence, artificial intelligence, artificial general intelligence, and cognitive science can then be mapped onto regions of this overall space, with the possibility of extending this mapping further as it proves useful.

The overall hope is that this analysis has brought some useful structure to how we think about intelligence and the different flavors of it, whether in terms of levels of intelligence and approximations to them or particular regions of the resulting space. In turn, the hope is that this helps us see through the existing thicket of conflicting definitions of both intelligence in general and artificial intelligence more specifically, toward a common understanding of the relevant space of possibilities. As mentioned at the beginning, this chapter is intended as only a first step in this direction rather than as a final word on the topic; so, if this does prove to be a useful direction, a variety of refinements and extensions would likely still be needed.

The more advanced topics covered here are also first steps at best. The analysis of the intelligence singularity raises questions about whether there is sufficient headroom above human-level intelligence for an exponential improvement over it. The analysis of generative AI attempts to identify where large language models fit within the intelligence space. There is uncertainty in this due to how little we—and I—still understand these models, but tentative conclusions can still be drawn in terms of their approximations to the three levels of intelligence. A similar analysis of Sigma enables a comparison of its positioning within the intelligence space with that of large language models.

The analysis of ethics has focused on questioning whether alignment is the right approach to AI safety versus developing AI systems that are responsible members of society; and on how rights and responsibilities might appropriately depend on aspects of full-spectrum intelligence. The analysis of intellectual property concludes that, due to their lack of interests, current AI systems do not by themselves generate intellectual property, yet it does not rule out the possibility that future systems might. The analysis of free will indicates that a lack of free will negates neither the importance of the concept nor the necessity of there being something like a belief in it.

Finally, it must be acknowledged that several of these analyses—most particularly of the second half of ethics and of intellectual property—are not only just first steps but also depend critically on the notion of something truly mattering, a concept in whose current definition I have at best only minimal confidence. The analyses of generative AI and Sigma also depend to some extent on this notion. Hopefully, though, at some point either confidence in the current definition can be increased or a better definition can be provided.

5 On Theories

5.1 Introduction

What is a *theory*, of what use are such theories, and how can they be evaluated? These are general questions that are explored in this chapter, but with an ultimate concern in each case for their implications on the architectural approach to theories of mind, whether natural or artificial. As discussed in section 2.2, the architectural approach to science and engineering decomposes the notion of a *system* into a fixed component—the *architecture*—plus *variable content*, according to the semi-formal equation *system* = *architecture* + *variable content*. *Cognitive architectures* (section 1.2), considered broadly across the cognitive sciences and the various names they take on in different fields—such as AI architectures, AGI architectures, and agent architectures—are fixed structures that, when paired with variable *knowledge and skills*, can yield *intelligent systems*, or as a variant equation: *intelligent systems* = *cognitive architectures* + *knowledge and skills*.

Cognitive architectures can be thought of as bare *minds*, independent of whatever content they might embody, although not simply blank slates ("Tabula rasa," 2024), as they do include a set of mechanisms plus possibly some innate knowledge and skills. The focus here will be on the common model of cognition and the architectures that played a central role in its initial development—ACT-R, Soar, and Sigma. But also mentioned will be other traditional cognitive architectures as appropriate, while gradually incorporating into the discussion other architectural approaches, such as ones based on neural networks (e.g., Park et al., 2023; Sumers et al., 2023) and reinforcement learning (e.g., Sutton & Barto, 2018).

Section 5.2 focuses on the first question of "What is a theory?," with a concentration on steps 3 and 4 of the specialization of the reflective method to essential analysis: identifying the essence of what a theory is and identifying its definitional space. It also includes some material on steps 2 and 5, of amassing and assorting exemplars. Step 1, of identifying the topic, is already presumed by the title of this chapter. This analysis does not attempt to be either exhaustive in identifying dimensions of variation or systematic in populating the space, but it does attempt to provide enough of both of these to be both comprehensible and supportive of answering the remaining two questions, concerning theory use and evaluation, which are the main focus of step 6. In the process the analysis yields not just a list of dimensions of variation, but a structure—in the form of a directed acyclic graph (DAG)—over them. Unexpectedly, the three dimensions at the first sublevel of this DAG—*fidelity*, *usability*, and *lawfulness*—have turned out to correspond, respectively, to the semantics, pragmatics, and syntax of theories, revealing a pleasingly systematic character at this level.

Section 5.3 moves on to the first critical set of implications by addressing the question of theory use. The term *use*, however, turns out to be overly ambiguous in this context, requiring distinguishing at least two potential aspects of the term. Here it refers to the *function* or *manner* in which a theory is used. For example, one theory may be used for prediction and another for explanation. In contrast, a *domain* is a body of phenomena to which the theory is relevant. Domains can range over the entirety of the universe and human experience and imagination. Domains may also be defined at a variety of levels of granularity. Each theory may support one or more uses across one or more domains.

Section 5.3.1 begins the exploration of theory use by merging multiple perspectives on it into a single broad yet coherent story. Some of this turns out to be driven by definitional aspects of the notion of a theory that are stripped away in section 5.2. Rather than yielding dimensions of the definitional space, as would be the expected output of such a process, these aspects characterize two ways theories are used—*prediction* and *explanation*—thus providing part of the foundation for section 5.3. Following this merging of perspectives, section 5.3.2 explores how individual theories might span multiple of these uses rather than, for example, only being usable for prediction or explanation.

Section 5.4 then explores a second set of critical implications by addressing the third question, concerning how to evaluate theories. This involves extending the DAG from section 5.2 to a set of interconnected

DAGs that include four new dimensions—*beauty*, *comprehensiveness*, *rigidity*, and *surprisingness*—that are relevant to theory evaluation. It also involves assuming polarity along the dimensions of these interacting DAGs, so for example higher fidelity is assumed to be better than lower fidelity.

Sections 5.2–5.4 each individually wrap up with a discussion of the implications of its general analysis for the more specific topic of cognitive architectures, with section 5.5 then concluding the overall analysis.

5.2 What Is a Theory?

In exploring what a theory is, steps 2–5 of essential analysis are applied. This starts with step 3, through examining several preexisting meanings of the term before settling on a simple, stripped-down essence (section 5.2.1). This then enables step 4, by yielding a set of dimensions along which theories can vary that correspond to the strictures that have been stripped out of standard definitions. This is followed by steps 2 and 5, with pedagogical examples amassed and assorted when appropriate (section 5.2.2). Section 5.2.3 considers implications for cognitive architectures by extending step 5 to identify where in the definitional space of variations they sit.

5.2.1 Defining *Theory*

There are many shades of meaning for the term *theory* and correspondingly many definitions. The ultimate concern here is with theories that take the form of cognitive architectures, and so the definitional focus can be narrowed down to versions that help shed insight on this topic. In particular, we'll concentrate on three specific definitions—from Newell (1990), Dictionary.com ("Theory," 2024a), and Wikipedia ("Theory," 2024b)—while acknowledging that there may be others that could also be relevant. We'll do this across multiple steps in the following subsections on aboutness, models, mechanisms, systems, a least common definition, and final words on the essence of theory.

5.2.1.1 Aboutness The first definition listed in Dictionary.com is substantive without being overwhelmingly detailed: "a coherent group of tested general propositions, commonly regarded as correct, that can be used as principles of explanation and prediction for a class of phenomena: Einstein's theory of relativity." As implied by the example embedded in this definition, it best fits theories from the physical sciences, where accepted theories expressed in mathematical terms—yielding coherent general propositions—are tested individually, evaluated experimentally for correctness as a whole, and used for prediction and explanation. However, even there, theories exist—such as string theory (Becker et al., 2006)—that are difficult to evaluate for correctness, and other theories—such as general relativity (Renn, 2005)—that required some time before they could be evaluated in such a manner. Theories in the other sciences can diverge even further from this physically ideal notion, with perhaps the social and cognitive sciences struggling the most to live up to it. In the cognitive sciences, prime examples where multiple definitional criteria are violated include cognitive architectures and—if they are to be thought of as theories—the kinds of models acquired by machine learning.

Viewed in another way, this definition can be seen as combining an essence concerned with *aboutness*, as expressed in terms of "for a class of phenomena," plus a starting point in understanding how theories are used—*explanation* and *prediction*—and what aspects may be relevant in specifying a definitional space: *general propositions* that are *coherent* and *tested*; serving as *principles* that are *accepted as correct* and that are *usable*. Further discussion of these latter two topics can be found in section 5.3 for theory use and section 5.2.2 for its definitional space. In addition, section 5.4 builds on the dimensions of the definitional space to explore how theories may be evaluated.

Stripping away both these uses and the now nondefinitional aspects, the claim here is that a theory at its core designates, represents, or is simply about some phenomena. Traditionally within AI, "an entity X designates an entity Y relative to a process P, if, when P takes X as input, its behavior depends on Y" (Newell, 1980). If we are considering, for example, entity X as a symbol in the mind of a human, P is human cognitive processing that operates on X in a manner that reflects Y. The notion of representation in AI also concerns aboutness, but it typically takes on a formal aspect, with entity X formulated as statements in logic and P consisting of theorem proving based on X that reflects Y.

Theories in the physical sciences are about the material world around us. Theories in the life sciences are about living organisms. Theories in the social sciences are about human behavior, whether individually or in groups. Theories in the cognitive sciences—when considered broadly—are about minds, typically human, but conceivably also belonging to other entities, whether natural or artificial, and even possibly about more abstract notions of mind (section 4.2.4).

This essence of aboutness roughly aligns with the definition found at the beginning of the "Theory" (2024b) definition: "a rational type of abstract thinking about a phenomenon, or the results of such thinking." This clearly focuses on aboutness, but in addition includes notions of *rationality* (a variant of coherent) and *abstractness* (a variant of general). While these two notions may be part of an ideal theory, they aren't part of the essence here, and so their further consideration is postponed until the discussion of the definitional space in section 5.2.2.

5.2.1.2 Models *Models* (section 4.1.3.2) share the notion of aboutness with theories, but there exist multiple, often inconsistent notions of how they differ, including level of abstraction or detail, a qualitative versus a quantitative nature, or type of use (see, e.g., "What is the difference between a model and a theory?," 2023). The notion of a theory considered here encompasses whatever might traditionally be considered as theories or models, with the potential differences between them available to further build out the definitional space around the concept of a theory.

Computational models have the distinction of being executable on computers, essentially combining Newell's entity X and process P into a mechanism that comprises both structures and processes (section 4.2.1.1). More generally, computational models can be thought of as raising the possibility of there being two different processes that need to be considered, and thus two different kinds of mechanisms that may be in play. One is the code that executes the model on a computer. The other is the understanding of the model by a human—the "abstract thinking" from the Wikipedia definition.

Mental models (section 4.1.3.2) are models that exist in the memories of agents. Large language models (LLMs) (e.g., Brown et al., 2020) are mental models that are about language usage—predicting the next word in a sequence given the previous words. They provide a fine example of the first kind of mechanism at play in a computational model, in terms of code that executes on a computer, but the structures by themselves—large neural networks with billions or trillions of learned parameters—are effectively black boxes that are uninterpretable by humans and so do not provide an example of the second kind of mechanism.

Traditional scientific theories, typically based on principles, are inherently instances of the second kind, whether directly computational or not.

Computational or mathematical underpinnings may be added to such theories, but even without this, the combination of the principles and human understanding can be seen as a mechanism of its own. Cognitive architectures typically support some form of both of these types of mechanisms, being accompanied by implementations that make them computational while still being transparent enough to enable some amount of human understanding.

5.2.1.3 Mechanisms Newell (1990), in his foundational book on how cognitive architectures can support *unified theories of cognition*, focused on the idea of a theory being "a body of explicit knowledge." As I understand it, *explicit* here is to at least connote that a theory is an object that is usable by arbitrary people, versus structures within a human mind that are usable only by that mind's associated human. This thus assumes that, in the context of a theory, although process P may be internal to a human—as with the second option earlier—entity X must be more broadly accessible. At the same time, entity X can in this view be expressed in terms of many different sorts of knowledge, not just in terms of those that may be considered propositions or principles. This makes particular sense for cognitive architectures as theories, as they often take the form of executable code.

Newell's full definition of a theory is a "body of explicit knowledge, from which answers can be obtained to questions by enquiries." Question-answering is an active process, which again implies the need for some combination of entity X and process P when considering a theory. Newell goes on to mention the kinds of questions he has in mind—predictions, explanations, and prescriptions for control. However, as with the uses embedded within the Dictionary.com definition, discussion of these uses is postponed until section 5.3, with discussion of theory usability in general found in section 5.2.2.

Thus, whether implicit or explicit, whether relying on humans or computers, or whether cast as a theory or a model, there must be processing that goes along with any structures to yield aboutness and thus to have a theory. Putting it all together, the essence of a theory becomes *a mechanism that is about some phenomena*. Still, when the process P is internal to a human, the conventional notion of a theory being just entity X, with process P implicit, can provide a useful shorthand. In such cases the acknowledged essence is essentially reduced to *structures about some phenomena*.

5.2.1.4 (Meta)Systems It is worth taking two additional steps beyond the essential formulation just articulated. The first generalizes from the notion of a mechanism to that of a *system*. As characterized in section 5.1, a system comprises an architecture plus variable content. Yet, from a different perspective a system can be viewed as just a set of mechanisms (section 4.2.1.1). Whereas a simple theory may implicate only a single mechanism, as in the essence just mentioned, more complex theories may require multiple mechanisms and, thus, be better considered as systems. This yields a generalization of the initial essence to *a system that is about some phenomena*, to leave open whether one or more mechanisms are involved, while also aligning with the definition of a *model* in that section. It further raises the possibility that not only whole systems but also architectures and variable content—that is, parts of systems according to the earlier decomposition—may be theories.

Second, if we consider what all may be encompassed by the phrase *some phenomena*, it should be clear that the subject of a theory can in general also be reconsidered as a system. This leaves us with a theory being, most generally, *a system about a system*, or perhaps *a metasystem*. The latter formulation omits the explicit mention of *about* while inherently including it as the heart of what it means to be *meta*. This is the final form we'll assume for the essence of a theory, centering on aboutness while permitting significant flexibility for both the nature of the theory (i.e., entity X in the company of process P) and the phenomena it is about (i.e., entity Y).

5.2.1.5 Contrast with a least common definition If we were to ask what a *least common definition*—the specialization of the notion of a least common core from section 1.1.2 to definitions—would be across the three definitions of a theory we have seen in this section, it would certainly focus on *aboutness*. It may, however, also include something like *lawfulness* (section 5.2.2) based on such phrases as *general propositions, principles, abstract thinking*, and *explicit knowledge* (according to some possible interpretations of this phrase), implying that the essence here minimizes beyond the least common definition by omitting lawfulness from it, allowing theories to be based on ground instances or even trillions of parameters—both of which can be found in various machine learning models—as well as, or even instead of, abstract/general laws. At the same time, taking clues from the definition of designation (section 3.1), as well as from the notion of abstract thinking, this essence explicitly incorporates the processes that combine with the structures to form mechanisms. The resulting notion of a theory as a system also then enables decomposing theories into architectures versus variable content.

5.2.1.6 Final words on the essence of *theory* The essence of *a metasystem* permits including within the notion of a theory a range of what otherwise might be considered only theorylike entities that transgress in one or more ways the strictures found in traditional definitions. A general question arises, however, with any such hypothesized essence of whether to invent a new label for its topic or to reuse an existing label whose traditional definition may be somewhat askew from this essence. With respect to the latter, there is also a question of which of several possible existing labels are most appropriate. In Rosenbloom (2012a), for example, the label *computing sciences* was used for the great scientific domain whose essence is information transformation, but other labels, such as *information sciences* and *mathematical sciences* were also considered. Here, the choice has been to go with *theory*, although other possibilities, including *model* or some new label, are also conceivable.[1]

5.2.2 Variations: From Strictures to a Definitional Space

The focus here is not on all possible sources of variations within the space of theories but specifically on those found in the definitions in section 5.2.1 and ones directly related to these that are relevant to theory evaluation (section 5.4). Thus, some of the possible distinctions between theories and models, such as level of detail and qualitative versus quantitative, are omitted. Although this does not yield an exhaustive set of sources or dimensions of variation, it does span what was considered important enough to be included in the three definitions of *theory*. In the process, it enables exploring where theories in the various sciences fall within this space (section 5.2.2.4) and yields a characterization of where cognitive architectures in particular fit (section 5.2.3).

The approach here begins with the now-nondefinitional strictures identified in section 5.2.1, but in a structured form rather than as a simple list or set. Given that the strictures arose in the context of multiple definitions, it is not surprising to find overlap among them that is reflected in the structure. There is also some structure within individual definitions that it is useful to tease out. Figure 5.1 is an attempt at such a structure, but given the vagueness of the definitions of the various terms involved, it is intended to be illustrative rather than definitive.

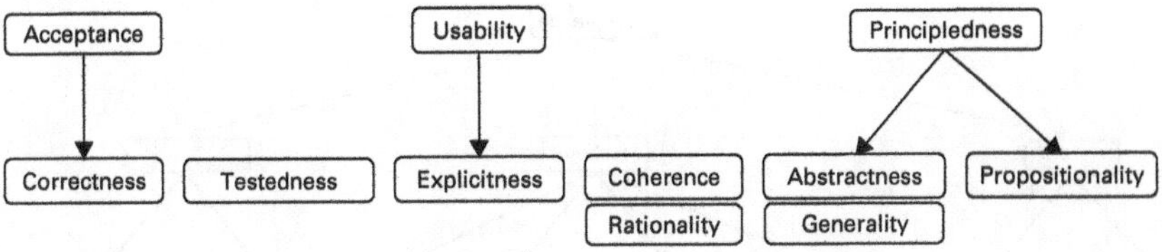

Figure 5.1
A beginning at a definitional space for theory based on its now nondefinitional strictures.

When two strictures are stacked in the figure, it denotes a rough equivalence between the two. Coherence and rationality are stacked because of a common concern with the logical relationship among the theory's elements. Abstractness and generality are in turn stacked because of a common concern with the breadth of applicability of the theory's elements. In chapter 4, generality referred to the breadth of applicability of an entire system, as is the normal usage in the context of discussing intelligence. Its usage here is different, however, focused as it is instead on individual elements of the theory. This can be confusing—and to add to the confusion, part II will switch back to the meaning from chapter 4—but it is done to stay compatible with how the terminology is traditionally used within the communities that are focused respectively on intelligence versus theories.

When instead one stricture points to another, it denotes a dependence of the former on the latter. Acceptance (of the theory by the scientific community) depends on correctness due to the phrase "commonly regarded as correct." Usability depends on explicitness, in Newell's sense of being accessible by many. Principledness is assumed to require both generality and propositionality, yielding a compound stricture.

Other ways of structuring these strictures are of course conceivable, with possibilities for adding and removing dimensions of variation as well as equivalences and dependencies among them. Figure 5.2 shows one way to significantly reengineer this structure into a single directed acyclic graph (DAG). The gray nodes provide additional dimensions of variation that (1) provide rough, but useful, equivalents to some of the existing dimensions; (2) tie together previously disjoint portions of the structure in figure 5.1; or (3) add some useful substructure. The original structure from figure 5.1 can still be seen here, although acceptance is now assumed to depend on more than it did in that figure.

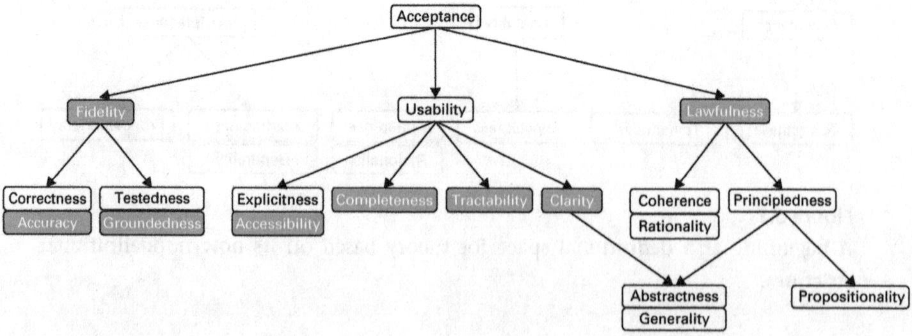

Figure 5.2
Augmented structure from figure 5.1 to include aspects for the definitional space of theory in light gray.

In figure 5.2, *groundedness* is added as a rough equivalent to testedness, making it more explicit that the testing applies to the elements of the theory rather than to the theory as a whole.[2] Correctness then does apply to the theory as a whole, but it assumes a hard criterion of correct-versus-incorrect, rather than the graded range of possibilities implied by *accuracy*. *Fidelity* then provides a new compound stricture based on the combination of these two that concerns how well the theory matches the phenomena. Usability depends on the explicitness—or, as it is termed here to reflect my understanding of Newell's usage of the term, the *accessibility*—of the theory. It has also been expanded to include a theory's *completeness*, *tractability*, and *clarity*, with clarity explicitly dependent on generality. This also makes usability indirectly dependent on generality, a connection that was not shown in figure 5.1. One additional compound stricture, *lawfulness*, combines coherence with principledness to concern the overall composition of the theory as a set of elements.

The most dramatic reconfiguration from figure 5.1 is shifting acceptance from merely depending on correctness, as implied by the Dictionary.com definition, to depend on fidelity, usability, and lawfulness. This broadens the notion of theory acceptance to include factors that may play a role in whether a community actually adopts a theory.[3] A correct theory that is intractable, for example, may be given lip service but largely ignored in practice. Likewise, a theory that is insufficiently lawful may be used when necessary while never quite being either respected or trusted.

Although figure 5.2 shows only one possible way of expanding on the structure in figure 5.1, the three new compound dimensions—fidelity, usability, and lawfulness—can yield a definitional space at a useful level of abstraction while remaining grounded in the three definitions found in section 5.2.1. This, in a sense, provides a bottom-up justification for these three new compound dimensions. Surprisingly—at least it was to me when I realized it quite far into this effort—there is also an appropriate top-down justification. Fidelity concerns the match of a theory to the phenomena it is about; usability concerns whether a theory can be used to answer questions concerning these phenomena; and lawfulness concerns the well-formedness of the theory. Thus, these three compound dimensions can be interpreted, respectively, in terms of the semantics, pragmatics, and syntax of theories.[4]

These dimensions are explored further in sections 5.2.2.1–5.2.2.3, before reappearing in section 5.4.1 in the more detailed discussion of evaluation criteria for theories. Additional criteria for evaluating theories that are not derived from these definitions, and which thus can be seen as proposals for further dimensions of the definitional space, are then explored in section 5.4.2.

5.2.2.1 Fidelity Fidelity combines the need for accuracy (or correctness) of the theory as a whole with groundedness (or testedness) of the individual elements of the theory. The need for accuracy and groundedness—and thus for fidelity—should be obvious. However, while ultimately desirable—and even required if a theory is to be applied successfully and safely—there can be many theories that have not yet reached such a status, or for which the path to such a status remains unclear. Two examples from physics, concerning string theory and general relativity, were already mentioned in section 5.2.1.1. This is also obviously true for LLMs (e.g., Brown et al., 2020), which amount to extraordinarily broad theories, but ones for which achieving groundedness, not to mention accuracy, is extremely challenging. Thus, fidelity as a whole, along with its two components, are considered sources of variation here, while also playing a key role in the later discussion of theory evaluation (section 5.4).

5.2.2.2 Usability A theory can fail to be usable in multiple ways. A lack of explicitness, or accessibility, has already been mentioned as one possibility. Three others worth considering are incompleteness, intractability, and lack of clarity.

Incompleteness implies the existence of gaps in the theory that impede answering questions of interest. However, it is not an all-or-none property, as it may vary depending on how the theory is used and on the problems to which it is applied. Completeness can vary with respect to use according to whether the need is for prediction, explanation, or other possibilities. For example, a procedurally encoded theory, such as various forms of symbolic AI systems and large neural networks, may be complete for prediction while missing much of what is needed to support explanation.

With respect to problems, some aspects of a domain may, for example, be predictable or explainable just based on an architecture, making the architecture complete for those aspects, while other phenomena may require a more complete system that also includes variable content. Returning to physics for a more concrete illustration of this point, Newton's law of universal gravitation is that $F=G(m_1m_2/r^2)$, with F a force of attraction, G a constant, and m_1, m_2, and r parameters corresponding to object masses and the distance between their centers. From the architecture here—that is, the equation plus either the specific value of G or simply the assumption that G is positive—it can be predicted that two masses will be attracted to each other. However, a quantitative prediction of the force for a particular situation requires a full system that includes specific values for all of the parameters. So, the architecture in this case is complete with respect to certain qualitative problems but not for quantitative ones.

An interesting twist on this can be found in recent work on the common model of cognition (section 1.2.3 and chapter 16). The common model is both abstract and incomplete. It also includes no ambition to be an implemented architecture. Yet it does specify a set of modules along with their functionalities and interconnections, plus sixteen assumptions about how all of this works. This might seem to provide a poor candidate for yielding even qualitative predications, yet once its modules were mapped onto functional circuits in the human brain its *connectome* could be compared quantitatively with traditional models of high-level brain architecture from neuroscience and shown to outperform them on a variety of human datasets (Stocco et al., 2021).

Intractability concerns mathematical, computational, or other pragmatic stumbling blocks that may prevent a theory from being used as intended, even if it is complete. Computationally, the notion of intractability considered here is more inclusive than the standard version from theoretical computer science, including not only problems that are in NP (for which solutions may be verified in polynomial time but for which finding such solutions may require exponential time) but also those that go beyond this to being incomputable (where no amount of computation can even in principle be sufficient to yield a correct answer), or simply require more computing power than is available.

For example, it is possible to provide a complete set of equations for the three-body problem in physics, but solving them analytically is difficult to impossible ("Three-body problem," 2024). Similarly, symbolic AI models may be intractable at scale due to the combinatorics involved in their processing, as can reinforcement learning algorithms that must be trained over combinatoric state spaces. Large neural networks, while not combinatoric themselves, were for many years intractable, and thus unusable, due simply to the amount of data and computing power required to learn their parameters.

Clarity as a factor in usability arises from the mechanistic nature of theories. In particular, it is a function of the extent to which the structures underlying the theory provide a clear path for the theory's processes to operate on them. This is most obviously an issue when the processes involve human understanding and application. A set of structures, no matter how complete and tractable, are unusable if the human who is to process the structures is incapable of comprehending and leveraging them. The same, however, can also be true for computational processing when, for example, the structures to be processed are ambiguous, such as sentences in natural language (which, as it turns out, can also be a problem for human processing of them).

5.2.2.3 Lawfulness Lawfulness was defined earlier in terms of the strictures of coherence/rationality and principledness. The need for a theory to be coherent should in principle be obvious. However, considering the essential, stripped-down notion of theory embraced in this chapter, it can't be taken as a given, yielding this first variation. For example, a model acquired via machine learning may comprise a neural network with billions, or even trillions (e.g., Fedus et al., 2022), of parameters. Such theories may be highly utile under some circumstances while maintaining little apparent coherence. Similarly, a theory may be self-contradictory, pushing the boundaries of rationality ("Theory," 2024b), while also being of potential use in, for example, generating some kinds of explanations.

The definitional requirements for "general propositions" ("Theory," 2024a) and "abstract thinking" ("Theory," 2024b) together raise two additional sources of variations that combine to yield principledness: whether the theory is built from general/abstract components rather than special cases, and whether it is encoded in terms of propositions. According to the standard notion, if something is not expressed in general terms—or does not at least have some general elements—it perhaps can be a model but not a theory. However, as viewed here, a theory can be composed of all general laws, as is typical of handcrafted theories; all special cases or individual parameters, as is typical of many theories developed via machine learning; any mixture of the two; or even potential intermediate forms. In the theories that result from *instance-based learning* (e.g., Aha et al., 1991), for example, phenomena are represented in terms of, possibly many, specific instances, each of which is easy to understand on its own but with a lack of abstractness that makes it difficult to discern any principles. Any generality exhibited by such theories is due to how the associated processes—whether in the form of case-based reasoning (e.g., Kolodner, 1992), analogy (e.g., Gentner & Forbus, 2011), blending (e.g., Gonzalez et al., 2003), or k-nearest neighbors (e.g., Taunk et al., 2019)—interpret the instances.

Propositionality in general concerns the form of encoding, which may be closely tied to the types of uses anticipated (section 5.3). A propositional encoding in terms of declarative statements does typically make explanation easier—particularly if it encodes *causal* knowledge—but if, for example, prediction is the only intended use, then a *procedural* encoding that may be executed but not directly examined may work just fine. From a theory-as-mechanism perspective, there are always structures and processes, but what are considered propositional encodings assume flexible, domain-general processes already familiar to humans—such as mathematical, logical, or verbal reasoning—implying that aboutness is concentrated in the structures and that humans are already prepared to understand these structures on their own. In contrast, procedural encodings imply that much of the aboutness may be embedded in less flexible, special-purpose processes and that humans in turn may need to understand a novel combination of both structures and processes.

As with generality, propositionality makes sense here as a dimension of variation rather than as a definitional property. It also fits naturally with the traditional controversy in AI concerning procedural versus declarative encodings in knowledge representation and reasoning (e.g., Winograd, 1975). As mentioned earlier, Newell's notion of a "body of explicit knowledge" might on the surface seem to fit here, as a call for propositional/declarative representations; however, he includes simulations—which are typically procedural—as explicit representations, so this does not actually appear to be what he intended.

As a final note, although generality and propositionality are often conflated, they are in fact distinct dimensions, with prominent examples that exhibit each without the other. Instance-based theories, for example, are canonical examples of propositional theories that lack generality, whereas there are many programs that embody general algorithms while lacking propositionality.

5.2.2.4 Positioning theories across the sciences Any individual theory may be situated at one specific location within the definitional space delimited by these three top-level dimensions or, especially for larger more complex theories, it may span a larger region of possibilities. For example, Newton's law just mentioned fully satisfies lawfulness. It also achieves fidelity, as long as we aren't in the quantum domain, and partially achieves usability. What is missing for the latter is completeness, requiring variable content in the form of parameter values to be added to enable quantitative predictions.

In general, theories in the physical sciences tend to be like Newton's law in satisfying fidelity and lawfulness. They may also be incomplete in the same sense of missing parameter values necessary for quantitative predictions. They do, however, vary in terms of their tractability and clarity. As already mentioned, the three-body problem provides one example of a physical theory that leads to intractability in use, but other such examples also exist. There may also be issues of clarity, but these most commonly involve opacity for humans not sufficiently facile with the relevant mathematics. Still, the status of physics with respect to these dimensions is likely one of the main reasons it is looked up to as the king or queen of sciences, beyond how it has also is assumed to underlie all other sciences.[5]

Theories in the life and social sciences—including the cognitive sciences—often strive for similar status with respect to these dimensions but are less frequently successful in doing so. Fidelity may fall short when phenomena are too opaque or complex to be accurately modeled. Usability may also fall short when theories are ambiguously described in natural language, or are too complex to be applied, or are missing important components. Lawfulness falls short when theories lack coherence or principledness, as is typical, for example, for any theory embodied as a learned neural network.

As argued previously, the nature of the theories that are possible for any body of phenomena is much more a property of the phenomena being modeled than of those theorizing about them, and if one is to study a body of phenomena, one must simply work with the best theories possible for it without being discouraged by negative comparisons with what is possible in other domains (Rosenbloom, 2012a). This very much holds for the cognitive sciences. Rosenbloom and Forbus (2019) not only argued for a more equal standing of theory with experimentation in cognitive science—as can be found, for example, in physics—but also for the importance of understanding the criteria for theory evaluation that are presently appropriate within the field.

5.2.3 Implications for Cognitive Architectures

The initial focus in this section is on traditional cognitive architectures and where they tend to be situated with respect to fidelity, usability, and lawfulness (section 5.2.3.1). This also includes a discussion of the variable content and systems enabled by such architectures. The focus is then shifted to machine learning architectures that are based on neural networks, along with their variable content and systems (section 5.2.3.2). Neural networks induce a different form of cognitive architecture that can yield a useful contrast to the more traditional form. Section 3.5 raised the question of whether neural networks should be considered as cognitive architectures; here we assume this but with a bit more discussion.

5.2.3.1 Traditional cognitive architectures Issues of fidelity tend to be much more challenging for traditional cognitive architectures than they are for other kinds of theories in the cognitive sciences, with both groundedness and accuracy often far from the physical ideal. Groundedness is challenging because of the internal complexity of the theories and the difficulty of evaluating the various pieces in isolation. Accuracy is, in its turn, challenging due to how it most often depends on not just the architecture itself but also on the wide range of variable content—that is, knowledge and skills—that may be required in building a system that can be compared with whatever reference data is relevant. Further affecting such positioning is the fact that many cognitive architectures tend to emphasize breadth of applicability (what is termed *power* in section 5.4.2.1) over accuracy in any one area—in a sense, they seek a single global optimization that does well overall while possibly sacrificing local optimality in any individual areas.

ACT-R (Anderson et al., 2004), with its strong basis in the modeling of human cognition, is likely the closest architecture to the physics end of the fidelity dimension. At the other end, AGI architectures (Goertzel, 2014) often have little scope to determine either groundedness or accuracy, although there are exceptions, such as AIXI (Hutter, 2005), where quantities such as asymptotic accuracy can be assessed even with no accompanying variable content. Most cognitive architectures fall somewhere in between these two ends. Evaluation of cognitive architectures—particularly with respect to accuracy—is a major, and rather unique, issue whose discussion becomes the entire topic of section 5.4.

Traditional cognitive architectures may support one or more uses but are always expected to be usable in many domains—that is, after all their raison d'être. Their accessibility is straightforward. With respect to completeness, their long-term aim is for extreme completeness in terms of the set of fixed mechanisms required for intelligent behavior while at the same time being radically incomplete in terms of the vast quantities of knowledge and skills that are also required to yield such behavior. With respect to tractability, cognitive architectures can be found across the spectrum. Some are incomputable (e.g., Hutter, 2005), others have tractability challenges in the classical computer science sense that they attempt to ameliorate in various ways (e.g., Tambe et al., 1990), while others are inherently tractable (e.g., Anderson et al., 2004). With respect to clarity, cognitive architectures specified in terms of executable code are clear for computational processing but not typically for humans reading them. So, such architectures may be accompanied by other material, such as manuals, tutorials, papers, and occasionally specifications, to assist with this. In contrast, cognitive architectures limited to on-paper designs, such as the common model of cognition, are typically—or, at least, hopefully—clear for humans reading them but not clear for computational processing.

With respect to lawfulness, traditional cognitive architectures, as with all human-devised theories in the cognitive sciences, tend to be coherent. They are also necessarily based on general mechanisms that, as implied earlier, are to apply across a wide range of problems and behaviors. Yet, although some aspects may be encoded descriptively, even mathematically or logically—and thus propositionally—typically cognitive architectures are encoded in terms of mechanisms, and thus procedurally. Beyond simply implying that they are nonpropositional, the need for procedural theories to be complete in the sense of being executable implies, as discussed briefly in section 2.5, that there are aspects included that complicate determining which portions are theoretical and which are merely atheoretical conveniences.

Some critiques of cognitive architectures as theories focus on this procedurality. Other critiques may focus on their fidelity—in particular, the difficulty of evaluating them with respect to fidelity—or the flexibility of the variable content that is required to make systems based on them executable. This flexibility can make them hard to pin down as theories that provide constrained answers to questions of interest. They thus clearly strain at multiple strictures while also spanning a number of the dimensions.

Considering the broader context in which cognitive architectures are a central component—expressed in section 5.1 as *intelligent systems = cognitive architectures + knowledge and skills*—it is also possible to consider bodies of knowledge and skills, as well as entire intelligent systems, as theories. A body of knowledge and skills will be about some phenomena or domain, but the domains themselves can range so widely that it is difficult to say anything more in general about what properties theories about them might have. Yet, theories of individual domains based on cognitive architectures, for example, a theory of language comprehension that is specified in terms of knowledge and skills, might be so analyzed.

Systems as a whole provide a major opportunity for completeness, in terms of being executable, but they also embody even more complexity, as the characteristics of any system derive from a combination of its architecture and variable content. The results may conceivably be like one or the other individually, or like some blend of the two, or even like something quite different that emerges from the interactions between the two.

5.2.3.2 Machine learning architectures based on artificial neural networks

A useful point of comparison overall for traditional cognitive architectures is the kinds of machine learning models, based on artificial neural networks, that are now so pervasive in artificial intelligence. Such systems minimally include an architecture that can represent nodes and links, learning weights on these links based on experience with many instances, and computing activations on output nodes given activations on their input nodes (e.g., figure 3.3). Yet, there is a plethora of examples where such architectures are extended with other capabilities commonly found in more traditional cognitive architectures, such as reinforcement learning (e.g., Francois-Lavet et al.,

2018; OpenAI, 2022), search (e.g., Silver et al., 2018), and declarative memories (e.g., Graves et al., 2014). Whether minimal or not, such an architecture can be conceived of as a theory of some kind of mind (e.g., LeCun, 2022), and thus as a form of cognitive architecture, if not as a brain architecture (e.g., O'Reilly & Munakata, 2000; Eliasmith, 2013).

Because of the simplicity and mathematical rigor typical of such architectures, particularly the minimal ones, they tend to be easier to test and determine to be correct—at least in terms of abstract standards of correctness if not in terms of match to human data—than other cognitive architectures (e.g., Dwivedi et al., 2022; Binz & Schulz, 2023). This is not meant to dismiss the significant efforts directed at testing and correctness for traditional cognitive architectures—particularly the long-standing ones such as ACT-R (Anderson et al., 2004), Clarion (Sun, 2016), EPIC (Kieras, 2017), and Soar (Laird, 2012)—but simply to note the inherent difference in the complexity of the problem.

Machine learning architectures based on neural networks may also be considered as propositional, given their clean mathematical expression. Still, they do possess many of the other attributes listed previously for cognitive architectures in general, including being radically incomplete in terms of being able to yield behavior on their own, and thus being far from usable for answering questions on their own.

An intelligent system based on such an architecture typically includes variable content in the form of hand-generated nodes and links and learned parameters (or *weights*). Classically—that is, with *discriminative models*—such variable content would amount to a theory of a single domain (e.g., Landhuis, 2020). However, the newer *generative models*, in the form of large language models, yield probability distributions over strings of words that effectively capture large swaths of what is known across the domains in which such language is used (e.g., Brown et al., 2020).

With respect to fidelity, accuracy tends to diverge between discriminative and generative models. It is a major trait of the more narrowly focused discriminative models, although there can be major problems with such assumptions (e.g., Szegedy et al., 2014). In contrast, accuracy is a significant challenge in generative models, leading to what are referred to as *hallucinations* (section 4.3.2.3). Still, a variety of benchmarks are being developed and deployed to attempt to both assess the accuracy of such models and to drive necessary improvements (e.g., "The Big Benchmarks Collection," 2024).

Neither form of model typically supports any significant level of groundedness, with the individual parts especially difficult to evaluate on their own.

With respect to usability, accessibility is normal, but there are distinct stories to be told with respect to completeness, tractability, and clarity. Discriminative models tend to strive for completeness with respect to an individual domain whereas generative models may strive for completeness across all of human knowledge and, to a lesser extent, human skills.[6] However, both are inherently limited by their training data and what they can extract from it, yielding gaps that can be both surprising and troublesome (e.g., Szegedy et al., 2014; Ji et al., 2022). The tractability of a neural network, in terms of the sheer amount of computing required, is directly related to how many parameters it has and the size of its training set and, thus, varies widely according to these two quantities. However, in contrast to traditional cognitive architectures, neural networks can readily take advantage of hardware acceleration of various sorts. Clarity for humans also decreases rapidly as neural networks get larger.

With respect to lawfulness, generality is typically toward the high end, at least to the extent that the training data supports it, as that is the implicit goal of the machine learning process at the heart of such systems—accuracy can be seen as the explicit goal. Coherence can be all over the board. As with completeness, it very much depends on the training data and what is learned from it. With respect to propositionality, neural network variable content and systems, in strong contrast to their architectures, are strongly procedural. Other than in rare cases where it is possible to analyze the "meaning" of a set of weights in such a network (e.g., Zhang et al., 2018), questions are only answerable by executing the networks.

5.3 Uses: Merging Perspectives and Spanning Uses

The traditional notion of *theory use* (section 5.1), in terms of such activities as prediction and explanation, is well captured by the Dictionary.com definition ("Theory," 2024a) found in section 5.2.1. Yet, this way of viewing the question reflects only one perspective worth considering. Section 5.3.1 starts with this traditional perspective before introducing several alternatives, along with an approach to merging all of them into a single, broader, yet more coherent story. This is followed in section 5.3.2 by a discussion of how theories can span multiple uses via *adaptation* and *combination* and, in section 5.3.3, by the implications this merged approach to the notion of theory use has for cognitive architectures.

5.3.1 Merging Multiple Perspectives

Starting with the traditional approach from philosophy of science, multiple perspectives on theory use are introduced and merged here into a single framework for it. These other perspectives come from machine learning, theoretical computer science, statistics, an analysis of the types of questions that theories might answer, and theory adaptation. We can start to develop this framework by noticing that a theory connects a *situation* with an *outcome*. Prediction essentially suggests outcomes given situations, whereas explanation tells us why a particular outcome occurs in a particular situation. In general, the theory's structures provide the potential to connect the situation with the outcome, whereas its processes actually realize this connection. The diversity of uses then turns out to arise from differences among how these structures are processed.

A key distinction made by the framework proposed here concerns *asymmetric* versus *symmetric* processes (see chapter 8). Asymmetric processes can be further differentiated according to whether they proceed *forward* from the situation to the outcome versus *backward* from the outcome to the situation. Processes can also be considered asymmetric, however, even when they proceed in both directions, as long as fundamentally different kinds of processing are used in the two directions—such as *deduction* (forward) versus *abduction* (backward). With symmetric processes, processing proceeds uniformly in both directions, essentially making the situation and outcome indistinguishable with respect to how processing occurs. Taking this to the extreme, a symmetric process can be viewed as eliminating any a priori distinction between which elements are part of the situation versus part of the outcome. Putting this all together, the framework proposes a tripartite distinction of forward, backward and symmetric theory uses.

5.3.1.1 Philosophy of science In philosophy of science, the outcome is the *explanandum* (or *explicandum*) whereas the combination of the theory and situation form the *explanans* (or *explicans*) (Hempel & Oppenheim, 1948). The Dictionary.com definition mentions two classical uses of theories that can be thought of in these terms: explanation and prediction. Prediction is a forward, deductive process that generates the explanans from the explanandum. In a sense, it doesn't distinguish the situation from the theory—in a logical sense, both are axioms—although it is still the theory that connects the situation to the outcome. Explanation is a rather broad notion, but in this context, it can be considered as a backward, abductive process that identifies the aspects of the explanans that yield the explanandum.

For example, given a billiard ball with an initial location, time, and velocity vector (the situation) and Newtonian mechanics (the theory) in combination as the explanans, explananda such as the time or location at which the ball will hit a cushion can be predicted in a forward direction and explained in a backward direction. Both prediction and explanation are thus asymmetric, with the difference simply being the direction of processing. This perspective does not focus on forms of symmetric processing, but that shows up next, with the introduction of the machine learning perspective.

Before proceeding, though, it is worth noting that at least one key aspect of the traditional notion of prediction is not captured simply by the direction of processing—that the explananda should be generated from the explanans prior to observing whether or not the explanans is true in the world. This temporal aspect of prediction—what may be termed *predictive power* ("Predictive power," 2024)—has always seemed awkward to me, and this awkwardness now has a name—*the paradox of predictivism* (Barnes, 2008). Oberauer and Lewandowsky (2019) rationalize this paradox by effectively including preprediction observations as part of a prediction's explanans, thus reducing the credit the theory's assumptions get—that is, their grounding—for making the prediction in such a circumstance.

Further discussion of grounding can be found in section 5.4.1, but what matters most here is that it is reasonable to assume that prediction can be considered in terms of its directionality without direct reference to its timing, with this timing instead being one factor in determining how much a match between a prediction and an observation should reflect on the theory's grounding.

5.3.1.2 Machine learning Machine learning (ML) assorts the three inputs to theory use in a different manner, lumping the situation and outcome into an *instance*, with the theory—or model, in ML terms—remaining distinct. Returning to the billiard ball example, the theory is just Newtonian mechanics, whereas what is to be predicted is when and/or where the ball hits a cushion given an initial situation, and what is to be explained is the instance of a billiard ball hitting a cushion at a specific location and time given the initial situation.

As was discussed in section 5.2.3.2, rather than distinguishing between prediction and explanation, the traditional distinction in machine learning is between discriminative and generative models. A discriminative model starts with a situation and produces an outcome. It thus maps directly onto forward, predictive processing. The distinction between whether the prediction is made pre- or postobservation is crucial here during the learning process, where one set of instances is used for training—leading prediction to occur postobservation for these instances—and another set is held aside for testing the accuracy of the theory's preobservation predictions. During performance, no observations are made; instead, predictions are used in lieu of the missing observations.

A generative model instead represents a distribution over the space of all possible instances. In probabilistic terms, discriminative models are based on conditional probabilities—P(situation | outcome)—whereas generative models are based on joint probabilities: P(situation, outcome). The latter thus induces a symmetric form of processing that naturally yields neither prediction nor explanation, although, as will be discussed shortly, it can be adapted to yield either form of behavior, with, for example, modern LLMs (e.g., Brown et al., 2020) adapted to predict what words are to come next.

As already noted, such a symmetric form of processing is missing from the standard philosophy of science story. However, if we shift from the statistical nature of the machine learning world to the deterministic nature of the traditional philosophy of science world, traditional *descriptive* theories cast in terms of mathematical equations are symmetric and can be used analogously to generative models for *verification*. A theory performs verification when, given a hypothetical instance, it responds with true or false as to whether that instance is accepted by the theory. This is simply a special case of returning the probability of a hypothetical instance via a generative model. Although verification is characterized here as a special case of generation, it will be used going forward as the generic label for this form of symmetric processing, as it is more analogous to the traditional terminology from philosophy of science.

The notion of backward, abductive explanation does not fit directly into the discriminative-versus-generative machine learning perspective, although it can be found as a topic in ML more broadly (e.g., Cai et al., 2021).[7] In fact, the importance of explaining the answers provided by deep neural networks—that is, *model transparency*—has led to considerable recent attention on this topic. Early approaches attempted post hoc analyses of learned networks (e.g., Samek et al., 2021), essentially attempting to adapt predictive models to yield explanations, although other approaches do exist, such as learning models directly for explanation (e.g., Situ et al., 2021).

Thus, both the previous perspective and this one directly include forward processing, whether it is called prediction or discrimination. At the same time, the top-level dichotomy for each is missing one of the other two forms of processing—symmetric for philosophy of science and backward for ML—although based on an understanding of the overall framework it is possible to extend each to include what it is nominally missing from it.

5.3.1.3 Theoretical computer science One of the major concerns in theoretical computer science is the time complexity of algorithms. As mentioned in section 5.2.2.2, solutions to problems in NP may be verified in polynomial time (P) but finding such solutions may require exponential time. Given that algorithms are simply mechanisms about certain classes of problems, with their inputs and outputs mappable to situations and outcomes, respectively, the distinction here can be thought of as being between the exponential computations required for forward prediction (or backward explanation) versus the polynomial computations possible for symmetric verification. In a sense, this can thus be seen as a reinstantiation of the types of theory use seen in machine learning but in the non-probabilistic context found in philosophy of science.

5.3.1.4 Statistics According to Shmueli (2010), in statistics a symmetric form of processing—known as *descriptive modeling*—is most commonly found. Yet, it is typically atheoretical, simply providing a succinct description in mathematical form of the data, with no distinction between situations and outcomes. She then goes on to provide a clear distinction between the two directions of asymmetric uses—predictive and explanatory modeling—with the former focused on predicting outcomes from so far unseen situations and the latter concerned with determining which aspects of the theory are relevant to known outcomes. All three directions of processing are thus included here, although with a largely theory-free version of symmetric processing and without structuring the three in terms of directionality.

5.3.1.5 Question types This perspective was foreshadowed by Newell's definitional focus on using theories for answering questions (section 5.2.1.3).

It focuses directly on the types of questions a theory may help in answering. To start, it is straightforward to map three kinds of standard question types onto the triad of prediction, explanation, and verification. In particular, *what*, in the form of "What is the outcome resulting from this situation?," corresponds to prediction; *why*, in the form of "Why does this outcome result from this situation?," corresponds to explanation; and *whether*, in the form of "Whether this outcome results from this situation?," corresponds to verification.

Other questions, of course, can also be considered. One of the most interesting is *how*, in the form of "How does this situation result in this outcome?" Although expressed here in the present tense, this can be a retrospective question, in the form of "How did this situation result in this outcome?," or a prospective question, in the form of "How can this situation result in this outcome?" *How* questions are asymmetric, and when prospective are typically considered as *problem-solving* if processed in a forward direction from the situation to the outcome and as *planning* if processed in a backward direction from the outcome to the situation. In either direction, prospective uses of *how* match Newell's (1990) proposed theory use of *prescription for control.*

Each question starts with some combination of the situation, theory, and outcome as input, and from this it returns a particular kind of answer. *What* returns outcomes (explananda). *Why* returns aspects of the situation and theory (parts of the explanans). *Whether* returns a Boolean truth value or probability. *How* returns a path produced by the theory that connects the situation to the outcome. Such paths may also be implicated in answering certain kinds of *why* questions.

5.3.1.6 Theory adaptation As with the perspective in section 5.3.1.2, this final perspective also stems from machine learning, but it concerns how theories are learned rather than the nature of the learned theories themselves. This form of *theory adaptation* amounts to a metatheory for how to improve theories. It thus focuses on the directionality of processing that underlies how such metatheories are used, which turns out to depend critically on the directionality of processing in the use of the theories being adapted.

Adapting forward theories is typically driven by backward processing of the same theory—that is, by some sort of explanation[8]—whether in the form of backpropagation in neural networks (Rumelhart et al., 1986), reinforcement learning in agent systems (Sutton & Barto, 2018), or the composition of rule firings in the common model of cognition (Laird et al., 2017). This general observation was reflected in assumption C2 of the common model.

Adapting-backward—that is, abductive—theories may occur when the explanans—that is, the description of the situation and the theory itself—is insufficient to explain the explanandum. It often involves hypothesizing new elements of the theory that bridge explanatory gaps in it (see, e.g., Fann, 1970), providing one tentative form of grounding for such elements. This form of *abductive theory extension* has shown up in the natural sciences, for example, with the hypothesis of the existence of *atoms*, with accompanying concerns as to whether they were in fact real or only a useful explanatory concept (Gardner, 1979). However, this assumption was ultimately grounded via more direct observation. Such theory extension generally occurs as part of the normal backward processing itself but may also incorporate some forward processing to assist in developing appropriate hypotheses. Adapting symmetric theories typically involves a symmetric learning mechanism, such as Hebbian learning (Hebb, 1949) or various adaptations of it to, for example, graphical models (e.g., Russell et al., 1995).

This theory adaptation perspective adds a metatheoretic aspect to theory use but otherwise follows the tripartite synthesis—of asymmetric (forward versus backward) versus symmetric uses—that underlies the merger of perspectives developed in this section.

5.3.2 Spanning Uses

Given any theory, it may be optimized for one or more uses; or it may be applicable to other uses, although possibly in an unoptimized form that requires significant *adaptation* (in a different sense from theory adaptation in section 5.3.1.6); or it may be completely unfit for other uses. We can get a crude, first-order sense of this by considering theories according to the tripartite distinction.

Forward theories tend to be optimized for prediction and discrimination, and for answering forms of *what* and *how* questions. But they may also be adaptable for (backward) explanation and answering forms of *why* questions by varieties of postprocessing (e.g., Samek et al., 2021) or inversion (e.g., Mitchell et al., 1986). For procedural theories, this may leverage how their sequences of actions yield implicit causal information.

Forward theories can also be adapted for (symmetric) verification, and thus *whether* questions, when situations are combined with predicted outcomes into theoretical instances that can be compared with those to be verified. However, when probabilities are involved, the conditional probabilities that yield predictions are insufficient by themselves to yield a full joint distribution without also knowing the probabilities of the situations. Mathematically, P(situation, outcome) = P(outcome | situation) × P(situation). Still, a forward probabilistic theory could be adapted into a symmetric generative model if the probabilities of individual situations can be factored in.

A symmetric theory built for verification can be used for discrimination and yield predictions by determining which outcomes are accepted for a situation of interest. To enable this, a *generate-and-test* approach can be used in which possible outcomes are generated independently of the theory with the theory then used to verify which outcome is correct for the situation of interest. When the theory is probabilistic, essentially providing a full generative model, rather than predicting a deterministic outcome a distribution would naturally be yielded over all of the potential outcomes given the situation, from which a most probable outcome can be chosen. Large language models essentially work in this manner, selecting the best next word from a distribution over the possible next words.

If a symmetric theory is propositional—such as a descriptive theory encoded as mathematical equations—and thus not built narrowly just for verification, adaptation of it for prediction can be greatly simplified if the terms can be rearranged so that the variable whose value is to be predicted is isolated on one side of an equation, while rearranging any remaining equations as necessary to yield the values required for the other variables in this equation. Extending such a theory by making it differentiable, and thus amenable to modern machine learning techniques, can further aid in this (e.g., Shoshan et al., 2024).

A symmetric theory can also yield explanations in a manner analogous to how this works for forward theories. However, ascription of causality to such an explanation is purely heuristic. A symmetric theory provides no statement of, nor means of inferring, the direction of causality. A mathematical equation is, for example, a canonical means of embodying a symmetric, descriptive theory. It may in principle reflect causality in one direction or the other, but it may also just reflect correlation, even if it is a "deterministic" correlation that is based on a coefficient of 1. The equation by itself does not enable inferring which of these possibilities actually holds.

A backward theory can yield predictions in a manner analogous to how this occurs with a symmetric theory, based on determining which out of all possible outcomes yields an explanation that is grounded in the situation of interest. The programming language *Prolog* (Colmerauer & Roussel, 1993) can be considered in these terms. Another possibility is directly inverting a backward theory to yield predictions, analogously to what has already been described in the reverse direction to enable forward theories to yield explanations. For example, this could happen by using a plan that has been developed in a backward manner to predict the future. In terms of adapting a backward theory to perform verification, an approach analogous to that taken for forward theories—in this case, of generating situations from outcomes and then using the resulting instance for verification—may work, although I have not been able to uncover any such approaches in the literature.

A simple theory may have only one native direction, with other uses being possible only through adaptation, whereas a complex theory may embody multiple native directions through *combination*. ACT-R (Anderson et al., 2004), for example, includes forward processing in procedural memory plus symmetric processing of propositions in its declarative memory and a symmetric/descriptive component in the form of equations that define activation processing. In this case, these three uses arise in complementary portions of the theory. In contrast, a generative adversarial network (GAN) (Goodfellow et al., 2014) combines a generative model with a discriminative one, but with the latter in service of producing a better generative model overall. A similar approach might combine a weak predictive theory with a stronger verification theory, in a generate-and-test manner, to yield better predictions than possible with just the predictive theory. This is akin to what was just described for adapting symmetric theories to prediction but goes beyond that by incorporating some theoretical aspects into the generator.

5.3.3 Implications for Cognitive Architectures

Section 5.2.3 contrasted traditional cognitive architectures with neural-network/machine-learning architectures while also shifting to include the latter as cognitive architectures. In this section, this shift is assumed, with the implications thus intended to apply fully across this broader class.

Not all cognitive architectures are intended to embody theories. Those intended to facilitate understanding certainly tend to do so, but those more focused on engineering purposes may or may not do so (chapter 2). Either way, when cognitive architectures are theoretical, their uses depend on three kinds of theories, according to the equation *system = architecture + variable content* introduced in section 5.1. A cognitive architecture is a theory about the fixed structures and processes that yield a mind. Variable content, encoded as skills and knowledge—even when in the form of trillions of parameters (e.g., Fedus et al., 2022)—yield content theories for particular domains; or as in the case of LLMs (e.g., Brown et al., 2020) and knowledge graphs (Ehrlinger & Wöß, 2016), broad knowledge across many domains. GPT-3, for example, is based on 45 terabytes of text from a broad range of areas (Brown et al., 2020) whereas the Google Knowledge Graph contains over 500 billion facts about five billion entities (Sullivan, 2020).

Neither architectures nor variable content typically provide theories that are complete enough to be usable on their own. That is the role for systems, which exist explicitly to provide usable theories of intelligent behavior across one or more domains. Yet, the kinds of uses intelligent systems can most naturally fulfill may depend critically on the nature of the cognitive architecture and the kinds of knowledge and skills it supports.

Cognitive architectures focused on forward processing, whether in the form of rules, action execution, discriminative models, heteroassociative networks, or action policies, are best at prediction, answering *what*—and, in some forms, *how*—questions, and comparable forms of processing. Architectures focused on backward processing, whether in the form of planning, abduction, or Prolog-like logic programming are best at explanation, answering *why*—and, also in some forms, *how*—questions, and comparable forms of processing. Architectures focused on symmetric processing, whether in the form of first-order logic, probabilistic graphical models, autoassociative networks, or generative models, are best at verification, generation, and comparable forms of processing.

Cognitive architectures that mix multiple of these forms of processing, whether in terms of a forward procedural memory and a symmetric declarative memory—as, for example, mentioned for ACT-R in section 5.3.2—or distinct modules for forward reactivity and backward planning (e.g., Bonasso et al., 1997), or a single formalism flexible enough to enable multiple forms of processing, as in Sigma, may naturally support multiple classes of uses. As with the discussion in section 5.3.2., it is also possible to adapt cognitive architectures that directly support one type of processing for other types. Part II of this book extensively explores these kinds of variations in architectural capabilities and the implications of spanning/combining them in various manners.

Cognitive architectures that adhere to the structure of the common model of cognition (section 1.2.3)—such as ACT-R, Soar, and Sigma—are built around a strong forward core, which is encapsulated as a procedural memory that directly supports *what* and *how* questions (and asymmetric learning) and a declarative memory that directly supports *whether* questions (and symmetric learning). In addition, such architectures are to be explanatory (supporting *why* questions) in the context of their procedural learning, which proceeds backward over the forward structures in procedural memory. This does not, however, naturally yield cognitively available answers to *why* questions that the system itself can articulate to others, although this could potentially be supported by leveraging the deliberative selection and execution of actions that is enabled by procedural memory and is at the center of the common model's cognitive cycle.

Irrespective of the directions of processing supported, cognitive architectures are rarely complete enough to be used for answering specific questions without the addition of relevant knowledge and skills. This is true for all of the uses just described. Yet, when such architectures are sufficiently principled, they may prove directly usable for answering more abstract questions. For example, a cognitive architecture may play a key role in explaining a whole class of coherent behaviors—what might be considered a *capability*, such as the control of behavior. This may occur due to that capability being explicitly built into the architecture, such as procedural memory for control of behavior, or because it is a consequence of the combination of multiple other capabilities, such as Sigma's demonstration of the basic reasoning processes required for distributed word representations—as found, for example, in Jones and Mewhort (2007)—via a combination of region tensors and the sum-product algorithm, plus the addition of the ability to compute *l2* normalization (Ustun et al., 2014).

In work on the Soar and Sigma cognitive architectures, a number of standard AI capabilities, such as declarative learning (Rosenbloom et al., 1989), reinforcement learning (Rosenbloom, 2012b), and neural networks (Rosenbloom et al., 2016b; Rosenbloom et al., 2017), have been *deconstructed* in terms of the existing architecture plus appropriate knowledge and skills and occasional architectural tweaks (as with the addition of *l2* normalization just mentioned). This can be thought of as providing an abductive explanation of the capability as a whole in terms of the existing architecture plus its content, with the tweaks then corresponding to the small amounts of abductive theory—or architecture/content—extension that are needed to complete the explanation. Perhaps a more traditionally grounded alternative to the term "deconstruction" would be "abduction with minimal extension."

Overall, these kinds of explanations bear a relationship to the evaluation criteria of *power* and *exuberance* to be discussed in section 5.4, and to the specific variant of functional elegance in Sigma, in that they expand the classes of phenomena that can be explained with the aid of at most minimal extensions to the overall theory.

5.4 Theory Evaluation

In section 5.2.1, an essential definition of a theory was provided that focused on *aboutness*, with an ultimate formulation in terms of *metasystems*. This was followed in section 5.2.2 with an attempt to characterize the definitional space for theories in terms, ultimately, of the structure in figure 5.2. By design, the dimensions of this space turn out to provide a good start in identifying the criteria relevant for evaluating theories.

But what does it mean to convert a dimension into an evaluation criterion? A dimension is a neutral axis of a space, whereas a criterion increases in value as you move from the less desired end toward the more desired end. Desirability itself can be a tricky concept. For some dimensions, such as accuracy, it is safe to say that higher accuracy should always be preferred, although this may be moderated if there are trade-offs with other dimensions that might compromise it to some extent. For example, large language models compromise on accuracy in service of breadth of applicability, a criterion that will be introduced shortly as *power*. For other dimensions, the more desired end may vary based on the intended use. For example, declarative theories may be more usable for explanation whereas procedural theories may be more usable for complex predictions. Similarly, theories based on general elements are more conducive for use in explanation whereas theories based on a multitude of instances may be as useful, or even better, for use in prediction.

It is also important to distinguish between absolute desirability and relative desirability (Rosenbloom, 2012a). Absolute desirability enables comparing theory quality across different domains of applicability; for example, using the same criteria for theories in physics and philosophy. Relative desirability focuses on comparing the quality of different theories for the same body of phenomena. Relative desirability is always relevant, but it is important to take more care with the application of absolute desirability. Some domains simply cannot support theories that are as desirable as those found in other domains, whether because of inherent issues of knowability (affecting fidelity among other possible criteria) or the special case nature of the phenomena (affecting generality among other possible criteria), or simply our current state of understanding of the phenomena (affecting many possible criteria). When absolute comparisons go beyond comparing the theories themselves to value judgments of either the inherent importance of the domain or the quality of the researchers working on it, this can yield false and counterproductive results.

The main limitation of figure 5.2 with respect to identifying appropriate evaluation criteria is its having been restricted to strictures eliminated from existing definitions when determining the essence of the notion of a theory—that is, those shown in figure 5.1—plus additional aspects in figure 5.2 that help in better understanding and structuring these strictures. This process provided no guarantee that the results would fully encompass the space of relevant evaluation criteria. Figure 5.3 thus starts with the structure in figure 5.2 and attempts to fill in the apparent gaps. It is based on what I have read and experienced over my research career, so it too may be incomplete, or even inappropriate in places, as well as providing just one possible perspective on this topic.[9] Still, in toto, it provides a rich space of evaluation criteria for theories that has not previously been assembled in one place.

This figure adds (1) substructure under clarity, including splitting up the direct dependence on generality shown in figure 5.2 via an intermediate dependence on minimality; (2) a new high-level compound criterion of *beauty* as part of acceptance, along with its substructure; and (3) three additional compound terms—*comprehensiveness*, *rigidity* and *surprisingness*—that at this point are not considered part of acceptance. The addition of these last three criteria converts the figure from a single DAG into four interconnected DAGs.

The remainder of this section explores the implications of the resulting structure for theory evaluation. Section 5.4.1 explores the three high-level criteria implied by figure 5.2—fidelity, usability, and lawfulness. Section 5.4.2 explores the four new high-level criteria of beauty, comprehensiveness,

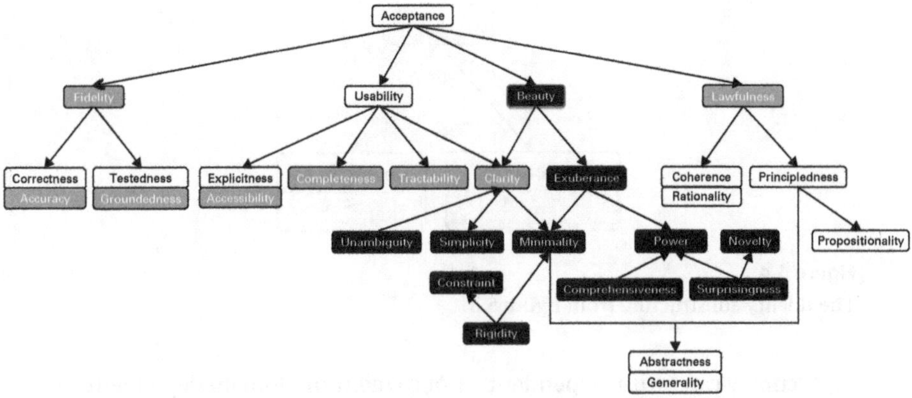

Figure 5.3
Augmented structure from figure 5.2 to include additional aspects for theory evaluation in dark gray.

rigidity, and surprisingness. Section 5.4.3 then examines the implications of all of this for evaluating cognitive architectures.

5.4.1 From Dimensions to Evaluation Criteria

This section considers the dimensions of fidelity, usability, and lawfulness—all three of which were previously definitional—as criteria for evaluating theories. Reformulated from section 5.2.2, fidelity evaluates how well a theory matches the phenomena it is about; usability evaluates how practical a theory is for answering questions concerning these phenomena; and lawfulness evaluates how well formed the theory is. As noted earlier, these three general evaluation criteria thus correspond, respectively, to the semantics, pragmatics, and syntax of theories.

5.4.1.1 Fidelity Fidelity (figure 5.4) begins with accuracy, which is at the very heart of theory evaluation. Without accuracy, theory use may sometimes prove fruitful, but it may just as easily prove ineffectual or even deleterious. For a theory in the sciences, or more generically the realm of understanding, accuracy typically is measured by comparing predictions against the reality of the phenomena to be understood. For a theory in engineering, or more generically the realm of applications, accuracy may instead be measured against either some abstract ideal or some concrete benchmark created to approximate the ideal.

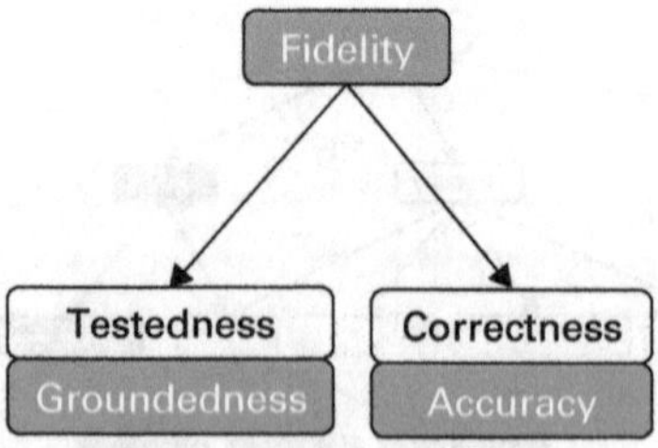

Figure 5.4
The fidelity substructure from figure 5.3.

Accuracy can be use dependent—not to mention domain dependent—as a single theory's accuracy may vary as a function of whether it is evaluated with respect to prediction—the canonical case—versus explanation, versus verification. The methodology of *controlled experiments* is geared toward determining accuracy, most often predictive accuracy but conceivably also explanatory or verificatory accuracy. This methodology also helps to bifurcate the notion of lack of accuracy into whether a theory is inaccurate—that is, does not do well in controlled experiments—versus of unknown accuracy due to a, possibly temporary, lack of appropriate experiments.

Despite the centrality of accuracy in evaluating theories, it is important to acknowledge that optimizing for accuracy in isolation is not always the best strategy in developing and using theories. In some cases, accuracy may need to be compromised to yield improved usability, such as by making an intractable theory tractable or an unclear theory clear for explanatory purposes. We still, for example, teach Newtonian mechanics despite its known inaccuracies in the quantum world. Some accuracy may also be sacrificed in favor of a theory that can apply more broadly (see the criterion of *power* in section 5.4.2). This comes up explicitly in section 5.4.3 for cognitive architectures, with the twist of it yielding a long-term research strategy that is akin to abstraction planning (e.g., Sacerdoti, 1974) for ultimately developing a theory that is both broadly applicable and accurate.

Groundedness can be viewed as applying the criterion of accuracy not to the theory as a whole but recursively to the individual elements out of which it is composed. This should be simpler in general than determining overall accuracy. However, it can be difficult if the elements are not easily separable. There are also cases where the pieces themselves may be as hard to evaluate as the theory as a whole, such as when an AI theory includes

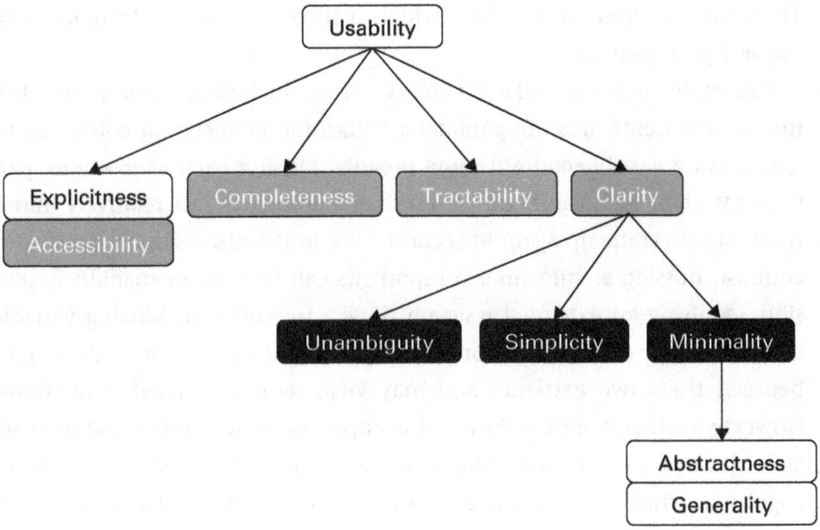

Figure 5.5
The usability substructure from figure 5.3.

components that are themselves *AI complete* (see, e.g., Shapiro, 1992). In other words, for such a case, success on the individual components requires a full solution to the general AI problem of producing intelligent behavior. Still, in principle, a well-grounded theory that is applied rationally (see section 5.4.1.3) should yield accurate results. As with accuracy, controlled experiments have a strong role to play in grounding. However, they play little to no role with respect to any of the remaining evaluation criteria.

Following up on section 5.3.1's discussion of the paradox of predictivism, when an observation occurs subsequent to a prediction—or, following the lead of Oberauer and Lewandowsky (2019), when the explanandum is not included as part of the explanans—it can contribute not just to assessments of the accuracy of the theory as a whole but also to the grounding of its pieces.

5.4.1.2 Usability As discussed in section 5.2.2.2, usability combines accessibility, completeness, tractability, and clarity (figure 5.5). Not much needs to be said here about accessibility, or explicitness in Newell's terminology. Clearly if a theory is only within the mind of a single individual, it is usable only for that one individual, whereas if it is more broadly accessible, it becomes potentially available for use—and criticism and revision—by the entire community of individuals concerned with the phenomena of interest.

Thus, this criterion impacts how widely a theory can be available for both use and improvement.

Incompleteness can take on many forms, including missing architectural components, missing parameter values for architectural components, and missing variable content more broadly. Missing parameter values, particularly when such values can be estimated during use, are relatively minor issues but do come up again in section 5.4.2 in the discussion of rigidity. In contrast, missing architectural components can be a major roadblock, possibly requiring an extended research project to work out. Missing variable content, when considered more broadly than just parameter values, falls between these two extremes and may yield more graduated evaluations. However, with any of these forms, it is important to remember that they are all with respect to intended usage. So, for example, the common model of cognition, which is incomplete with respect to full humanlike intelligence in terms of both architectural mechanisms and variable content may still be sufficiently complete with respect to its intended use as a theory that reflects the best current consensus across the field—a use that can be considered as a variation on explanation.

Intractability is more subtle. A theory may be complete in principle with respect to a use while still not supporting its intended purpose. Neural-network learning was intractable in this sense for much of its early history, as were more traditional AI-based game-playing programs (Levy, 1988) and reinforcement learning systems (Sutton & Barto, 2018)—the first because of the sizes of the networks needed and the amount of data needing to be processed, the second because of the combinatorics of the search required, and the third because of the number of experiences that must be processed.

AIXI (Hutter, 2005), an attempt at a simple mathematical approach to general intelligence, combines reinforcement learning with universal induction (Solomonoff, 1964), an approach to induction that yields key asymptotic guarantees while failing miserably on tractability—it is in fact incomputable. Although there may still be some uses of it as it is, it is possible to consider deep reinforcement learning (Francois-Lavet et al., 2018) as substituting neural-network learning for universal induction to yield a tractable approximation for a much wider variety of uses. It thus sacrifices some accuracy, even if only theoretical or asymptotic, for usability.

Clarity gets at the interpretability of a theory by both man and machine. This combines lack of ambiguity with simplicity and minimality. Expressing a theory in a formal language, such as mathematics or logic, is one way to minimize ambiguity. The well-defined processes associated with such languages tightly constrain the interpretations of structures expressed in them, yet without constraining how they map to the phenomena of interest. Embodying a theory in executable computer code provides a comparable constraint on interpretation. However, this may also introduce an additional ambiguity, as often seen with cognitive architectures, of which aspects of the code express the theory versus which exist only to make the theory executable (section 2.5). Further, when clarity with respect to a human is considered, how clearly the code is written and how well it is documented also play large roles. Theories expressed in natural language can be quite ambiguous, with those expressed in semi-formal languages existing at intermediate points along this dimension.

Simplicity concerns how easy it is to interpret the theory, independent of either any ambiguity or the sheer size of the theory (i.e., its lack of minimality). For example, a theory expressed obscurely in natural language would lack simplicity, whereas other more straightforward ways of expressing the same theory in natural language may be quite simple. It can also be the case that expressing a theory in a formal language, while reducing its ambiguity may reduce its simplicity, yielding a trade-off within the criterion of clarity itself.

Minimality may at first seem confusable with simplicity, but instead of concerning the form in which the theory is expressed, it concerns the theory's size, essentially in terms of the number of elements involved in defining it. Measuring theory size may be tricky, and may be done in multiple ways, but it is the core notion behind traditional criteria such as Occam's razor. Minimality depends on generality because the more general the theory's elements are the fewer of them that are required.

5.4.1.3 Lawfulness Figure 5.6 shows the structure of lawfulness, although much of what needs to be said about it with respect to evaluation has already been sufficiently foreshadowed by the discussion of it as a dimension of variability in section 5.2.2.3. In brief, coherence and rationality both concern how well a theory holds together, whereas principledness concerns whether the elements take the form of principles; that is, whether they are stated as abstract/general propositions.

Handcrafted theories automatically tend to do well when evaluated according to coherence/rationality and abstraction/generality, whereas

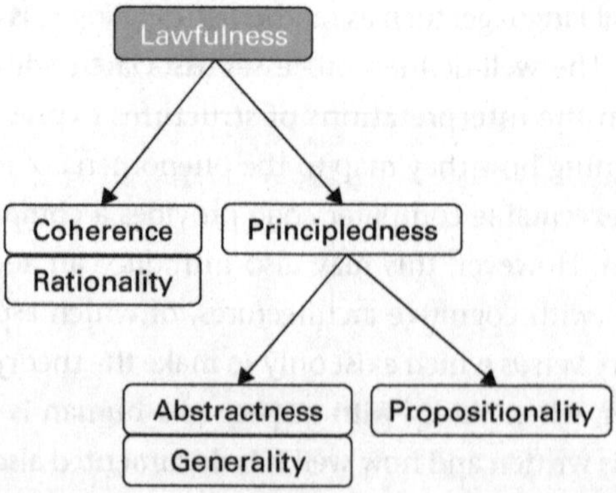

Figure 5.6
The lawfulness substructure from figure 5.3.

many theories created via machine learning—particularly those that are either instance based or based on many parameters—do not. Yet, it is worth noting that these criteria matter much more for some uses than for others. They are clearly, for example, much more critical for explanation than for either prediction or verification; although, even for these latter uses a coherent set of abstract rules may be much easier to test/ground than massive sets of either instances or parameters. Thus, there are good reasons to prefer theories that are more coherent/rational and abstract/general, while acknowledging that these criteria may not always be essential.

Propositionality can be thought of in terms of declarative versus procedural encodings. Many computational theories are encoded procedurally, and although they may be very general and effective for prediction, they can leave much to be desired for either explanation or verification. As discussed in section 5.3.2, there may be ways to adapt such theories to these latter uses, but it is typically still easier to adapt propositional theories.

5.4.2 Additional Evaluation Criteria

This section considers four criteria for evaluating theories that were not derived from definitional strictures: beauty, comprehensiveness, rigidity, and surprisingness. Beauty concerns the aesthetics of a theory. Comprehensiveness concerns whether the range of phenomena covered by a theory is exhaustive with respect to some natural maximal domain. Rigidity, in a sense, concerns a theory's necessity as an integrated whole. Surprisingness concerns how a theory yields an important form of value added, in terms of providing more than what was expected of it.

5.4.2.1 Beauty As hinted at earlier, beauty can be seen as an extension of the triumvirate of syntax, semantics, and pragmatics to include aesthetics. However, it can also be seen as embodying an assumption—or heuristic—that the truth of how the universe works should be beautiful; what Ritz and Vaidyanathan (2025) refer to as *useful beauty*. This assumption seems to have its origins in physics, where many of the theories are clearly beautiful, although the use of this criterion remains controversial even there (Hossenfelder, 2018). Beauty is a notion that is hard to define in general, but with respect to theories it may be possible to think of it as some combination of clarity and *exuberance* (figure 5.7).

Clarity has already been discussed sufficiently in section 5.4.1.2. Exuberance (Wilczek, 2016), or *explanations with reach* (Deutsch, 2011), concerns a theory's breadth of applicability—that is, its *power*—plus how much must be stated in the theory to yield these predictions (i.e., its minimality, to double-dip into an aspect of clarity and its dependence on generality). It can in fact be considered as a composite criterion based on the ratio of the former to the latter.[10] It is a criterion I have personally felt driven to use, although informally, with respect to cognitive architectures, in the form of functional

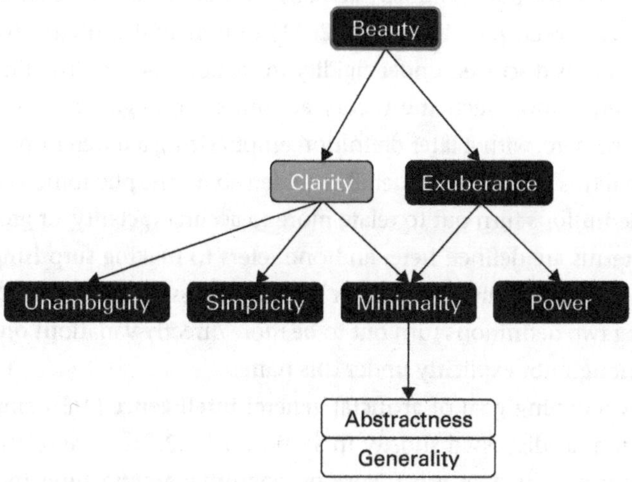

Figure 5.7
The beauty substructure from figure 5.3.

elegance (section 1.2.2, chapter 14, and section 15.4), although it is even further from being commonly accepted in that community. Occam's razor has already been mentioned under minimality, but a version of it can also be considered as a special case of exuberance—when the power is constant across multiple theories, exuberance prefers the one with the smallest statement. Exuberance also can be seen to align with approaches concerned with the level of compression embodied by a theory (Wolff, 2013).

As just mentioned, power—or more technically *explanatory power*—concerns the breadth of applicability of a theory (i.e., what was referred to as generality in chapter 4). Because of the inherent ambiguity in the use of English terms, power may seem confusable with both completeness and generality (as this term has been defined in this chapter). However, as used here, each has a distinct meaning. Power refers to the range of phenomena covered by the theory. Completeness refers to a lack of missing elements that are necessary to answer questions within this scope. Generality refers to the scope of the theory's individual elements rather than to the scope of the theory as a whole. In analogy to what was seen for fidelity, power is to generality as accuracy is to groundedness.

Power can conceivably be measured either in terms of the number of phenomena covered or the number of classes of phenomena covered, an approach known as *consilience* (Whewell, 1840; Thagard, 1993). Wikipedia ("Explanatory power," 2023) actually lists eight distinct definitions of explanatory power, plus Deutsch's (2011) notion that a theory should be hard to vary (as discussed under rigidity in section 5.4.2.3). The first definition concerns how much the theory accounts for, aligning with how the term is used here, with a later definition emphasizing a different perspective on this in terms of the level of detail provided about the phenomena covered. Several definitions turn out to relate more to accuracy, clarity, or grounding, as these terms are defined here; and one refers to making surprising facts a matter of course (as discussed under surprisingness in section 5.4.2.4). The remaining two definitions turn out to be more directly variations on exuberance, although not explicitly under this name.

The overarching goal of artificial general intelligence (AGI) emphasizes power—and, as discussed shortly in section 5.4.2.2, its close relative comprehensiveness—as does most work on cognitive architectures in general. AIXI, as introduced in section 5.4.1.2, can be seen as an exuberant and even beautiful attempt at a simple mathematical approach to AGI, even with its issues with respect to tractability.

It might typically be assumed that power comparisons between theories only matter for theories of comparable accuracy; however, as hinted at in sections 5.2.3.1 and 5.4.1.1, an argument can be made that as a route to an ultimate theory that maximizes both accuracy and power, it may be more effective to prefer powerful theories early on, even when less accurate, as it may be easier to improve their accuracy by small tweaks than it is to improve the power of more accurate theories by comparable changes.[11]

In being partly based on clarity, beauty's definition here may overlap with how it is used in the arts, but it can also diverge in significant ways—ambiguity, size, and complexity can all conceivably play significant roles in the beauty of artistic creations. Exuberance is also a term that can be relevant in the arts, as, for example, found in the baroque period that spanned parts of the seventeenth and eighteenth centuries. But the meaning here specifically reflects how much phenomena can be spanned—that is, how much power is achievable—by how little theory. Power too is a term found in the arts, but there it instead concerns the impact on the audience, or "user."

5.4.2.2 Comprehensiveness As shown in figure 5.8, comprehensiveness depends on power, but it also depends on a notion of what the maximal domain is that isn't itself considered a criterion here. When physicists talk about a "theory of everything" (2024), they have in mind a comprehensive theory of the universe. More on point, unified theories of cognition (Newell, 1990) and AGI (Goertzel, 2014) strive for comprehensive theories of natural or artificial intelligence. Comprehensive cognitive architectures contain all of the mechanisms required to yield general intelligence, whether the goal is human(like) or artificial intelligence. Comprehensive systems combine

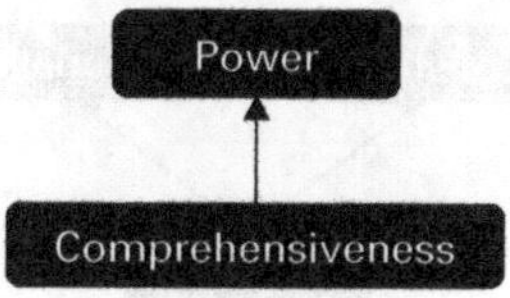

Figure 5.8
The comprehensiveness substructure from figure 5.3.

comprehensive architectures with all of the variable content required to perform intelligently across, at least, the domains of human performance.

5.4.2.3 Rigidity *Rigidity* (figure 5.9) can take on two different forms. The dominant one concerns the latitude enabled by a theory's variable content, with the classical version simply being a preference for fewer parameters. This thus involves a form of minimality, an aspect of clarity already mentioned in section 5.4.1.2 but here focused not on clarity per se but on minimizing the flexibility enabled by the variable content. As discussed in section 5.4.3.5, this notion gets quite murky for cognitive architectures, where the variable content must by definition yield significant flexibility.

The second form of rigidity concerns how constrained the theory is to take on a specific form. In other words, even small changes to a rigid theory can alter it for the worse. Deutsch (2011) focused on a theory being *hard to vary* in the sense of any changes that are made to it altering its predictions, and thus potentially reducing its accuracy. In a complex theory within which the elements mutually constrain each other, changing any one of them may also make the theory as a whole inoperable, essentially damaging its completeness to where it can no longer answer questions that it previously could. This form of rigidity may be considered desirable for the architectural portion of a theory—ensuring some sense of inherent "rightness" or optimality for the theory as stated; however, nonrigidity is the essence of variable content. Moreover, if the variable content is flexible enough, assessments of this form of rigidity with respect to the architecture can be complicated when changes to the architecture can be compensated for by changes to the variable content.

5.4.2.4 Surprisingness *Surprisingness* at its core relates to novelty, but this can come in at least three forms. The first occurs when there are predictions from the theory that come as a surprise. This form of surprisingness is shown

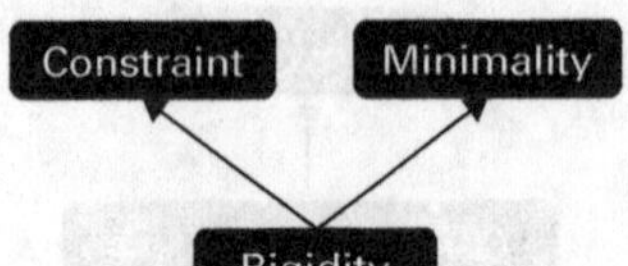

Figure 5.9
The rigidity substructure from figure 5.3.

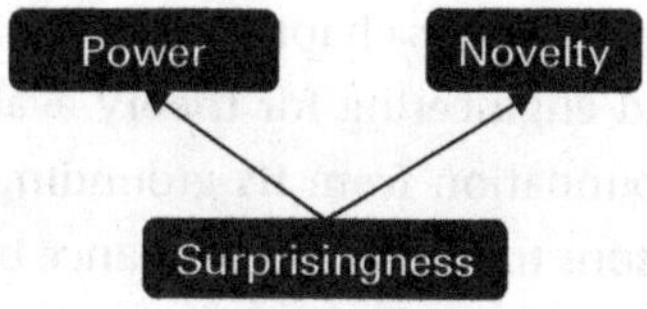

Figure 5.10
The surprisingness substructure from figure 5.3.

in figure 5.10 via dependencies on both *novelty* and power, with the latter signifying that surprisingness arises from what phenomena are covered by the theory. If validated, this form of surprisingness should provide strong support for a theory's accuracy and grounding that goes beyond controlled experiments that simply validate anticipated results. It is a major factor that is absent when the results are known before the theory is developed.

The second form of surprisingness occurs when the theory itself is surprising; that is, it is not a straightforward follow-on from what has come before. In figure 5.10 this depends just on novelty. This form of surprisingness can be considered as more an evaluation of the inventiveness of the theorist than of any inherent property of the theory itself. A third form relates more to the nature of the phenomena than either the theory or the theorist. It corresponds to one of Wikipedia's definitions of "Explanatory power" (2023), as to whether the theory reduces surprising phenomena to normality. If phenomena are surprising because they are not predicted by existing theories, then this says something about the novelty of a new theory that does predict them, as well as the power of the new theory in covering such phenomena. It also, however, raises the paradox of predictivism (sections 5.3.1 and 5.4.1.1), as the phenomena must already be known before the new theory is created for them to have been surprising.

5.4.3 Implications for Cognitive Architectures

For years, how to evaluate the large complex theories provided by traditional cognitive architectures has been a major concern in the field, with whole events devoted to the topic (e.g., Kaminka & Burghart, 2007) as well as other more individual attempts to cope with the issue (e.g., Varma, 2011). This section focuses on evaluation of these types of architectures, but with mentions of both neural-network and reinforcement learning architectures when especially relevant.

Of the criteria discussed in this chapter, accuracy is dominant in general across both science and engineering for theory evaluation. Cognitive science is built on this foundation from its grounding in experimental psychology, with comparisons to human performance being at the center (see, e.g., Bower & Clapper, 1989). In artificial intelligence the dominant criterion at this point is also accuracy, but here measured against standardized benchmarks (see, e.g., Olson et al., 2017).

For cognitive architectures, power (i.e., applying across a broad range of domains), completeness (i.e., embodying a broad range of humanlike cognitive capabilities, such as memory, learning, and natural language), and comprehensiveness (i.e., yielding general intelligence) come demandingly into the foreground, leading to a tussle for dominance between accuracy on one side and these other three criteria on the other. In cognitive science, the accuracy of cognitive architectures typically remains preeminent, but power, completeness, and comprehensiveness also remain critical. The Newell test (section 4.2.1) and its ilk (e.g., Sun, 2004; Langley et al., 2009), for example, concern lists of functionalities that cognitive architectures should support. In artificial intelligence, the balance for cognitive architectures is tilted more toward power, completeness, and comprehensiveness (e.g., Langley et al., 2009; Kotseruba & Tsotsos, 2020). In artificial general intelligence, comprehensiveness is essentially definitional, but in service of this, lists of implicated functionalities can also be found (e.g., Goertzel, 2014). Accuracy tends to be even further in the background here.

Beyond overall disciplinary priorities, it is possible to understand much about individual architectures, as well as their similarities and differences, from the criteria that they foreground, whether explicitly or implicitly. This is explicit with Sigma from the set of four desiderata that have driven its development (section 1.2.2). Grand unification is all about completeness and comprehensiveness, in extending the scope of cognitive architectures beyond traditional high-level cognitive processing to such things as perception and motor control. Generic cognition can be seen as being about a form of power, in the sense of modeling both human and machine cognition, but also a form of minimality in assuming that both can result from a common set of mechanisms. Functional elegance most closely aligns with exuberance, while in the process relating to power, minimality, beauty, and generality. Sufficient efficiency relates to usability, particularly in the form of tractability.

Overall, completeness and comprehensiveness can be seen as the dominant criteria in Sigma's desiderata but with nods also to generality, minimality, power, exuberance, beauty, and tractability. The absence of fidelity from this list is notable. This does not mean that accuracy (or groundedness) does not matter, but that it is not part of the driving force of architectural development, and thus while in focus at times it is not pursued pervasively. Lawfulness is also not foregrounded as it might be in an architecture that is based on logic (e.g., Genesereth & Nilsson, 1987). Still, generality does play a direct role, and coherence/rationality indirectly takes on a key role through the construction of Sigma's cognitive architecture on top of a graphical architecture that is an adaptation of the clean formalization of factor graphs (Rosenbloom et al., 2016b). With respect to usability, tractability is foregrounded and there is an unstated requirement to have a functioning computational system that implies a need for both accessibility and completeness. Clarity does play a role, but it is definitely in the background, to the detriment of potential users.

The two leading cognitive architectures—ACT-R and Soar—are less explicit in their desiderata, but clearly foreground power, completeness, comprehensiveness, and accuracy, while differing in terms of how much of this emphasis concerns human versus machine phenomena. They also place a much greater emphasis on usability overall—with significant attention to clarity—given that each supports an active community of users. As such, acceptance may be considered as foregrounded by both as well. There is almost always a desire for an architecture to be accepted by the community, but explicitly attracting and supporting a large user group is a considerable step beyond this in foregrounding acceptance. Tractability has been a key focus in Soar over the years (e.g., Tambe et al., 1990; Derbinsky et al., 2012), but with it now generally fast enough for its current ambitions. Both ACT-R and Soar do seem to place less emphasis than Sigma on beauty, exuberance, and minimality. Minimality and exuberance were foregrounded in the early version of Soar but became less central to the late version as more memory and learning mechanisms were added (section 1.2.1).

Together, these three architectures helped to inspire the common model of cognition. Although both power and fidelity are important for the common model, its dominant criterion is its overall acceptance. It is to include only those aspects of cognitive architectures for humanlike cognition about which a consensus can be reached within the overall community. It thus willingly sacrifices completeness—it is an abstract, on-paper design not intended to be implemented as is—and comprehensiveness to limit itself to just those aspects for which a consensus can be reached.

In section 2.6 a somewhat ad hoc set of four evaluation criteria were introduced for cognitive architectures:

(1) whether an architecture functions as intended,
(2) how broadly the architecture models and/or produces the phenomena of interest,
(3) how simple and elegant the architecture is,
(4) and how much insight the architecture inspires and captures, whether into anticipated implications or wholly unanticipated ones.

Starting more broadly and systematically from the notion of theories in general, a different perspective on evaluation criteria has been developed here that centers on acceptance and its four main pieces—fidelity, usability, lawfulness, and beauty—plus comprehensiveness, rigidity, and surprisingness. Still, the four prior criteria can be understood in terms of those identified here. The first maps onto fidelity; the second onto power and comprehensiveness; and the third onto aspects of simplicity and minimality, while extending up from there to clarity and beauty. A combination of the second and third yield exuberance, which also then extends up to beauty. The fourth roughly maps onto aspects of surprisingness, either in terms of making what is already known but perhaps insufficiently explained normal or in yielding new surprises.

The remainder of this section explores in some detail the implications of fidelity, usability, lawfulness, beauty, comprehensiveness, rigidity, and surprisingness for evaluating cognitive architectures in general. Acceptance by a particular research community, or beyond this to a broader scientific community, is assumed to flow from some combination of the first four of these criteria, with the details varying by architecture and community. Acceptance will not be explicitly addressed further here, beyond what has already been said about it.

5.4.3.1 Fidelity How to evaluate—or, perhaps more bluntly, whether it is possible to evaluate—the accuracy of a traditional cognitive architecture has been a long-term concern on which the field unfortunately has made little progress despite considerable effort in the form of workshops and papers (e.g., Kaminka & Burghart, 2007; Varma, 2011). This is partly due to the complexity of such architectures, but a larger share is likely due to the lack of rigidity resulting from the variable content, and thus in any resulting system that is able to produce results that can be evaluated, whether in comparison to human performance or a synthetic benchmark. But, without evaluations of accuracy, it is hard to tell when progress on this critical criterion is being made with respect to either individual architectures or the field as a whole.

The field has neither standardized benchmarks nor competitions, with a key reason being that each architecture falls short of comprehensiveness in its own way—in the sense of which architectural capabilities are missing from the total of what is needed for a complete mind that can operate across many domains when possessing the appropriate knowledge and skills—and so it is hard to create either benchmarks or competitions that are not biased toward one architecture or another. Yet, lacking such standardized means of evaluation deprives the field of both the concrete benefits of being able to measure such progress and the more abstract intellectual/academic respect that comes from being able to perform such comparisons. This, in turn, likely reduces both the availability of funding and the participation by graduate students who consider the ability to excel in such contexts essential for building the kind of scientific career they seek.

I am not going to propose an overall solution here, as I have no better idea about how to do this than anyone else, but I do want to mention two types of exceptions and two relatively new possibilities. As to exceptions, there appear to be at least two general classes: (1) minimal architectures where the focus is on how well they support the vast amounts of appropriate types of variable content necessary for high accuracy, and (2) bare-bones tasks—or those that Newell would have characterized as being "close to the architecture" (Rosenbloom & Laird, 1993)—where the variable content plays a minimal role in yielding accuracy.

The first class typically consists of simple AI architectures with learning, such as artificial neural-networks and reinforcement learning, but may also include slightly more complex architectures, such as those found in deep reinforcement learning (Francois-Lavet et al., 2018) and neural-network-based game-playing programs (e.g., Silver et al., 2018). The second class typically focuses on architectures from cognitive science that are used to model phenomena at such short timescales that the architecture shows through what little variable content is required. ACT-R has been especially good at this approach over the years (Anderson et al., 2004). Other architectures, such as Clarion (Sun, 2016) and EPIC (Kieras, 2017), may also sit within this class; and Soar has also played in this space at times (e.g., Newell, 1990).

One relatively new possibility worth considering is leveraging an abstract consensus, such as the common model of cognition. With such a consensus it may be possible to evaluate abstractly both (1) how accurate the common model is with respect to humanlike cognition, and (2) how accurate any cognitive architecture is with respect to the common model. The first form of evaluation enables talking about progress in the field as a whole. The second form enables talking about progress on individual architectures (although with the caveat that at times individual architectures may be more accurate than the common model).

As mentioned in section 1.2.3, and explained in more detail in Laird, Lebiere, and Rosenbloom (2017), it is now possible to talk about how both ACT-R and Soar differed from the common model back in the 1990s in terms of its sixteen assumptions, as well as how they have both converged today to be much more well aligned with it. This analysis also enables comparing the two architectures with each other based on these same sixteen assumptions. More generally, it may prove possible to create benchmarks and contests around the assumptions embodied by the common model, while ignoring those not required by it, to evaluate architectures quantitatively in terms of how well they reflect the consensus.

The other relatively new possibility is to, in effect, transform traditional cognitive architectures into members of the first exception class, by forcing nearly all of their variable content to be learned from a sequence of experiences that can be replicated across architectures. *Interactive task learning* (Gluck & Laird, 2018)—where systems learn from a combination of exploration in the domain of interest and interaction with an expert human—provides one such path. A slightly different variant of this approach might measure how well different architectures can do on various problems when initialized with whatever content can be extracted from a particular knowledge graph (e.g., Dancy, 2022) or large language model (e.g., Ribeiro & Forbus, 2021).

Grounding for cognitive architectures considers the support existing for its individual components. As discussed in section 5.4.1.1, this effectively involves recursively applying the criterion of accuracy to the individual components of the theory. Thus, much of what has just been said about accuracy applies here, although grounding can be somewhat simpler for complex architectures because it decomposes them into pieces that may be evaluated separately, at least to the extent that their impact can be divorced from the effects of the other pieces.

When considering the accuracy of not just individual pieces, but of combinations of them, things are inherently more complex. Unexpected interactions can yield novel predictions—that is, the first flavor of surprisingness—one of the more interesting things that can come out of a theory. In addition to surprisingness, this also relates to exuberance.

Two relevant examples with which I have been closely involved are the range of learning behaviors that came out of adding a simple procedural learning mechanism (i.e., chunking) to Soar (e.g., Steier et al., 1987); and (2) the emergence of *base-level activation* (Anderson et al., 2004) as a temporal prior when episodic learning was developed in Sigma (Rosenbloom, 2014). Some of these unexpected interactions may prove "right," in terms of however accuracy is measured, and thus act as compelling further evidence for the theory. The addition of chunking to Soar, for example, still appears to be "right," with the resulting incorporation of an abstraction of it, and of the comparable mechanism from ACT-R, into the common model. Further evidence is provided by how an analog of it can now also be found in the world of generative AI (Yu et al., 2024), although with no apparent awareness of this earlier work in cognitive architectures and the common model.

Other unexpected interactions may prove "wrong" and thus implicate theory revision or even abandonment. A good example of this involves declarative learning and memory in Soar. These capabilities were first found to emerge as an interaction between chunking, problem-solving, and bits of appropriate content (Rosenbloom et al., 1989). But this approach ultimately proved to be too cumbersome—in particular, requiring too much disruptive deliberation to learn facts while pursuing other tasks—and was therefore replaced by new purpose-built architectural mechanisms in the late version of Soar.

It was argued in section 2.6 that exploration is a central activity in investigating cognitive architectures. For cognitive architectures, exploration partly concerns individual pieces of the architecture and partly concerns attempts to apply the entire architecture. However, it also implies a significant concern with the interactions among the architecture's various pieces. Evaluation of these interactions is an activity that seems to fall on a dimension that spans from grounding and its focus on individual pieces to accuracy and its focus on the entire system.

This concern with interactions may not be as obvious for the kinds of mathematical theories traditionally found in disciplines such as physics, as it might be expected that if each of the individual equations is accurate that their combination must be so as well. This doesn't rule out the possibility of surprising interactions, which may be touted as major discoveries—or at least hypotheses—but does seem to presume that they are relatively rare. However, for architectures, the complex pieces out of which they are built and the diversity of ways in which they can be combined yield a rich space of possible interactions that cries out for exploration and accompanying evaluation. Whether this is considered as an extended form of grounding, or as an aspect of accuracy, or perhaps as something in between, its importance is clear.

5.4.3.2 Usability There is nothing specific worth saying about accessibility here. With respect to completeness, much has already been said with respect to cognitive architectures, including their dependence on large amounts of variable content to yield a full system. But they also offer a possible midpoint, where they start incomplete but utilize learning to become self-completing over time. This could amount to simply learning a few architectural parameters—including what is known as *hyperparameter optimization* when the parameters affect the learning algorithm itself (Feurer & Hutter, 2019)—or entire bodies of knowledge and skills. Evaluations of completeness may also be relevant with respect to the capabilities embodied in such architectures, whether the issue is with the abstractness of the specification of such capabilities—as in the common model—or even their existence.

With respect to tractability, one of the earliest issues to arise with traditional cognitive architectures was their potential for generating combinatorics. One main source of combinatorics involves searching through possible sequences of actions and states when the choice of the right action isn't clear (e.g., Laird et al., 1987). This can be approached via either handwritten or learned knowledge that helps to control such search (e.g., Laird et la., 1987; Botvinick et al., 2019), but with the warning that adding such knowledge may lead to its own tractability problems (Minton, 1988). Intractability can also result from combinatorics within the architecture itself, particularly given learned content (Tambe et al., 1990), although in this case, optimization of the Soar architecture has largely reached the point where problematic combinatorics within it is rarely, if ever, a dominant concern.

In neural-network architectures, even without inherent combinatorics, tractability of the learning process can be a critical issue for evaluation.

GPT-3, for example, required more than 300 zettaflops to learn its largest model (Brown et al., 2020). This may be due to combinatorics in the space of examples over which learning is to occur, tying it into the issue of combinatoric spaces just discussed, or to some other cause.

Reaching beyond combinatorics, it has already been mentioned that some architectures, AIXI in particular, can be incomputable. In the reverse direction, the simple question of the efficiency of the algorithms used in an architecture, and of how they are implemented, can be a constant challenge. With respect to the two architectures with which I was most involved—(early) Soar and Sigma—efficiency was ever an issue to be worked on.

Clarity, as a qualitative evaluation criterion, can be problematic for cognitive architectures. Beyond their sheer complexity, two other factors that play a critical role here are their procedurality and the typical necessity of including code that isn't part of the theory but is required to make the theory executable, and thus operational. The addition of such code obscures what is actually part of the theory (section 2.5).

One of the major advantages of both neural-network and reinforcement learning algorithms is that their architectures are quite simple and easily expressed mathematically. Systems that are based on these architectures may be completely opaque—leading to an intense demand for explainability (e.g., Samek et al., 2021; Situ et al., 2021) and concern about the edge cases where they may fail (e.g., Heaven, 2019)—but their architectures themselves are strikingly simple and clear, at least for humans with a sufficient mathematical background. In contrast, lack of clarity for complex cognitive architectures can be a significant impediment for entry into the field by new students and researchers. This is on top of the lack of pull yielded by clear benchmarks and the lack of commercial appeal from benchmarks that reflect valuable capabilities.

Attempts to make complex cognitive architectures clearer include creation of documentation in the form of user/reference manuals (e.g., Bothell, 2017; Laird et al., 2024), specifications (e.g., Milnes et al., 1992; Cooper et al., 1996), and tutorials (e.g., Taatgen & van Rijn, 2007; Rosenbloom & Ustun, 2016; Soar, 2017). It is hoped that the common model might also help here, assuming individual architectures can be described as adhering to the common model but with a relatively small and clear set of differences. Still, clarity remains an issue, and one that is at present only measurable in a qualitative manner.

5.4.3.3 Lawfulness As designed artifacts, cognitive architectures tend to be coherent. By their very nature they also tend to be built from general components, attempting as they do to explain the full scope of intelligence and to operate across many different domains. Classical architectures are typically encoded procedurally—as large bodies of code—and are thus not propositional, although mathematically expressed architectures, such as those based on neural networks and reinforcement learning, are more propositional. Parts of classical architectures may also be encoded in this manner, with ACT-R's equations concerning such things as activation providing prime examples. AIXI also falls within this subclass.

This story can, however, be quite different for the variable content supported by a cognitive architecture. Depending on the domain of interest, and how the content about that domain is acquired and represented, it can be coherent or not, general or not, and propositional or not. In architectures aligned with the common model, for example, the contents of working memory and long-term declarative memory are propositional, whereas the contents of long-term procedural memory are nominally not.[12] In some sense this is exactly the distinction between the two types of long-term memory (but see part II for a deeper analysis of this distinction). Generality can still go either way, with, for example, rules and facts about the world on one side and instance-based memories on the other side. When parameter learning is the focus, as in neural-network and reinforcement-learning systems, coherence and generality are little more than a crapshoot, and propositionality is nonexistent.

5.4.3.4 Beauty Beauty decomposes into clarity and exuberance. Clarity in its turn reduces to unambiguity, simplicity, and minimality. As concluded generically in section 5.4.1.2 for theories that are expressed via computer code, cognitive architectures tend to be unambiguous in interpretation, although there may be ambiguity in terms of which parts of the code are actually part of the theory. A slightly different divergence is found for simplicity, where cognitive architectures are simple for computers to interpret while possibly quite difficult for humans to do so.

The situation with respect to minimality is quite different, as there is not even a consensus as to whether smaller is necessarily better than larger. Competing perspectives can be found. A physics perspective, and Occam's razor in general, suggests that smaller is better. A competing perspective from biology is that evolution implies an extended sequence of hacks that more likely than not will fail to yield a simple and neat result (e.g., Levin & Yuste, 2022). Architectures exist that take the physics stance, either explicitly or implicitly, including AIXI, deep reinforcement learning, LLMs, and Sigma, with the early version of Soar also relevant. Architectures more compatible with the biological stance include CogPrime (Goertzel et al., 2014) and the Society of Mind (Minsky, 1986).

Exuberance decomposes into minimality and power, essentially being defined as power over size (i.e., lack of minimality). With minimality just covered, only power remains to be explored. Given that it is one of the driving concerns of cognitive architectures, it is clear that more is better with respect to power. How to define and measure the power of a cognitive architecture remains somewhat obscure in general, although one such attempt created a list of "application categories" and evaluated which cognitive architectures supported each category (Kotseruba & Tsotsos, 2020).

As discussed in section 5.4.3, for cognitive architectures there is an acceptance of the notion that power may trump accuracy, especially during the early—and possibly even later—stages of architecture development. This does not completely dismiss the importance of accuracy but suggests that pursuing breadth given approximate accuracy may be a more powerful long-term strategy than attempting to prematurely search for local maxima with respect to accuracy. In a sense, it is a strategy akin to abstraction planning (e.g., Sacerdoti, 1974), where a rough sketch of the entire solution is created before attempts are made to dig further into the details of any one part of the overall plan. This can place architectural work at a disadvantage in narrowly defined competitions, but it is part of the price exacted by this type of approach.

Both of the major architectures with which I have been closely associated—(early) Soar and Sigma—have had exuberant characters. As discussed in section 1.2.2, early Soar can be seen as rather minimal, based as it was on a single representation for long-term knowledge, one way to make decisions, one way to learn, and one way to reflect on its own behavior. In work on Sigma, exuberance has traditionally been referred to as functional elegance,[13] which prefers achieving completeness—and thus, indirectly, power—from interactions among a small number of general mechanisms rather than a large number of specialized ones. From an extended form of factor graphs plus a simple local learning mechanism, it evinces probabilistic graphical models, rules, neural networks, sum-product networks, procedural and declarative memories, reinforcement learning, and more.

Other notions—such as the *free energy principle* (Friston, 2010)—that *attention is all you need* (Vaswani et al., 2017), or that *reward is enough* (Silver et al., 2021)—also clearly embrace exuberance. Whereas functional elegance is a generic call for exuberance, each of these notions advances a specific technology for achieving it, akin to how Sigma focuses on extended factor graphs.

The tension between the physics and biology perspectives for minimality does imply that there is no consensus in the cognitive architecture world on exuberance—as opposed to power—being an appropriate evaluation criterion for cognitive architectures in general. Perhaps, it is only appropriately applied to architectures based on the physics perspective. Yet even architectures more aligned with the biological stance do expect to yield via interactions among their components much more than is directly specified.

Given that beauty is partially based on exuberance, the same qualifiers concerning its use as an evaluation criterion for cognitive architectures apply. Although minimality also plays a role in clarity, lack of consensus about its importance doesn't raise comparable questions about the role of clarity, since its two other components—unambiguity and simplicity—can stand on their own as part of both clarity and beauty.

I have implicitly striven for beauty in my own work, but with a combination of significant exuberance and a relative lack of clarity, the reality is definitely mixed. As mentioned in section 5.4.2, AIXI, which is extraordinarily exuberant and mathematically rather simple, can be considered quite beautiful, although it suffers along other criteria, such as tractability given its incomputability. Possible claims concerning beauty can also be made for the architectures of deep reinforcement learning and LLMs, even if not necessarily for the systems these architectures enable.

5.4.3.5 Comprehensiveness Both cognitive and neural-network architectures fall short with respect to comprehensiveness, although in different ways. One major example for standard neural-network architectures, whether discriminative or generative, stems from how they are simply models of memory (and learning), thus being effectively limited to System 1 behavior. Game-playing programs such as AlphaZero (Silver et al., 2018) do add limited forms of System 2 behavior, as does the inclusion of reinforcement learning from human feedback (Ziegler et al., 2019) or chain-of-thought prompting (Wei et al., 2022). More ambitious attempts to build broader architectures around neural networks are also in progress (LeCun, 2022; Park et al., 2023; Sumers et al., 2023).

Traditional cognitive architectures are typically more comprehensive in this sense than neural networks, including significant System 1 and 2 capabilities. However, as mentioned in section 5.4.3.1, they still do fall short, with each architecture tending to fall short in its own unique way, yielding a key reason there has been essentially no success over the years in developing useful benchmarks for progress with cognitive architectures.

When the focus is shifted to systems, those based on generative neural networks, particularly large language models (LLMs), tend to be the most comprehensive due to the vast amounts and varieties of data from which they learn. This is also likely why benchmarks have found a role in evaluating such systems (e.g., Evidently AI Team, 2024), as it can be assumed that they are all comprehensive enough to be compared with each other via the same benchmarks. With additional tuning they may also compare well with more focused systems (Binz et al., 2024)

Cognitive architectures that leverage knowledge graphs (Rytting, 2000; Forbus & Hinrichs, 2006; Douglass et al., 2009) reach partway toward this, with ones that incorporate generative AIs as modules (e.g., Romero et al., 2023) approaching even closer. Otherwise, systems based on both cognitive architectures and discriminative neural networks tend to span only limited classes of tasks.

5.4.3.6 Rigidity It was argued in section 5.2.3.1 that flexibility in terms of variable content is key for cognitive architectures, which should in turn make rigidity of the dominant form—of minimizing the variable content—not a relevant evaluation criterion in general for systems based on them. However, as discussed in section 5.4.3.1, when the addition of knowledge and skills is restricted wholly, or at least mostly, to occur via learning, aspects of this form of rigidity can become relevant. There may still be vast quantities of variable content, but the architectural component of the theory would restrict what it contains. This is already typical for systems based on neural-network and reinforcement learning architectures, but systems based on traditional cognitive architectures are still far from being where rigidity can be a relevant evaluation criterion for them.

When considering the architecture itself as a theory, there are at least two kinds of circumstances in which rigidity of this dominant form may be relevant as an evaluation criterion. The first concerns architectural parameters that may be set in distinct ways for modeling different domains. In a sense, such parameters can be viewed as just one form of variable content, subsuming this case under the previous one. But, in a different sense, they can also be viewed as a less general part of the architecture itself. In the latter case, evaluating the architecture in terms of the number of such parameters, or whether their values are constrained by learning, can be relevant. The second kind of circumstance arises when the architecture takes on the flavor of a toolkit (section 2.2), with many different approaches available for instantiating each type of capability. Evaluating such architectures in terms of their (lack of) minimality, and thus (lack of) rigidity (and beauty), can make sense.

The second form of rigidity, which focuses on how constrained the theory is to a particular form, can again be problematic as an evaluation criterion for systems based on cognitive architectures. Humans can clearly do the same thing in multiple ways—based possibly on quite different combinations of knowledge and skill—and cognitive systems must maintain this same level of flexibility. With respect to the architectures themselves, this form of rigidity harkens back to the notion of *identifiability* in cognitive modeling; for example, from behavioral measures is it possible to identify whether processing is serial or parallel versus there being some variants of serial and parallel models that are indistinguishable with respect to that measure (e.g., Townsend, 1972).

When I began work on Sigma, it was unclear whether what would result would be an architecture or simply a toolkit for developing architectures based on graphical models. The effort received a name when it became clear that the constraints imposed by using graphical models to replicate many of the functionalities found in Soar, and to a lesser extent ACT-R—while also attempting to go beyond both of them in critical ways—was leading to a single architecture rather than to a space of possible architectures. There were choices concerning which capabilities to add to the architecture, but ultimately the choices about how to implement those capabilities was most often constrained by Sigma's underlying graphical architecture and the accompanying need for the new capabilities to work smoothly with the existing ones. This doesn't mean that there wasn't flexibility in various choices, but this notion of how the various parts of an architecture can be mutually constraining does seem to relate to this form of rigidity; that is, in such circumstances, it is hard to vary how these capabilities are instantiated.

5.4.3.7 Surprisingness The first form of surprisingness from section 5.4.2.4 typically results from unanticipated interactions among an architecture's basic elements. In this sense it is closely related to exuberance, but it focuses on the unexpected consequences of element combination. Examples with which I am directly familiar stem from the Soar and Sigma architectures. In Soar, this included how a universal weak method could arise from the ability to search plus small amounts of knowledge in rule form (Laird & Newell, 1983); how universal subgoaling—yielding a broad set of reflective/metacognitive abilities—could arise from recursion on a small variety of impasses in decision-making (Laird, 1986); how combining these two capabilities with a simple skill-acquisition mechanism (chunking) could yield a wide variety of learning behaviors (Steier et al., 1987), including knowledge acquisition (Rosenbloom et al., 1989) and plan acquisition (Laird & Rosenbloom, 1990).

In Sigma, this included how procedural, semantic, and episodic memories (Rosenbloom, 2010a; Rosenbloom, 2014)—plus distributed vector word representations (Ustun et al., 2014), and neural-network inference (Rosenbloom et al., 2016b)—could arise from a uniform base of extended graphical models; how combining this with a simple gradient descent learning mechanism could yield a wide variety of learning capabilities (Rosenbloom et al., 2016a), including reinforcement learning (Rosenbloom, 2012b); and how further tweaking of this combination could yield backpropagation for neural networks (Rosenbloom et al., 2017). Other examples from Sigma included the emergence of base-level activation (Anderson et al., 2004) as a temporal prior on episodic memory (Rosenbloom, 2014) and how mental imagery could arise from further tweaking of extended graphical models based on region tensors (Rosenbloom, 2011a; Rosenbloom, 2012c).

Nearly all of these—with base-level activation being the main exception—resulted from efforts to deconstruct in terms of what was already in the architecture a capability familiar in the field that was not already explicitly part of the architecture, rather than something that was wholly unanticipated. The surprise in these cases was that it was possible to yield these capabilities without explicit architectural mechanisms for them. Some of the examples in Sigma did require abductive architectural extensions, as introduced in section 5.3.3, but there was still a surprise that only relatively small architectural tweaks were required rather than whole new modules.

The recent work on LLMs (e.g., Brown et al., 2020) has also yielded a strong sense of surprisingness, at least for me. From an architectural perspective, it is surprising that such a simple combination of mechanisms—descended from even simpler models for assigning probabilities to sequences of words—even when combined with vast amounts of training data, can yield such sophisticated, nuanced, and (at least apparently) novel responses, whether correct or not, across such a wide range of domains. This would seem to combine versions of both the first and third flavors of surprisingness, as in addition to being surprising in its own right, it reduces a surprising phenomenon—in this case, that a computational system can produce such a body of responses—to normality.

There are also undoubtedly examples of the second form of surprisingness with respect to cognitive architectures. I personally have been surprised by several approaches to cognitive architectures that have appeared over the years, with AIXI probably yielding the biggest surprise, but in general this may be better as an evaluation factor employed by a community rather than by individuals.

While surprisingness may typically not be included in the evaluation criteria for cognitive architectures, the functional elegance I've searched for in Sigma has led to a focus on it, as inherited from earlier experiences with Soar that did help attract the field's attention to it and to the idea that there might be something unexpectedly worthwhile in considering cognitive architectures and integrative approaches to AI more generally.

5.5 Conclusion

In this chapter, the methodology of essential analysis, as recast in terms of a specialization of reflective analysis, has been applied to better understand the nuances involved in the notion of a theory. A particular focus has been on the implications of these nuances for cognitive architectures as theories of the fixed aspects of a mind that form part of complete cognitive systems according to the semi-formal equation *intelligent systems = cognitive architectures + knowledge and skills*. This has involved exploring what a theory is, how such theories are used, and how they are evaluated.

To start, a stripped-down essence for the concept of a theory has been identified that focuses on aboutness—or the more technical notion of metasystems—along with dimensions of a definitional space that encompasses it. Starting with the strictures that have been stripped out of three existing definitions of a theory, a directed acyclic graph (DAG)—which is almost simply a tree—of dimensions has been identified. At the top is whether the theory is accepted. Just below this are three high-level dimensions upon which acceptance depends: fidelity, which turns out to map to the semantics of the theory; usability, which turns out to map to its pragmatics; and lawfulness, which turns out to map to its syntax. Each of these high-level dimensions can also be decomposed further, ultimately yielding the full DAG. Within the resulting definitional space, it is possible to roughly identify regions within which traditional theories across the sciences can be found, and in which more particularly both cognitive architectures and their variable content can typically be found.

Consider first traditional cognitive architectures. With respect to fidelity, they have distinct challenges due to their breadth of applicability, complexity, and lack of variable content. Yet, different architectures still fall in various places along this dimension. With respect to usability, cognitive architectures strive for completeness with respect to the fixed mechanisms required for intelligent behavior but are radically incomplete by themselves with respect to answering most questions due to their lack of variable content. They can also be all over the board with respect to tractability and typically have significant issues with respect to clarity. With respect to lawfulness, they are typically coherent but not propositional.

Due to their simple mathematical formulations, machine learning architectures based on neural networks tend to fare better in general with respect to fidelity, clarity, and lawfulness. With respect to the rest of usability, they exhibit the same level of incompleteness due to a lack of variable content on their own but typically exhibit even more incompleteness with respect to the full set of fixed mechanisms required for intelligent behavior. However, in their favor, by focusing on learning they provide a direct means of countering that first source of incompleteness. Like traditional cognitive architectures, these architectures also exhibit significant variability with respect to tractability.

The variable content for both types of architectures, when viewed as theories, can vary with respect to fidelity. Neural networks are often situated at one or the other of the two extremes along this dimension, with discriminative networks tending toward the high end and generative networks tending toward the low end. In contrast, traditional cognitive architectures tend to be more in the middle. Neural-network content tends to suffer in comparison with traditional architectural content in terms of clarity, coherence, and propositionality.

Following this essential analysis, multiple perspectives on theory use have been merged into a simple yet coherent perspective that maps a variety of distinct uses onto three possible directions of processing:

1. Forward processing: prediction, discrimination, some problems in NP, *what* questions, certain *how* questions, and prescription.
2. Backward processing: explanation, other problems in NP, other *how* questions, and certain *why* questions.
3. Symmetric processing: verification, generation, some problems in P, and other *why* questions.

Ways of spanning multiple of these uses via single theories have also been discussed.

An exploration of theory evaluation began with the DAG of dimensions generated by essential analysis and expanded from there to a set of four interconnected DAGs that now also include beauty, comprehensiveness, rigidity, and surprisingness, each with its own substructures. These additional criteria have not been derived from definitions, but they can all be relevant to questions of theory quality.

Following a general exploration of the application of these DAGs to theory evaluation, the focus shifted to their more specific implications for evaluating cognitive architectures. How to apply these criteria to cognitive architectures, as well as to the variable content and systems associated with them, is a complex question. Evaluating accuracy—and, to some extent, grounding as well—remains a difficult challenge, although two classes of exceptions and two potential new paths are identified. The remaining criteria further flesh out the picture, with the criteria of power, completeness, and comprehensiveness turning out to be especially significant for cognitive architectures, to the point where accuracy may often be traded off for them.

II Dissections and Mappings: Dichotomic Analysis

6 Introduction to Part II

When defining artificial intelligence (AI) in section 4.2.3, the focus was on the overarching goal of achieving intelligence in artificial agents. In this part, the focus shifts to understanding AI in terms of the space of technologies and systems that have been developed in service of achieving this goal. From either perspective, there is a strong overlap between AI and both cognitive science and artificial general intelligence (AGI). Although AI technologies are at the heart of this part, as they have often been in the development of intelligent systems, these related fields also play a role when there is sufficient overlap or near adjacency. From these technologies, each discipline tends to create its own systems, whether simple or complex. The phrase *integrated systems*—or sometimes just the term *systems*—will be used throughout this part to refer to complex AI and AGI systems that combine multiple technologies as well as to complex models in cognitive science that similarly combine multiple technologies.[1]

AI is presently dominated by technologies based on machine learning, particularly in the form of *deep learning* over very large, multilayer neural networks (Goodfellow et al., 2016). Yet, a much wider variety of technologies has been investigated since the field's beginnings, many of which will likely still play critical roles in building complete systems that are capable of human-level intelligence. The spotlight here is on the range technologies that are at the heart of representation, reasoning, and learning, all construed quite broadly—even neural networks are assumed to afford forms of representation (via edges and weights) and reasoning (via computing output from input vectors) beyond their obvious usage in (machine) learning. These core technologies may also support other functionalities—such as planning, natural language processing, and perception—but there is not a specific concern here for any of the more special-purpose technologies that may also be associated with these other functionalities.

A variety of both symbolic and neural technologies are included in this analysis but given my more extensive experience with the former, there will almost certainly be an imbalance in their favor. Rather than attempting to explain each technology as it is introduced in the main body of the text, appendix A summarizes the range of technologies included and provides informal descriptions of each.

The specific aim of this part is to improve our understanding of the broad space of AI technologies via a combination of *dissection*[2] and *mapping*. To dissect is "to cut apart (an animal body, plant, etc.) to examine the structure, relation of parts, or the like" and "to examine minutely part by part; analyze: to dissect an idea" ("Dissect," 2024). To map is to "to represent or delineate on or as if on a map" ("Map," 2025). Together this combination provides a path toward understanding and representing the space of AI technologies, along with their similarities and differences.

One possible source of joints for dissection is disciplinary boundaries. As already mentioned, AI has an affinity with both AGI and cognitive science. It also typically has its home in computer science and overlaps further with data science, economics, neuroscience, operations research, psychology, statistics, and more. In earlier work, an interdisciplinary perspective was taken on *computing*, considered quite broadly as a *great scientific domain* (Rosenbloom, 2012a). As part of this, AI was abstracted down to a simple relationship, captured in the metascience expression (ME) language, between the social (S) and computing (C) sciences—S/C—that involves implementing, or simulating, the social domain of human cognition—which includes psychology, cognitive science, and other social sciences—on top of a computational substrate.

It is possible to elaborate further upon this core relationship, such as by adding the life (L) sciences when the brain, and thus neuroscience, is involved to yield S/L/C—where a neural substrate is simulated on a computer, which then yields cognition at the next layer up—but, even if such an analysis could be refined to where it identified the distinct role of each discipline relevant to AI, leveraging such disciplinary boundaries for dissection only yields coarse-grained maps that would likely provide little in the way of further insight into technologies and integrated systems.

A second possible source of joints is boundaries between distinct functionalities, much like how dissection in natural systems may follow boundaries among organ systems. AI textbooks (e.g., Russell & Norvig, 2010), for example, traditionally align chapters along the boundaries of high-level cognitive functionalities such as search, reasoning, planning, learning, perception, and language processing. At a slightly more granular level, the common model of cognition (section 1.2.3) focuses on core memories plus a cognitive decision cycle that, working together with appropriate knowledge and skills, are to yield multiple higher-level functionalities, such as planning and natural language processing.

Whichever granularity is chosen, such an approach can potentially yield insight into both the field as a whole and integrated systems that combine multiple such components. However, carving up AI according to functionalities tends to yield a rather ambiguous basis for the mapping of technologies, as many of the latter may be relevant to multiple of the former. For example, although search algorithms may in general be assigned to a problem-solving functionality, they may also be critical to reasoning, planning, and other functionalities. Such an approach also doesn't provide a direct path toward understanding the similarities, differences, and trade-offs among technologies.

A third possible source of joints is *dichotomic boundaries*, which are the basis for the *dichotomic analysis* central to this part. A dichotomy—such as procedural versus declarative or symbolic versus subsymbolic—implicitly defines a joint between its poles. A cross product of such dichotomies then defines a multidimensional space that includes a joint within each dimension, thus yielding multidimensional *dichotomic maps*, although even a single dichotomy can be thought of as yielding such a map, albeit in only one dimension.

When combined with *evaluation criteria* for integrated systems capable of intelligent behavior—or, alternatively, well-functioning minds—plus an understanding of how these criteria apply to locations within these maps, and thus to individual technologies, trade-offs become discernible, both across locations within single maps and between multiple maps. These criteria—which, as mentioned in chapter 1, can be characterized at a high level as *accuracy*, *efficiency*, *generality*, *simplicity*, *improvability*, *usability*, and *operability*—are introduced in chapter 7, along with a brief discussion of how they relate to the evaluation criteria already introduced for theories in section 5.4. The criteria here are intended to help in understanding, evaluating, and comparing integrated systems composed of different technologies. They may ultimately also help guide technology selection for new systems that are to possess particular combinations of advantages. To facilitate this, the *criterial advantage (CA)* language is incrementally developed and applied in this part as a succinct semiformal means of capturing and comparing the potential advantages of individual technologies based on their locations as well as of integrated systems that combine technologies that span multiple locations.

This approach leads to the four phases of dichotomic analysis introduced in section 1.1.1: (1) identifying the three core dichotomies (section 6.1); (2) exploring evaluation criteria relevant to dichotomies, technologies, and integrated systems (chapter 7); (3) analyzing the core dichotomies individually (chapters 8–10); and (4) exploring interactions among the core dichotomies (chapter 11). The remainder of this chapter, in addition to identifying the three core dichotomies (section 6.1), provides background on how each of them is to be analyzed in chapters 8–10 (section 6.2) and the critical role that interactions among them are to play in chapter 11 (section 6.3).

6.1 Dichotomies and Dichotomic Maps

Dichotomies have long played a role in attempting to understand AI, although typically a rather contentious one in which arguments are propounded for or against the opposing sides. The traditional procedural-versus-declarative controversy (e.g., Winograd, 1975) epitomizes this. Yet, dichotomic maps can provide an uncontroversial and relatively unambiguous means of assigning technologies to locations in a map based on pole-specific correspondences among the technologies—such as assigning both rules and feedforward neural networks to the asymmetric pole of the corresponding dichotomy—with adjacencies across locations providing additional insight into similarities and differences among them.

This ultimately yields two kinds of maps. A *disciplinary map* reflects the field as a whole. Rosenbloom, Joshi, and Ustun (2019a), for example, present a 3D map that stretches from symbolic logic and statistical relational AI in one corner to feedforward and recurrent neural networks in the opposite corner, while breaking down the traditional but coarse distinction between symbolic and neural technologies across three finer distinctions—symmetric versus asymmetric, symbolic versus subsymbolic, and combinatory versus noncombinatory. Other locations in the map, based on other combinations of these distinctions, correspond to such technologies as rules, analogical mapping, constraint propagation, and Boltzmann machines.

Disciplinary maps of this sort hold out the potential to yield a novel organization over the field that situates its technologies within a systematically generated overall structure. In considering trade-offs across dichotomies, as well as interactions among them, a variant of these disciplinary maps will also be seen in which locations are populated by expressions in the CA language that capture *criterial advantages*—that is, criteria that favor the technologies at these locations—rather than by technologies themselves.

An *architectural map,* in contrast, characterizes the fixed structures and processes that are at the heart of integrated systems—whether considered as cognitive architectures (chapters 1–2), artificial general intelligence (AGI) architectures (Goertzel, 2014), agent architectures (Müller, 1998; Bordini et al., 2006; Kakkasageri et al., 2012; Girardi & Leite, 2013; Chin et al., 2014; Silva et al., 2020) or intelligent robot architectures (Medeiros, 1998; Kortenkamp & Simmons, 2008; Kumova & Heye, 2017; Ingrand & Ghallab, 2017). Rosenbloom (2019a), for example, presents a 2D map based on the same distinction between symmetry and asymmetry but this time crossed with monotonicity versus nonmonotonicity to map combinations of key technologies found in many integrated systems, such as procedural and declarative memory, and reactive, deliberative, and reflective reasoning. Architectural maps of this sort hold out the potential for understanding not only why particular combinations of technologies are found in existing systems but possibly even what combinations may be appropriate for envisioned integrated systems of particular sorts.

Each location in a disciplinary map may include a variety of technologies from across the field, whereas each architectural map focuses more narrowly on just those technologies included in the architecture of interest. Either type of map may usefully be described at multiple levels of abstraction.

Given the similarities among technologies that map to the same pole of a dichotomy and the differences among those that map to different poles, two technologies can be considered as *identical with respect to a map* if they share all of the same poles in the map, and *opposite with respect to a map* if they differ along each dimension. Degrees of similarity or difference then depend on the number of shared or unshared poles, yielding a *rectilinear, Manhattan,* or *L1 metric* for the map that is calculated for any two locations as the length of the minimum path through adjoining locations that connects them.

Two integrated systems may in turn be considered as identical with respect to a map if their architectural maps span the same locations. It is not clear that there is a useful notion of opposition for such systems, although some form of a distance metric might be based on lack of overlap in locations. Instead, one other useful notion for integrated systems is that one may be considered *complete with respect to a disciplinary map* if its architectural map includes at least one technology in all locations in which the disciplinary map includes at least one.

Section 4.1.4.1 introduced one way of scoping the *intelligence space*—that is, the space of types of intelligences that may be found across both natural and artificial agents—in terms of the trilevel definition of intelligence and approximations to its three levels. The combination of disciplinary and architectural maps introduced here turns out to provide a complementary approach to scoping this space. A disciplinary map structures the range of technologies that may be combined into integrated systems for intelligence, whereas each corresponding architectural map specifies a particular such system. Together they thus can be seen to scope and populate an intelligence space. One key qualification though is that both the dimensions of the disciplinary maps and how fully fleshed out they are with relevant technologies determine how complete such an approach can be with respect to the full range of natural and artificial intelligence.

When augmented with relevant evaluation criteria, such as accuracy, efficiency, and generality, a disciplinary map can potentially facilitate understanding the functional trade-offs that arise across differing locations in it, and thus what combination of locations are needed in integrated systems for particular purposes. From there, a down selection of individual technologies from within the included locations can potentially yield an architectural map for the system of interest. Although, such an approach doesn't provide guidance about how to do the down selection in a manner that yields a compatible set of technologies, nor about how to wire the selected technologies together—as is the focus of, for example, the common model of cognition—it can potentially yield justifiable combinations of locations from within which to choose the technologies.

The two prior efforts (Rosenbloom, 2019a; Rosenbloom et al., 2019a) explored four basic dichotomies between them as a means of understanding AI technologies: symmetry versus asymmetry—aka *(a)symmetry*—which concerns whether processing is uniform in all directions across structures; monotonicity versus nonmonotonicity—aka *(non)monotonicity*—which concerns whether processing only accumulates results or can alter or remove them; symbolic versus sub/nonsymbolic—aka *(sub)symbolic*—which concerns whether meaning is derived from data or metadata; and combinatory versus noncombinatory—aka *(non)combinatory*—which concerns the flexibility with which structures can be combined. However, this resulted in a rather disjointed overall approach with distinct 3D disciplinary and 2D architectural maps rather than a single 4D map that could be projected down to lower dimensions in creating both disciplinary and architectural maps.

One of the initial goals of the work that led to this part was to synthesize what was learned from these two prior efforts into a single coherent 4D disciplinary map that could yield further insights into the structure of AI broadly while also serving as the basis for useful architectural maps of integrated systems. Yet, as the work progressed, it became increasing clear that a simpler approach might be preferable at this point that conflates (non)monotonicity and (non)combinatory into a single new dichotomy that can be labeled *(un) controlled*. Technologies that are either nonmonotonic or combinatory may need to be controlled, whereas technologies that are monotonic and noncombinatory may be able to proceed in an uncontrolled fashion. Leveraging this underlying commonality, and replacing (sub)symbolic with (meta) data, has led to the trio of core dichotomies used here: (a)symmetric, (un) controlled, and (meta)data.

These three can each be understood from a single generic perspective that considers how technologies, as forms of mechanisms (section 4.1.1.1), comprise specific combinations of structures and processes. From this perspective, (a)symmetric concerns whether processing proceeds in a uniform manner in all directions across the relevant structures; (un)controlled concerns whether decisions precede individual actions versus their proceeding automatically on their own impetus; and (meta)data assumes that the structures are based on symbolic data with quantitative metadata, as discussed in chapters 1 and 3, while then considering whether meaning primarily resides in the data or the metadata.

One of the more unexpected outcomes from Rosenbloom (2019a) was the discovery that many previously articulated dichotomies from across multiple subfields could be mapped onto its two dichotomies—such as the mapping of rules versus logic, procedural versus declarative memory, and heteroassociative versus autoassociative neural networks[3] all onto (a)symmetry—enabling closer ties to be made across multiple subfields while reducing the cacophony of terminology these dichotomies span. In what is to come, these mappings are extended to include additional preexisting dichotomies, including some that may have only been implicit rather than explicitly stated. This core idea is also extended to all three dichotomies of interest. In the process, a deeper dive is taken into how to think about the three sets of dichotomies that result.

One component of this deeper dive is a rethinking of each of these dichotomies as a *cluster*, with a central element (centroid)—or *label*—onto which all the other elements in the cluster can be mapped. Each element within a cluster shares a core aspect that is critical to the nature of that cluster, but each may also differ from its cluster mates in other ways. For example, although both are (a)symmetric dichotomies, rules versus logic concerns symbolic approaches to intelligence, whereas heteroassociative versus autoassociative neural networks concern what are traditionally considered subsymbolic approaches. Some of these differences can be explained by appealing directly to other dichotomies explored here, such as (meta)data for the difference here, but no systematic effort has yet been attempted to identify and explain all such differences. As explored in a bit more detail in chapter 12, such an effort would be an interesting candidate for further follow-up.

In a sense, any of the elements within a cluster could serve as its label; however, focusing on a centroid onto which all the other elements can directly map increases the likelihood of ending up with a coherent cluster. As a result, (a)symmetry is the one label out of the original four that has been retained here—although now referred to as *(a)symmetric*—because the otherwise disparate dichotomies within its cluster, such as the three already mentioned, can all map directly onto it. In contrast, none of these other three dichotomies map so directly onto each other; for example, there is no direct mapping between rules versus logic and heteroassociative versus autoassociative networks. Using centroids of this sort ensures that there is a single essential conceptualization for each cluster, rather than risking the conceptual drift that could result if a chain of mappings was allowed in which the cluster's meaning might wander according to those aspects of the dichotomies that are mapped at each step of the chain.

During this exploration of dichotomies, it has turned out that, in general, it is much easier to map disparate dichotomies onto computationally defined labels, such as (a)symmetric, that inherently capture the essential commonality of the cluster rather than onto labels that focus on the divergent functionality of its poles, such as procedural versus declarative memory. Such labels also facilitate identifying criterial trade-offs across the poles. So, ascertaining essential computational commonalities among related dichotomies has been the focus here. For conciseness and consistency across dichotomies, there has also been a bias toward labels that take the form *(inverter) term*, such as (a)symmetric, with mappings of the other variants within each of the respective clusters onto these essential computational ideas.

This is, of course, not the first attempt to identify sets of dichotomies relevant to AI. In addition to the directly preceding work on my own part, there have, for example, been several recent analyses and surveys of cognitive architectures that involve dichotomies to some extent. A mammoth survey of cognitive architectures (Kotseruba & Tsotsos, 2020) categorized them according to whether they were symbolic versus emergent versus hybrid, which roughly corresponds to the (meta)data dichotomy that appears here, along with approaches to spanning it that yield hybrids. But the other dimensions of categorization in that work were based either on approaches to implementing a range of functional capabilities or the kinds of applications for which they have been used.

Langley et al. (2009) earlier focused on the capabilities required in integrated systems; the internal properties that enable these capabilities, all concerned with aspects of how knowledge is handled; and oppositions—aka dichotomies—relevant to these properties. Many of these oppositions correspond to elements of the clusters considered here, but there was no attempt in this earlier work to seek deeper relationships among these elements nor to use them systematically in creating maps.

Goertzel (2014), focused more on AGI, expanded on the trio of symbolic, emergent(ist), and hybrid forms of categorization to add a fourth, universalist approach that further diverges from the notion of a dichotomy toward a set of competing paradigms. There have also been several attempts to use a small number of dichotomies to understand the structure of particular integrated systems (Anderson & Lebiere, 1998; Sun, 2016; Thomsen, 2018); in particular, they constructed architectural maps for these systems. All three of these latter efforts exploited variations on (meta)data, with the first two also using variations on (a)symmetric, although mostly under different names. The third also included several additional dichotomies, such as top-down versus bottom-up and early versus late.

Other efforts have assessed what is common among integrated systems in terms of functionalities or modules and their interconnections (Samsonovich, 2011; Thórisson & Helgasson, 2012; Lucentini & Gudwin, 2015; Laird et al., 2017; Kotseruba & Tsotsos, 2020). Within AI more broadly, approaches are often organized in terms of functionalities—as in AI textbooks—or paradigms. Such approaches relate to the idea of using functional boundaries for dissection; however, other than attempts to relate functionalities to integrated systems at some level of abstraction, little interest has been shown in digging deeper into the overall structure of the technologies in the field.

Newell (1973a; 1983) generated two lists of dichotomies, one for psychology and the other for AI. The first list was described as "a partial list of binary oppositions in psychology." However, the concern there was with how these oppositions were piling up rather than on how to reduce piles of them down to just a few, as his point was to argue for the futility of pursuing an understanding of the human mind given such a cacophony without overall guidance from a *control structure*, a term he used at the time for what we would now call a cognitive architecture. Many of these oppositions—such as nature versus nurture and existence versus nonexistence of subliminal perception—are not terribly relevant to AI technologies, but some of them do make an appearance as variations in the coming chapters.

The second list provided a timeline of "intellectual issues" in AI, cast as dichotomies, reaching back to 1640 with the dichotomy of mechanism versus teleology, an issue he felt was resolved in the mid-1940s with the development of cybernetics (Rosenblueth et al., 1943). He also did some clustering of these issues in support of distinguishing among various disciplines. For example, he distinguished AI from cybernetics via a cluster comprising symbolic versus continuous systems, problem-solving versus recognition, psychology versus neuroscience, performance versus learning, and serial versus parallel. An echo of two of these dichotomies can be found in variants of (un)controlled and (meta)data (chapters 9 and 10) that today help distinguish symbolic from neural technologies.

In contrast to these efforts at enunciating and elucidating such dichotomies, Minsky (2012) argued against their use in the modeling of minds as too simplistic—referring to them as "dumb-bell" models—and argued instead for more options. Although the context here is a bit different, as highlighted in maxim 23 (appendix B), the approach here has been to explore whether there are deeper underlying commonalities that decompose trichotomies and other higher-order -otomies into cross products of simpler dichotomies, and even to see if preexisting dichotomies can be better understood in terms of sets of simpler dichotomies that typically, but don't necessarily, align. Examples of the former will be seen in section 8.1, where several existing trichotomies—for control, emotion, and causality—are reconceptualized as L-shaped patterns in a 2D map; and of the latter, more broadly, where forms of the coarse neural versus symbolic dichotomy align all three of the core dichotomies considered here.

Chapters 8–10 focus on the three core dichotomies individually, each of which is shown to subsume many other dichotomies. Are these the only dichotomies that matter in understanding AI technologies and systems? The answer is almost certainly no. However, the assumption here is that these three capture enough of interest both individually and in combination that limiting the consideration to just them provides an important enough step in what ultimately might be a larger and more comprehensive endeavor.

6.2 Analyzing Dichotomies

Analyzing a dichotomy turns out to involve a complex multistep process that is common to the three core dichotomies discussed in chapters 8–10. The first step in each of these analyses is to construe its core dichotomy in terms of how it evolved from one of those found in Rosenbloom, Joshi, and Ustun (2019a) or Rosenbloom (2019a). A space of variations on it is then explored that subsumes key distinctions across a range of preexisting dichotomies from across the field.

The next step examines the dichotomy's overall relationship to learning. Although most of the dichotomic variants directly concern representation or reasoning rather than learning, some do directly relate to learning, and many of the others have indirect implications for it that are worth considering. Additional learning technologies may also be considered during this step that help to flesh out the relevant maps.

There are at least three generic approaches to grounding learning in a dichotomy that is not itself focused on learning: (1) the learning process itself may reflect the dichotomy; (2) the structures that are input to the learning process, such as examples and experiences, may reflect the dichotomy; or (3) the structures produced by the learning process—that is, what is learned—may reflect the dichotomy. If both the input and output structures reflect the dichotomy, there is the added question of (4) whether they align—for example, asymmetric inputs yielding asymmetric outputs and symmetric inputs yielding symmetric outputs—or whether learning converts one into the other.

Following this consideration of learning, both sides of the chapter's core dichotomy are evaluated in terms of the criteria identified in chapter 7.

Each dichotomy polarizes along at least some of these criteria, implying that each also embodies trade-offs to be considered. For example, as will be seen in section 8.3, (a)symmetric implicates high-level trade-offs involving accuracy, efficiency, generality, improvability, and operability. With respect to accuracy, asymmetric approaches may have an advantage with respect to precision whereas symmetric approaches may do so with respect to correctness and robustness (see chapter 7 for definitions of these three criteria). Similarly, asymmetric approaches may be more efficient in finding a single solution to a problem whereas symmetric approaches may be more efficient when seeking many, or all, solutions. With respect to generality, there may be problems solvable at each pole of this dichotomy that aren't solvable at the other one, with improvability—in the form of learnability—following along similar lines. With respect to operability, asymmetric approaches tend to advantage reactivity, and the overall ability to affect the environment, whereas symmetric approaches tend to advantage safety.

The process of analyzing trade-offs has a subjective character to it at this point, although the hope is that further analysis and experimentation might make this more objective. Still, it is these trade-offs that help to explain why the dichotomies were of interest in the first place and why there have so often been arguments about why one side or the other should be preferred. Rather than a focus on taking sides, what comes across most strongly in the analyses to come is how valuable spanning these dichotomies can be in enabling the best of both worlds across a broad range of situations. This notion has been central to work on hybrids between symbolic and emergent(ist)/neural approaches to AI—which maps most closely to the (meta)data dichotomy here—but it can be extended to the other two dichotomies as well.

The next step in each chapter then explores just how its dichotomy might be spanned. Rosenbloom (2011b) discussed two generic strategies for spanning dichotomies in cognitive architectures that is adapted for use here. *Addition* incorporates distinct modules for each side of the dichotomy, with each to be used when appropriate. *Reduction* finds a commonality across the two sides that enables both to be supported via a single module.[4] Three variations on reduction were considered in this earlier work but these are reduced to two here: (1) one side of a dichotomy is implemented natively while the other is implemented, or simulated, on top of it; and (2) a native generalization enables direct implementation of both sides.

6.3 Interactions Among Dichotomies

The analysis of the individual dichotomies in chapters 8–10 is followed in chapter 11 by consideration of interactions among them. This is where mapping particularly comes to the fore, with dichotomies supplying dimensions along which the maps extend. Whereas a single dichotomy yields a 1D map, cross products among dichotomies yield 2D and 3D maps. Table 6.1 shows an abstract 2D map that crosses (a)symmetric with (un)controlled. It is an adaptation of maps found in Rosenbloom (2019a) but with (non)monotonicity replaced with (un)controlled.

Procedural and declarative memory are well-established notions in cognitive science, as seen, for example, in both the common model of cognition and the architectures that map onto it, with procedural memory concerned with the control of cognitive behavior and declarative memory concerned with representing, and possibly reasoning with, knowledge about the world. Reflection is not quite as standardized, as witnessed by its continuing absence from the common model, but it roughly amounts to the ability of an integrated system to step back in a way that lets it observe, reason about, and modify its own cognitive behavior.

Deliberation has a more ambiguous status, not in terms of its presence but in terms of how it is conceived and even labeled. It clearly refers to thinking of some sort, but under various perspectives it can be considered merely inference over the contents of declarative memory, or the selection and application of an action, or the routine selection of a sequence of actions, or even some form of reflection. One specific meaning is intended here—the sequential selection and execution of one or more actions without the necessity of reflection. In the common model this corresponds to a sequence of one or more cognitive cycles in which memory—most particularly, procedural memory—is accessed to select and or execute actions. Such a capability could alternatively be called (sequential) action selection and execution, or routine behavior, but deliberation is the label used here.

Table 6.1
Abstract 2D map of the cross product between (a)symmetric and (un)controlled

	Asymmetric	Symmetric
Uncontrolled	Procedural Memory	Declarative Memory
Controlled	Deliberation	Reflection

Table 6.1 can be viewed as an abstract architectural map that spans both dichotomies and thus is complete with respect to the corresponding disciplinary map. In this map, the uncontrolled row concerns memory access and the controlled row concerns behavioral control. Refocusing on the columns yields directed activity versus the pondering of a more coherent overall picture, somewhat like the well-known caricature of the difference between surgical and medical practitioners. Cognitive architectures such as Soar and Sigma can be seen as more concrete instantiations of this abstract architectural map. So too can be seen the common model of cognition and neural-network game-playing systems such as AlphaZero (Silver et al., 2018), although each of these latter two lacks one of the four components—reflection for the common model and declarative memory for AlphaZero. 2D versions of these four architectural maps can be seen in tables 6.2–6.5. Extensions of these to 3D maps can be found in section 11.4.

Such maps highlight similarities among technologies whenever they align along a dimension and differences whenever they contrast. So, for example, rules and feedforward neural networks align along (a)symmetric and (un)controlled—at least for parallel rule systems such as the one in Soar—but differ along (meta)data, with rules focused on symbolic data and feedforward neural networks focused on quantitative metadata.

Table 6.6 shows how both of these technologies can be viewed as forms of procedural memory, jointly nestling within the top-left corner. Similarly, monotonic logics—such as first-order logic—and Boltzmann machines can be viewed as forms of declarative memory, sharing uncontrolled with rules

Table 6.2
2D architectural map of Soar (adapted from Rosenbloom, 2019a)

	Asymmetric	Symmetric
Uncontrolled	Parallel Rules	Facts and Episodes
Controlled	Selection and Execution	Subgoaling

Table 6.3
2D architectural map of Sigma (adapted from Rosenbloom, 2019a)

	Asymmetric	Symmetric
Uncontrolled	Asymmetric Graphs	Probabilistic Graphical Models
Controlled	Selection and Execution	Subgoaling

Table 6.4
2D architectural map of the common model of cognition (adapted from Rosenbloom, 2019a)

	Asymmetric	Symmetric
Uncontrolled	Procedural Memory	Declarative Memory
Controlled	Selection and Execution	

Table 6.5
2D architectural map of AlphaZero (adapted from Rosenbloom, 2019a)

	Asymmetric	Symmetric
Uncontrolled	Feedforward NNs	
Inconsistent	Game Moves	Monte Carlo Tree Search

and neural networks, but being symmetric in their inference procedures. Moving down from there, search, planning, and subgoaling provide examples of reflection that share the symmetric aspect of monotonic logic and Boltzmann machines, in being able to consider whole patterns of actions but being controlled in how they proceed. Moving left from there yields the selection and execution of individual actions, a technology category that shares the controlled aspect of planning but has the asymmetry characterizing rules and neural networks.

As an architectural map, table 6.6 would make sense only for a system that actually includes all nine technologies found in the map. Sigma comes close, although the symmetric uncontrolled technologies listed here have not been explored with respect to it, and they thus may or may not be within its scope. Alternatively, this table may be considered as a partially fleshed-out disciplinary map, with table 11.2 later providing a more fully fleshed-out version. Table 6.1 can also be considered as a disciplinary map, albeit a very abstract one, in addition to its use as an architectural map.

From a dual perspective, each location, row, or column in a map can be considered as having its own unique signature, determined by where it falls along each dichotomic boundary. This approach leverages the ordering of the dichotomies introduced earlier—(a)symmetric, (un)controlled, and (meta)data—and the first letter of each side of the dichotomies, while using *X* to signify either a lack of commitment along a dichotomy or a hybrid of its poles. Clockwise from the top left in table 6.6 this yields signatures of *AUX*,

Table 6.6
A sampling of more concrete possibilities for the locations within table 6.1

	Asymmetric	Symmetric
Uncontrolled	Rules; Feedforward Neural Networks	Monotonic Logics; Boltzmann Machines
Controlled	Action Selection and Execution	Search; Planning; Subgoaling

SUX, *SCX*, and *ACX* for the four locations. Each adjacent pair of signatures, including the loopback from the last to the first, differs in one position. This can be thought of as similar to a *word ladder*—a word game originally invented by Lewis Carroll ("Word ladder," 2024)—that loops back on itself.

Both rules and feedforward neural networks share the signature of *AUX* in table 6.6, but they can still be distinguished via (meta)data, as *AUD* versus *AUM*. Similarly, monotonic logics can be distinguished from Boltzmann machines via *SUD* versus *SUM*. In ACT-R (Anderson et al., 2004), as opposed to Soar and Sigma, rules are considered to be controlled (*ACX*), with one selected to fire each cognitive cycle. It would thus be possible to consider rules as occupying the entire first column in table 6.6 (*AXX*), although the discussion of (un)controlled variants in section 9.1 takes a different tack, of considering rules in general as uncontrolled and referring to ACT-R's controlled asymmetric (*ACX*) structures as operators.

Tapping a bit into cognitive science, these signatures can be considered as personality types for technologies that are analogous to those for humans. Chapter 13 mentions how I test out as INTJ—Introverted (rather than Extraverted), iNtuitive (rather than Sensing), Thinking (rather than Feeling), and Judging (rather than Perceiving)—by a Myers–Briggs personality test ("Myers–Briggs Type Indicator," 2024). The more academically respectable five-factor model of personality (McCrae & John, 1992) can also provide a comparable analog. Either way, we can thus talk about the personality of an AI technology in terms of a three-factor signature that distinguishes whether it is *S* versus *A*, *U* versus *C*, and *M* versus *D*.

7 Criteria

The criteria that guide the development and evaluation of work within any intellectual discipline may be stated explicitly within the discipline or may instead be more of an implicit part of its culture. The goal of this chapter is to be explicit with respect to many of the criteria central to AI. The approach is to consider general criteria found across computer science, AI, and cognitive science for evaluating technologies and integrated systems. The seven high-level criteria that were foreshadowed in chapters 1 and 6—*accuracy*, *efficiency*, *generality*, *simplicity*, *improvability*, *usability*, and *operability*—are considered in some detail here, including how each decomposes into multiple finer-grained criteria.

Accuracy and efficiency are the dominant criteria within computer science, of which AI is typically considered a part. Accuracy, in terms of match to human data, is also a dominant criterion in cognitive science. Generality—which here aligns with chapter 4's focus on the breadth of applicability of a system rather than chapter 5's focus on the breadth of applicability of the individual elements of a system—is at best of second-order importance across these fields broadly, but it is central for integrated systems that are intended to yield human-level intelligence. Simplicity relates to engineerability in both computer science and AI and to Occam's razor in cognitive science. Improvability concerns whether a system can improve its own capabilities without direct human intervention. Learnability, whether in machine or natural form, is a key component of improvability. Usability—the ability to support human use of intelligent systems—is relevant for any system that is to support human activities. Operability—the ability for systems to be effectively deployed in external environments—matters for any system that is to behave in such environments. It is critical for robots operating in the physical world but is also of significant concern for systems that operate in other real-world and online environments.

Lists of criteria, particularly for integrated systems, can be found in a number of existing sources. Langley, Laird, and Rogers (2009) provided one of the most thorough prior discussions of criteria for integrated systems intended to support human-level cognition. Their list hits on aspects of six out of these seven classes, just missing simplicity. Anderson and Lebiere (2003) merged Newell's (1990) list of constraints on mind with an earlier such list (Newell, 1980) and then edited down the result to yield what they called the Newell test. As discussed in section 4.2.1, it comprises twelve functional criteria for a mind. With respect to the seven classes of criteria mentioned here, this list turns out to distribute across four of them: accuracy, efficiency, generality, and improvability. Other discussions of criteria for integrated systems can be found, for example, in Sun (2004) and Varma (2011).

There is also obviously a relationship to the criteria for theory evaluation found in section 5.4. As discussed in section 2.5, theoretical computational architectures combine aspects of both systems and theories. But, even without such a combination, there is a significant overlap in criteria, although with lawfulness playing much less of a role for systems and operability much less of a role for theories. The different traditions that feed into the two sets of criteria also lead to differences in both focus and terminology that I mostly won't attempt to disentangle here, leaving the development of a more complete and consistent ontology for evaluation across theories and systems to the future. However, as we just saw for generality, there are instances where identical terms are used in ways that are sufficiently different across the two traditions to require some clarification.

The remainder of this chapter considers the seven classes of criteria listed earlier, differentiating each further into usefully distinct subclasses, and even into subsubclasses on occasion, before wrapping up with a summary table (7.1) and a winnowing of this down to a subset of the criteria that will be used in the coming analyses. Once explicitly identified, these criteria may seem obvious, but having them all in one place, each accompanied by a discussion of its meaning, is important for supporting the trade-off analyses to come.

7.1 Accuracy

In theoretical computer science, proofs of correctness play a central role (Sipser, 2012). In experimental computer science, systems are built and tested to determine how accurately they produce the anticipated results (Desprez et al., 2012). To distinguish these two aspects of accuracy, they will be referred to respectively as *correctness* and *precision* (in section 5.2.2, what is called precision here was given the broader label of accuracy). In other words, correctness is a qualitative criterion concerning a lack of errors, whereas precision is a quantitative criterion concerning how close the results come to an ideal. *Robustness* can also be considered an aspect of accuracy, concerned as it is with guaranteeing that even in the worst case—in the presence of errors, attempts to mislead, or other exceptional situations—the results will not be so far off as to be disastrous (Cristian, 1982).

Evaluation of accuracy requires some standard for comparison, whether in the form of an abstract specification of correct behavior or a test set assumed to characterize precise performance. As part of computer science, AI adheres to these same standards (Cohen, 1995; Li et al., 2020). Cognitive science does as well, although given that its test sets are derived from human performance, experimental precision is much more central than either provable correctness or worst-case robustness (Polk & Seifert, 2002; Sun, 2008).

For integrated systems, the same standards also matter in principle, but the story is considerably messier given the number of parts and variations on them that may be relevant to producing any particular results (Langley & Messina, 2004; Goertzel, 2014). The situation is even worse for evaluating the fixed architectures underlying integrated systems, where any evaluation of behavior may be impacted by whatever assumptions are made about the variable knowledge and skills that sit atop the architecture (section 5.4.3.1).

7.2 Efficiency

Efficiency can concern the usage of *time*, *space*, *energy*, and other possible resources. Temporal efficiency can itself be decomposed into abstract scaling effects—or *combinatorics*—for algorithms and more concrete *real-time* effects for systems. The former concerns the functional relationship between increases in problem size, according to some appropriate measure of this size, and increases in time. Growth of time with problem size may, for example, be absent—yielding constant time algorithms—linear, polynomial, exponential, or even super-exponential. Algorithms in the first three growth classes are considered tractable, whereas those in the latter two are intractable (Sipser, 2012). Real-time efficiency matters when systems must operate effectively in environments where time passes, and where the utility of results depends to some significant extent on how soon they become available.

Within AI, temporal efficiency can matter for both performance and learning, with the former determining whether the system can do the job it has been developed for and the latter determining whether—or at least, how effectively or at what cost—it can be trained for such a job. In cognitive science, combinatorics has been less of a focus, except for theoretical concerns about whether a proposed algorithm or system could model human processing if it cannot scale to the problems humans tackle. However, real-time efficiency still matters in terms of whether a model can run quickly enough to enable whatever experimentation it is to support. When the focus is on integrated models of human cognition, there is also the question of whether the core cycle time of the model can match in real time, rather than just simulated time, the ~50 msec cycle found in humans (Card et al., 1983; Anderson, 2007; Kieras & Meyer, 1997; Laird, 2012; Rosenbloom, 2012d; Laird et al., 2017).

Spatial efficiency has traditionally been of much less importance than temporal efficiency in AI and integrated systems. However, in an era of mobile computing and the Internet of Things (Greengard, 2021), where there may be limited local storage, it can be critical. It can also be critical at the opposite end of the spectrum, where massive amounts of data or knowledge are necessary, in turn likely leading to massive models. In between these extremes, it can still be relevant to evaluating knowledge representations. Consider, for example, the difference between propositional and first-order logic, where the former requires individual representation of each instance of interest, whereas the latter may get by with a single quantified element that subsumes many such instances.

Energy efficiency follows some of the same considerations as spatial efficiency at the extremes, mattering particularly for small platforms with little power availability and for massive forms of computing, such as deep learning over huge datasets (Toews, 2020). It does not, however, seem to have attracted much concern in between these extremes. Energy efficiency will not be a focus in the dichotomic analyses to be found in chapters 8–11, and spatial efficiency will play a much lesser role than temporal efficiency.

When talking in general about the efficiency of a technology, questions of how the technology is to be applied can, of course, make a difference. The discussion of generality in the next section gets more explicitly into how to think about the breadth of ways that technologies can be used. However, one example here, comparing propositional with first-order logic, will set expectations for how efficiency will be considered in the rest of this part. If there is a need to reason about only one instance, propositional logic is simpler and can be slightly more efficient than first-order logic. Yet, if the intent is to reason about many similar instances, first-order logic quickly becomes much more efficient due to its ability to use quantifiers. Although we won't be directly comparing propositional and first-order logic in what is to come, this general distinction between working with single instances versus multiple related instances does play a role.

7.3 Generality

Generality is itself a very general concept, even when considered in the specific context of technologies for intelligent systems. It can refer to the range of problems that can be approached, or to the range of problem domains—that is, distinct types of problems—that can be approached, or to the range of environments within which these problems can be approached. From a more internal perspective, it can refer to the range of functionalities embodied within a system—such as reasoning, learning, and language processing (section 4.1.3)—to which the technology is relevant,[1] or to the range of temporal bands within the system (section 2.1) at which the technology can be applied. It can also refer to the range of kinds of entities to which it is applicable, whether humans—and for humans, whether child versus adult, or novice versus expert, or "normal" versus different in some way—or other natural organisms, or artificial entities not found in nature. Abstractly, these three classes of generality can be referred to as *problem*, *functional*, and *entity* generality. Problem generality is typically the dominant concern in intelligent systems and is also therefore the focus here, although with acknowledgment of entity generality. Functional generality shows up in a variety of aspects of full-spectrum intelligence (section 4.1.3), but its use in evaluation is beyond what will be attempted here.

In section 5.4, a distinction was made between generality and (explanatory) power with respect to criteria for theory evaluation. Generality there referred to the scope of a theory's individual elements and power to the breadth of applicability of the theory as a whole. A similar distinction can be made here, for example, between the generality of a single statement in propositional logic or first-order logic versus the overall power of the two forms of logic. However, in the context of technologies, the latter is what would typically be referred to as its generality, which is how—as mentioned at the beginning of this chapter—the term will be used here. How much can be expressed by a single statement can impact a technology's efficiency, most directly its spatial efficiency but potentially also its temporal and energy efficiency. In what follows, the expressibility of individual statements will be considered only as an aspect of efficiency.

For integrated systems, such as those that strive for humanlike intelligence or artificial general intelligence (AGI), generality—particularly problem and functional generality—is often the dominant overall evaluation criterion, demoting accuracy and efficiency either explicitly or implicitly to a lesser importance (Langley et al., 2009; Kotseruba & Tsotsos, 2020). One way to think about this is that such integrated systems strive for a global optimum in terms of their ability to do something reasonable for any human-level (at least) problem or functionality while being prepared to sacrifice some amount of accuracy or efficiency locally for this, as opposed to achieving more local optimality in terms of accuracy or efficiency within a narrower range of applicability in exchange for limiting applicability to just that range (Newell, 1990). These approaches have also been contrasted as general versus narrow AI (Goertzel, 2014) or strong versus weak AI, although this latter contrast has evolved from an earlier meaning that distinguished between systems that were actually intelligent versus ones that could only act as if they are so (Searle, 1980).

7.4 Simplicity

Simplicity relates to engineerability in AI—and computer science more broadly—and to Occam's razor in cognitive science. In the former, simplicity might be measured in terms of the number of entities and connections among them, or the number of kinds of entities and connections. In cognitive science, it may also, or alternatively, concern the number of free parameters that can be set in modeling human data. In section 5.4.1.2, simplicity referred to the ease of interpretability of a theory, whereas minimality referred to the size of a theory. Here, simplicity refers to the size of a theory.

Philosophers of science break down simplicity into *ontological simplicity* (or *parsimony*) versus *syntactic simplicity* (or *elegance*), with the former indicating the number of kinds of entities postulated and the latter indicating the number and conciseness of basic principles (Baker, 2016). Parsimony directly maps onto the corresponding aspect of the pragmatic criteria above.

Elegance would seem to relate to the number of parameters involved but also to go beyond this to a full measure of the principles involved.

One relatively simplistic mapping of this to the more pragmatic usages is to consider the number of entities and connections as measures of parsimony and the number of kinds of entities and connections as measures of elegance. The reduction approach to the spanning of dichotomies—particularly when grounded in a native generalization that directly implements both poles of a dichotomy—clearly focuses on improving elegance in comparison to the addition approach, by sharing of kinds of entities and connections across capabilities, while not necessarily saying anything about parsimony.

Simplicity has at best been a secondary or tertiary criterion across the relevant fields, at least for integrated systems, as witnessed by its omission from the criterial lists discussed earlier. However, it has been central to my own work on cognitive architectures. This was mostly implicit in the early days working on the Soar architecture, which had only one of each kind of mechanism—one long-term memory, one learning mechanism, one way of making decisions, and so on—but has been more explicit under the banner of *functional elegance* in the work on the Sigma architecture. In chapter 5, this was mapped to exuberance, which itself was partly based on minimality there, and thus on simplicity here. Simplicity will not be explicitly considered in the analyses to come here, other than a brief mention with respect to Sigma in section 11.4.

7.5 Improvability

The Newell test includes three criteria concerning model origins: *learning*, *development*, and *evolution*. Given a partitioning of an integrated system into a fixed architecture plus the variable knowledge and skills that it supports, learning can be thought of as (relatively) rapid modification of the variable aspects, whereas development is slower, perhaps genetically driven, modification of what might otherwise appear to be fixed aspects. Evolution is an even slower modification of the fixed, architectural aspects across generations of systems, rather than any form of modification within a single system.

As so expressed, these three appear more directly as capabilities than criteria and, thus, potentially more relevant to functional generality than to improvability. Yet they can be viewed as criteria when thought of in terms of whether knowledge and skills are learnable, whether aspects of an architecture can develop, and whether an integrated system could have resulted from evolution. Only learnability will be in focus in chapters 8–11.

7.6 Usability

The seven interrelated criteria that impact overall usability reflect the relationship between technologies and the human users of them: *ease of use*, *modularity*, *programmability*, *instructability*, *taskability*, *understandability*, *explainability*, and *trustability*. The first concerns simply how easy the technology is to use by humans. The second concerns whether a technology is decomposable in a manner that constrains the interactions among the resulting parts. These first two criteria could also be considered as aspects of simplicity, but ultimately it does not matter greatly to which of the two categories they are nominally assigned. The next two are human-in-the-loop variations on learnability, with programming involving a human directly updating an internal memory and instructability involving more indirect modifications based on learning from a combination of interaction in natural language and demonstrations of solutions (Gluck & Laird, 2018). Taskability involves telling a system what to do, typically by giving it goals to accomplish, in contrast to programming or instructing it how to achieve goals in general.

The remaining three criteria have received significant recent play in engineered AI and integrated systems (e.g., Arrieta et al., 2020; Thiebes et al., 2020), particularly when the core content is learned, and even more when it is learned in the form of neural networks that by default are too much like black boxes on their own. Understandability and explainability also apply in the scientific context of integrated models in cognitive science, in terms of the utility of the models both for the scientists who developed them and for others to whom they want to communicate them. These two criteria, plus ease of use and programmability, are the only ones that will appear in the coming chapters.

7.7 Operability

The complete list of criteria provided by Langley et al. (2009) comprised the following groups of criteria: generality (focused on problem generality), versatility (another aspect of problem generality concerned with how difficult it is to apply an architecture to problems within the domains to which it is relevant), and taskability (an aspect of usability); rationality (i.e., correctness given what the system knows) and optimality (i.e., precision); efficiency (in particular, real-time efficiency) and scalability (i.e., combinatoric efficiency); reactivity (real-time efficiency with an ability to respond directly to what is changing) and persistence (continuing as appropriate on problems even as the world changes); improvability (i.e., learnability); and autonomy (not requiring constant external interventions) and extended operation (i.e., long life). The two groups in their list that are not already sufficiently covered in this chapter are *reactivity* and *persistence*, and *autonomy* and *extended operation*. These four criteria can be seen as aspects of effective operability of integrated systems in real environments and have thus been added as such. The criterion of autonomy here follows its traditional usage in AI rather than how it is defined in section 4.1.3.9.

Along these same lines, three additional criteria may also be mentioned. The first can be termed *effectivity*, which concerns whether the system can actually change the external environment in the manner needed, as opposed to being merely a brain in a box. The second is *safety*, which—as in section 4.3.4.7—concerns avoiding (unintended) harm that might be caused by the system. The third is *security*, which as with all computing systems concerns the ability of a system to continue behaving without undesired external intrusions.

Of the seven operability subcriteria introduced here, only effectivity, reactivity, and safety will be considered in what is to follow.

7.8 List of Criteria

In summary, the criteria considered in this chapter can be found in table 7.1. The first column provides a name for each criterion, with a parenthetical abbreviation that provides the first increment in the development of the CA language. The second column provides brief characterizations of each of the criteria.

Although it is useful to bring all of these criteria together, as has been mentioned in the previous sections, only a subset of them—those that are italicized in the table—are actually used in chapters 8–11. The remaining criteria can, and do, matter in various circumstances, but they are beyond what I have been prepared to deal with here. Still, the ones that are used will aid in evaluating both individual locations—and thus technologies—as well as combinations of locations, most particularly entire architectural maps and the integrated systems they represent.

Table 7.1
Criteria relevant to developing and evaluating work in AI, including nearby criteria in computer and cognitive science, that have been considered here

Criteria	Description
Accuracy (A)	Match to some standard
Correctness (AC)	Fully meeting standard
Precision (AP)	Degree to which standard is met
Robustness (AR)	Guarantee of minimal degree to which standard is met
Efficiency (E)	Amount of resource required
Time (ET)	Amount of time required
Combinatoric (ETC)	Rate of growth of time with problem size
Real-time (ETR)	Ability to complete quickly enough
Space (ES)	Amount of space required
Energy (EE)	Amount of energy required
Generality (G)	Breadth of overall system
Problem (GP)	Breadth of applicability across problems, domains, and environments
Functional (GF)	Breadth of functionality that can be exhibited
Entity (GE)	Breadth of organisms or systems that can be modeled or built
Simplicity (S)	Lack of size and complexity
Ontological (or Parsimony) (SP)	Number of entities and connections hypothesized
Syntactic (or Elegance) (SE)	Number of kinds of entities and connections hypothesized
Improvability (I)	Ability to improve without human intervention
Learnability (IL)	Ability to rapidly improve "variable" aspects of entities
Developability (ID)	Ability to slowly improve "relatively fixed" aspects of entities
Evolvability (IE)	Ability to improve "fixed aspects" across generations of entities
Usability (U)	Ability to support human use of technology
Ease of use (UEa)	Ability for human to easily use the technology
Modularity (UM)	Ability of system to be decomposed into modules with limited interactions
Programmability (UP)	Ability for human to directly encode behaviors in system

Criteria	Description
Instructability (UI)	Ability for system to learn in context from instruction
Taskability (UTa)	Ability for system to accept tasks to be performed
Understandability (UU)	Ability for human to understand system behavior
Explainability (UEx)	Ability for system to explain itself to humans
Trustability (UTr)	Ability for system to be trusted by humans
Operability (O)	Ability to operate in external environments
Effectivity (OEf)	Ability to alter the external environment appropriately
Reactivity (OR)	Ability to respond immediately to environmental change
Persistence (OP)	Ability to continue working on a task even when environment changes
Autonomy (OA)	Ability to proceed without constant human intervention
Extended operation (OEx)	Ability to continue operation over long term
Safety (OSa)	Ability to operate without unintended harmful consequences
Security (OSe)	Ability to operate without undesired external intrusions

Those criteria used in understanding trade-offs in the rest of this part are highlighted in italics.

It is worth noting that *optimality* does not appear in this table, as it rightly refers to the degree to which a criterion is met rather than being a criterion on its own. For example, it is possible to talk about optimality with respect to various forms of either accuracy or efficiency. Optimality is also what is typically sacrificed when considering trade-offs, whether within or across criteria.

One simple example of a within-criterion trade-off that we have already seen occurs when an asymmetric approach can be more efficient in generating a single answer whereas a symmetric approach can be more efficient in generating all answers. A comparable across-criterion trade-off can occur, for example, between temporal efficiency and precision, whether expressed in terms of anytime algorithms in AI (Zilberstein, 1996) or speed-accuracy trade-offs in psychology (Wickelgren, 1977).

8 (A)Symmetric

The (a)symmetric dichotomy is the only one that played a central role in both of the prior analyses and is not reformulated here. Variations of it in other terms pervade much of AI and cognitive science, and even computer science more broadly, but these analyses were the first to explicitly identify it in this manner and to explore why this might be the correct form to serve as the basis for unifying a broad range of prior variants. The central distinction in (a)symmetric is whether processing proceeds in only a single direction over the structures that define it versus bidirectionally—or omnidirectionally, as the case may be—over these same structures.

More generally, there may be processing in multiple directions with asymmetry, but if so, then it is not uniform in nature across the different directions. Consider, for example, the symmetry of processing that occurs in both propositional logic and Boltzmann machines (Ackley et al., 1985) versus the asymmetry found in rules and feedforward neural networks. With the latter pair, forward processing yields inference, whereas backward processing yields something more like abduction or learning.

Table 8.1 shows a range of AI technologies across a 1D map as they divide between asymmetric versus symmetric. The headers here, as with all such tables moving forward, are annotated with the respective personality traits for this dimension, of *A* versus *S*. As introduced in section 6.1, each asymmetric technology is identical with respect to this map as is each symmetric technology. Likewise, each asymmetric technology is opposite with respect to this map from each symmetric technology.

It should be clear from this table that when listing technologies, they can be at many different levels of abstraction. For example, under symmetric, the technology of self-attention networks is relatively concrete and

Table 8.1
1D map of technologies partitioned according to the (a)symmetric dichotomy

Asymmetric (A)	Symmetric (S)
Action-Centered; Action Selection and Execution; Arithmetic Circuits and Sum-Product Networks; Attention Networks; Causal Association; Causal Intervention; Decision Trees (and Learning); Deliberation; Doing; Functions; Heteroassociative Neural Networks (and Backpropagation); Hidden Markov Models; Implicit Memory; Induction; Operators; Policies; Procedural Memory (and Learning); Procedural Programs; Reaction; Reinforcement Learning; Rules (and Composition); Shallow Reasoning; Skills (and Acquisition)	Autoassociative Neural Networks (and Hebbian Learning); Causal Counterfactuals; Constraint Satisfaction; Declarative Memory (and Learning); Declarative Programs; Deep Reasoning; Dynamical Systems; Explicit Memory; Knowledge (and Acquisition); Logic; Non-Action-Centered; Planning, Probabilistic Graphical Models and Statistical Relational AI; Reasoning; Reflection; Search; Self-Attention Networks; Thinking; Truth Maintenance Systems

specific, whereas the technology of thinking is a much broader and more abstract notion. Largely due to such abstraction, some of the terms used in this table can be ambiguous in isolation, and the astute reader may be able think of variants of some of them that belong in the other column, such as forms of backward planning that are more obviously asymmetric. This is partly dealt with via the addition of brief explanations of each in appendix A and partly by focusing on canonical or generic versions of each rather than on all possible ways of further instantiating them.

8.1 Variants

The thirty-one variants of this dichotomy identified in table 8.2 (adapted and expanded from Rosenbloom, 2019a) stem from knowledge representation, reasoning, decision-making, neural networks, memory, learning, programming, compiling, emotion, and causality. There are many differences among them, but what they all have in common is the underlying distinction between asymmetric versus symmetric processing. In the discussion that follows, attempts will be made to include one or two references relevant to each variant, although this relevance is stretched at times to include a reference that bears at all on the variant, and at other times there is nothing at all of relevance that has been found to be included.

Table 8.2
Identified variants of the (a)symmetric dichotomy

	Asymmetric (A)	Symmetric (S)
1	Rules	Logic
2	Rule-Based Reasoning	First-Principles Reasoning
3	Shallow Reasoning	Deep Reasoning
4	Reaction	Reflection
5	Deliberation	Reflection
6	Greedy Search	Exhaustive Search
7	Policies	Planning
8	Arithmetic Circuits and Sum-Product Networks	Probabilistic Graphical Models
9	Function-Based Reasoning	Model-Based Reasoning
10	Heteroassociative Neural Networks	Autoassociative Neural Networks
11	Attention Networks	Self-Attention Networks
12	Procedural Memory	Declarative Memory
13	Implicit Memory	Explicit Memory
14	Action-Centered	Non-Action-Centered
15	Doing	Thinking
16	Skill	Knowledge
17	Procedural Learning	Declarative Learning
18	Skill Acquisition	Knowledge Acquisition
19	Macro/Composition Formation	Fact Storage
20	Backpropagation	Hebbian Learning
21	Classification	Clustering
22	Supervised Learning	Unsupervised Learning
23	Discriminative Learning	Generative Learning
24	Genetic Programming	Inductive Logic Programming
25	Control (Reinforcement) Learning	Model Learning
26	Machine Learning	Knowledge Representation and Reasoning
27	Program	Data
28	Procedural Programming	Declarative Programming
29	Executable Code	Source Code
30	Primitive Emotions	Cognitively Elaborated Emotions
31	Causal Intervention	Causal Counterfactuals

Variants 1–3 are from traditional symbolic AI and its application in expert systems, although the asymmetric if-then rules mentioned in the first two actually showed up earlier in work in cognitive science (Newell, 1973b), and even earlier in theory of computation (Post, 1943). In variant 1 (rules versus logic), from the world of symbolic AI (Davis et al., 1993), rules authorize conclusions only in the forward direction; for example, given the rule $A \rightarrow B$, you can conclude that if A is present, then B should be generated. Such a rule does not license similar conclusions in the reverse direction, although it may be used asymmetrically to generate abductive hypotheses, such as that if B is present, then A may also be present. In contrast, a logical statement is deductively valid in any direction. From the implication $A \supset B$, you cannot only conclude that B is true if A is true but also, given that $A \supset B$ is logically equivalent to $\neg B \supset \neg A$, that if B is not true, then A is not true.

Variant 2 (rule-based versus first-principles reasoning) is a slight generalization of variant 1, from the world of expert systems, in which the second element of the pair is generalized from just logic to any approach that allows reasoning based on general principles that underlie the domain rather than simple associations between elements of it (Davis, 1983). Such a set of principles might be expressed in terms of logical statements, but they could also be stated in other forms, such as via qualitative models of physical processes (Forbus, 2019). Variant 3 (shallow versus deep reasoning) is a further abstraction of variant 2, where any form of simple associative reasoning is contrasted with any form of reasoning based on a deeper model of how the domain works (Karp & Wilkins, 1989; Steels, 1989).

Variant 4 (reaction versus. reflection) can be seen as another level of abstraction over variants 1–3, where simple asymmetric/associative processing of any sort yields reaction, immediately suggesting outputs in response to inputs. The contrast here is with reflection—or metacognition (Cox, 2005; Kralik et al., 2018)—where it is possible to step back from the choice being made to view it as a declarative structure over which general reasoning is possible in arbitrary directions. It is possible to approach reflection in an asymmetric manner, such as via rule-based reasoning; however, reflection itself is inherently symmetric.

Variant 5 (deliberation versus reflection) is a slight shift from variant 4. The symmetric pole is still reflection, but the asymmetric pole has shifted from reaction to deliberation. Deliberation, as defined in section 6.3, can be viewed as the iterative selection and execution of one or more actions that in combination yield sequential thought or behavior. Reaction, deliberation and reflection together can be seen as the three tiers in a 3T robotic architecture (Gat, 1998; Kortenkamp & Simmons, 2008) or as the trilevel control structure found in some cognitive architectures (Laird, 2012; Rosenbloom et al., 2016a). Variant 6 (greedy versus exhaustive search) can then be viewed as a special case of variant 5. Greedy search—aka hill climbing or gradient ascent/descent—performs a sequence of actions that each lead to immediate improvements in an objective function, ultimately stopping at a local maximum or minimum, whereas exhaustive search—such as in best-first search—can more flexibly explore the space in arbitrary directions for paths that reach a global maximum or minimum (Pearl, 1984).

Variant 7 (policies versus planning) specializes variant 4 (reaction versus reflection) in the context of planning problems, which consider how to control behavior over a sequence of situations. Policies, as used for example in Markov decision processes (Bellman, 1957) and reinforcement learning (RL; Sutton & Barto, 2018), provide tables that directly associate action valuations with states, enabling direct selection among actions that are available without any deeper understanding of the domain. Universal plans (Schoppers, 1987; Ginsberg, 1989) are similar but are less concerned with valuations and more concerned with direct specifications of which actions to select in which situations. Deep reinforcement learning (Wang et al., 2020) is also similar, except that the policy is encoded as a deep neural network rather than as a table or a universal plan.

Planning instead uses a model of the domain, in terms of states that represent situations and operators that represent actions, to search for paths that achieve goals given initial situations. In prosecuting these searches, planning may use these models in forward or backward directions or in both directions at once. Although operators are similar to rules, which are asymmetric, moving backward across an operator corresponds to applying the operator in the reverse direction rather than to abduction per se. This distinction between policies and planning can also be seen as a different specialization of variant 3 (shallow versus deep reasoning), although it comes out of quite a different tradition (Blythe, 1999).

Variant 8 (arithmetic circuits and sum-product networks versus probabilistic graphical models) comes from yet a different tradition within AI, that of probabilistic reasoning. The canonical technology for probabilistic reasoning is Bayesian networks, which capture probabilistic dependencies among items (Pearl, 1988). Although the dependencies are themselves expressed in an asymmetric manner, as with the asymmetrically stated implications in logic, they are processed symmetrically, albeit via message passing or sampling algorithms rather than via logical inference rules. Probabilistic graphical models are a generalization that includes both Bayesian networks and more explicitly symmetric forms such as Markov networks and factor graphs (Kschischang et al., 2001; Koller & Friedman, 2009). All of these forms are processed in a symmetric fashion by the same classes of algorithms. Arithmetic circuits (Darwiche, 2002) and sum-product networks (Poon & Domingos, 2011) are inherently asymmetric representations for probabilistic problems that reduce the number of problems that can be solved given any individual network/graph in exchange for solving those that can be solved more efficiently. There are improvements in real-time efficiency across the board and in some cases—such as probabilistic parsing—there can be combinatoric improvements as well (Poon & Domingos, 2011).

Variant 9 (function-based versus model-based reasoning) also arose out of probabilistic reasoning, as a means of extending the contrast embodied by variant 8 to include feedforward neural networks among the asymmetric possibilities, and to further understand the improvements in accuracy they achieve over their asymmetric probabilistic counterparts, such as arithmetic circuits (Choi & Darwiche, 2018). In particular, function-based reasoning computes a single value in one direction, whereas model-based reasoning can compute omnidirectionally to yield multiple answers from the same structures. Although derived from concerns with probabilistic reasoning and its relationship to neural networks, this ends up being similar to variants 3 (shallow versus deep reasoning) and 4 (reaction versus reflection). As will be discussed further under variant 29 (executable versus source code), all three of these traditions have also considered how efficiency can be improved by compiling asymmetric variants from symmetric ones (Rosenbloom et al., 1985; Darwiche, 2003).

Variant 10 (heteroassociative versus autoassociative neural networks) stems from work in neural networks, distinguishing asymmetric from symmetric flavors of them. The core asymmetric, or heteroassociative, form is the feedforward network (Rumelhart et al., 1986), including such extensions as convolutional neural networks (LeCun et al., 1989; Waibel et al., 1989), recursive neural networks (Goller & Kuchler, 1996; Frasconi et al., 1998), encoder-decoder networks (Sutskever et al., 2014), and generative adversarial networks (GANs; Goodfellow et al., 2014). Additional forms, such as recurrent neural networks (Elman, 1990)—including long short-term memories (LSTMs) (Hochreiter & Schmidhuber, 1997) and gated recurrent units (GRUs; Cho et al., 2014)—and attention networks (Bahdanau et al., 2015) are likewise asymmetric.

Symmetric, or autoassociative, forms include Hopfield networks (Hopfield, 1982), Boltzmann machines (Ackley et al., 1985), and the kinds of self-attention networks found in transformers (Vaswani et al., 2017). These examples can each be viewed as a relational neural network that starts with a fully connected graph among its elements and learns parameters that reflect relationships among them (Santoro et al., 2017).

Autoassociative neural networks that directly embody structural relationships by limiting initial connectivity—such as local connectionist networks (McClelland & Rumelhart, 1981; Feldman & Ballard, 1982), tensor product representations (Smolensky & Legendre, 2006), and graph neural networks (GNNs; Scarselli et al., 2009; Zhou et al., 2019)—are also symmetric. So too are autoencoders (Kingma & Welling, 2019), although as discussed further in section 8.4, they are encoded by restricting asymmetric encoder-decoder networks so that the input and output units of the overall network represent the same elements. The notion of a symmetric autoassociative network in general turns out to dovetail neatly with earlier analyses that have shown how some forms of autoassociative networks are equivalent to probabilistic graphical models (Jordan & Sejnowski, 2001).

Variant 11 drills down into an important special case of variant 10, contrasting attention networks with self-attention networks. Both poles involve replacing the normal Markovian assumption in neural networks, that layers in standard feedforward networks and stages in recurrent networks prior to the immediately preceding one can be ignored when learning to predict values in the next layer. They differ, however, in whether they do this by adding asymmetric connections from earlier layers/stages to later ones versus replacing the original feedforward structure across the layers/stages with bidirectional connections—based on dot products—among all of them.

Variants 12–14 stem from cognitive science, with variant 12 (procedural versus declarative memory) based on the evidence for distinct long-term memory systems in humans for procedural skills versus declarative knowledge (Squire, 1987). There is a broad sense in which everything stored in a cognitive memory can be considered knowledge; however, the narrower sense used here—that knowledge is just what is stored in declarative memory—provides a useful contrast with the notion that it is skills that are stored in procedural memory. Even in this narrower sense, knowledge may comprise, for example, simple facts, instances (Aha et al., 1991), frames (Minsky, 1975; Fikes & Kehler, 1985), cases (Kolodner, 1992; Leake, 1996), ontologies (Arp et al., 2015), knowledge graphs (Kejriwal et al., 2021), logical sentences, constraints, or the contents of autoassociative networks (Hopfield, 1982; Ackley et al., 1985).

Declarative memory itself may be partitioned into a separate semantic memory for general knowledge versus an episodic memory for experienced events (Tulving, 1972), but this further distinction remains less settled and is not necessary for the argument here. What matters instead is that procedural memory contains code that is run—often modeled as rules but potentially comprising feedforward neural networks or other asymmetric structures—while not otherwise being examinable, and is therefore asymmetric and implicit, whereas declarative memory contains knowledge that can be examined and used in arbitrary ways and is therefore symmetric and explicit.

Variant 13 directly concerns the examinability of memory, in the form of implicit versus explicit memory (Graf & Schacter, 1985). This variant is in general a difficult one to pin down precisely. We will, for example, see two additional interpretations of it under (un)controlled (section 9.1) and (meta)data (section 10.1). Its status as a variant of (a)symmetric is also somewhat iffy. Rather than relying on a direct argument about its mapping to (a)symmetric—although it may be possible to create a vague such mapping—it relies on the fact that it can chain to (a)symmetric not just through one of the other variants but through many of them, at least including numbers 3 (shallow versus deep reasoning), 4 (reaction versus reflection), 7 (policies versus planning), 9 (function-based versus model-based reasoning), 12 (procedural versus declarative memory), 15 (doing versus thinking), 25 (control versus model learning), and 27 (program versus data). This is thus a bit of a stretch according to the direct-mapping principle, but the close relationship to variant 12 makes it important to view the former as akin to the latter along the (a)symmetric dichotomy.

An even further stretch would be to include (un)conscious processing in the list of variants, as it would best be accommodated by chaining through implicit versus explicit. Morsella and Poehlman (2013), for example, described a connection between procedural versus declarative, implicit versus explicit, and unconscious versus conscious. They also included other dichotomies from psychology that have not been of concern in this chapter, such as subliminal versus supraliminal, preattentive versus attentive (Treisman & Gelade, 1980), and motor programming versus voluntary action, plus one additional dichotomy that corresponds to variant 4 of the (un)controlled dichotomy (automatic versus controlled processing). Although (un)conscious has not been included here, it can be found under both (un)controlled (chapter 9) and (meta)data (chapter 10).

Variant 14 (action-centered versus non-action-centered) stems exclusively from the Clarion cognitive architecture (Sun, 2016). Although both of these subsystems are implemented via asymmetric technology, the former functions as an asymmetric procedural memory and the latter as a symmetric declarative memory. More will be said about this in section 8.4, concerned with spanning this dichotomy. Variant 15 (doing versus thinking) is then a generalization of variant 14 beyond Clarion to an everyday distinction between people who prefer action versus those who prefer mulling things over.

Like variants 12–14, variant 16 (skill versus knowledge) is based on the notion that skills are asymmetric structures represented in procedural memory whereas knowledge comprises symmetric structures represented in declarative memory. This variant will also come back up in a different form under (meta)data (chapter 10) and will be discussed further there.

Variants 17–25 all stem from work on learning, whether in humans or machines. Some of these variants focus on (a)symmetry in the structures learned, while others concern (a)symmetry in the learning process itself. Variants 17–18 (procedural versus declarative learning; skill versus knowledge acquisition) follow directly from variants 12 (procedural versus declarative memory) and 16 (skill versus knowledge), concerned as they are with acquisition of their respective forms of asymmetric versus symmetric structures (Cohen et al., 1985); although, as with skill versus knowledge there is also a (meta)data aspect to variant 18 that we will return to in chapter 10. Variant 19 (macro/composition formation versus fact storage) is a specialization of these two prior variants: integrated systems typically include forms of macro(operator) formation—often in the form of rule composition (Neves & Anderson, 1981; Laird et al., 1986a)—that combine into single procedural structures whole strings or trees of existing such structures (Korf, 1985; Alphonse, 2004) versus fact storage that simply stores facts that are transiently present in working memory into long-term declarative memory (Laird et al., 2017).

In Variant 20 (backpropagation versus Hebbian learning), backpropagation (Rumelhart et al., 1986) retraces the paths that were followed in a forward direction during inference in feedforward neural networks, but in a backward direction based on qualitatively different information concerning errors and derivatives. In contrast, Hebbian learning treats the neurons situated both before and after a synapse symmetrically (Hebb, 1949).

In Variant 21 (classification versus clustering), classification asymmetrically produces one output—a choice of the class to which an example belongs—given the features of an example as input, learning a function from inputs to outputs. In contrast, clustering simply groups examples according to the similarity among their features, effectively learning a model of the space of examples. Variant 22 (supervised versus unsupervised learning) then maps onto (a)symmetric in a manner that is isomorphic to variant 21, although here the asymmetry focuses on singling out the supervisory input.

In variant 23, discriminative learning determines decision boundaries between classes whereas generative learning acquires models of the input data (Ng & Jordan, 2001). In probabilistic terms, a discriminative model learns (asymmetric) conditional probabilities of classes given example features whereas a generative model learns the (symmetric) joint probability distribution, with the latter supporting a full probabilistic graphical model and the former only supporting choice of the class given the features. As discussed in section 5.3.2, although conditional probabilities are in principle reversable via Bayes' law, this reversal requires the ancillary use of symmetric prior probabilities, thus limiting conditional probabilities to being asymmetric when considered on their own.

Genetic programming (Koza, 1992) and inductive logic programming (Muggleton & de Raedt, 1994), as contrasted in variant 24, may not normally be considered as in opposition, but while both search over a space of programs given fitness measures, the former is defined over trees that represent asymmetric procedural programs whereas the latter is defined over symmetric logical theories. In variant 25 (control versus model learning), control learning, of which reinforcement learning (RL) is the dominant form, acquires asymmetric procedural knowledge about action choices in particular situations—that is, policies—whereas model learning in its purest form acquires symmetric declarative knowledge about the task environment (Shapiro, 1981; Choi & Darwiche, 2018).

While many of the variants seen so far contrast forms of learning or forms of reasoning, variant 26 contrasts machine learning more broadly with knowledge representation and reasoning. Part of this is a (meta)data contrast that will come up in section 10.1. However, a significant part of it also contrasts the asymmetry of feedforward neural networks—as the current standard in machine learning—with the symmetry of logical representation and reasoning, while still acknowledging that there are indeed symmetric forms of machine learning and asymmetric forms of knowledge representation and reasoning.

Variants 27–29 stem from computer science, with variant 27 distinguishing code from data, under the standard assumption of asymmetric procedural code and symmetric declarative data. Variant 28 distinguishes traditional asymmetric procedural—or imperative—programs from symmetric declarative programs, where the former directly specify how to compute results while the latter only characterize the results to be computed (Lloyd, 1994). One significant form of declarative programming that will also play a role in other dichotomies is constraint programming (Rossi et al., 2006), or constraint satisfaction (Dechter, 2003), where programs are defined in terms of constraints on values of variables rather than by specifying directly how to compute specific values for those variables.

Variant 29 distinguishes executable code from source code. For this variant, the assumption is that executable code is what is run directly by the computer, asymmetrically one step after another, whereas source code is what is written by, and thus examinable and usable arbitrarily by, the programmer. Source code can of course be compiled into executable code, but here it is viewed as procedural information that is expressed in declarative form.

As was mentioned earlier, a comparable notion of the compilability of symmetric knowledge into asymmetric skill can also be found in several of the other variants. For example, in the context of variant 12, proceduralization can combine facts in declarative memory into rules in procedural memory (Anderson, 1995). Similar results can be found in the context of variant 8, with the compilation of Bayesian networks into arithmetic circuits (Darwiche, 2003); in the context of variants 3 and 4, with the compilation of deep knowledge and reflective processing into shallow knowledge and reactive processing (Rosenbloom et al., 1985); and in the context of variant 7, with the compilation of planning into a form of policy known as a universal plan (Schoppers, 1987).

Variants 30–31 are from trilevel emotional and causal hierarchies that were reconsidered in Rosenbloom (2019a) as L-shaped paths in 2D maps. The trilevel emotional hierarchy originated in cognitive science (Ortony et al., 2005), with primitive emotions occurring via routine behavior there, or what is referred to as deliberation here, and cognitively elaborated emotions occurring via reflection. Variant 30 (primitive versus cognitively elaborated emotions) can thus be seen as a specialization to emotional processing of variant 5 (deliberation versus reflection). The trilevel causal hierarchy originated in AI (Pearl & Mackenzie, 2018), with causal intervention occurring via action selection and execution—thus mapping onto deliberation—and causal counterfactuals occurring via reflection. As with variant 30, variant 31 (causal intervention versus causal counterfactuals) too can thus be seen as a specialization—in this case to causal reasoning—of variant 5.

8.2 Learning

Variants 17–25 directly partition learning into pairs of asymmetric versus symmetric learning capabilities, providing one way of tabulating the kinds of learning that may be implicated overall. There may be more such capabilities required in a full integrated system, but these provide a reasonable start in including procedural and declarative learning; skill versus knowledge acquisition; macro/composition formation versus fact storage; backpropagation versus Hebbian learning; classification versus clustering; supervised versus unsupervised learning; discriminative versus generative learning; genetic versus inductive logic programming; and control/reinforcement versus model learning.

These variants do reveal that learning interacts strongly with the (a)symmetric dichotomy, and that approaches (1)–(3) from chapter 6 predominantly align along poles of this dichotomy. For example, backpropagation is an asymmetric mechanism that transmits information backward through heteroassociative neural networks in service of learning parameters from asymmetric (classified) examples for use in heteroassociative neural networks, whereas Hebbian learning is symmetric in how it treats neurons before and after synapses in service of learning parameters from unclassified examples for symmetric neural networks.

With respect to approach (4), although mostly the inputs and outputs of learning align in terms of the (a)symmetric dichotomy, the conversion of symmetric declarative structures that require reasoning or interpretation into asymmetric procedural ones (Anderson, 1982) is a counterexample. Such transformations are typically motivated by real-time efficiency but may under some circumstances support improvements in combinatoric efficiency or precision, just as such benefits can be found in the trade-offs for asymmetric structures. It is less common to find work focusing on the other direction—of converting procedural structures into declarative ones. However, work on extracting explanations of behavior from feedforward neural networks (Townsend et al., 2020) could be said to fit within this category, although some of these techniques translate from metadata to data while still yielding asymmetric structures, such as rules, as their output.

8.3 Trade-Offs

For some of the variants listed in section 8.1, such as program versus data, it has always been assumed that the two poles provide distinct enough strengths that both are required in a complete system. For others, such as discriminative versus generative learning, controversies have raged over which one is the more appropriate one to use. For yet others, such as shallow versus deep reasoning, an initial controversary ultimately settled down to an implicit agreement that both are necessary. No matter which path any variant has taken, the trade-offs that are of interest to us here involve accuracy, efficiency, generality, improvability, and operability.

Let's start with accuracy. The discriminative-versus-generative and function-versus-model-based variants of the dichotomy both reveal how the asymmetric pole can yield more precision when comparing observed versus predicted values (Ng & Jordan, 2001; Choi & Darwiche, 2018). However, in the other direction, symmetric encodings—such as logics and probabilistic graphical models—tend to be mathematically more elegant and thus lend themselves more directly to proofs of correctness. Symmetric encodings, by inherently encoding solutions to whole classes of problems that may be covered in only a piecemeal fashion by whatever asymmetric structures exist, also tend to be more robust than their asymmetric equivalents, although there are technologies such as universal plans that attempt to provide an asymmetric equivalent for every possible symmetric output (Schoppers, 1987).

An important point to make at this juncture is that all of these relative advantages are a matter of potential rather than necessity, as captured by the use of such terms as "can" and "tend." They thus are best interpreted as rough upper bounds on what is possible rather than as what is always to be expected. This is a theme that will resonate through all of the discussions of criterial advantages whether stated explicitly or not. It will also show up in additional forms when considering the possibility of spanning dichotomies, as in the next section, or of integrated systems that combine multiple locations across multidimensional maps, as will be seen in chapter 11.

With respect to efficiency, by sharing structures and processing across multiple similar problems, symmetry typically yields better temporal and spatial efficiency than is possible via the set of asymmetric structures that would be required for solving the same problems, although algorithms do exist for at least partial sharing among similar asymmetric structures (e.g., Forgy, 1982). In contrast asymmetric structures may find solutions to individual problems in a more temporally efficient manner, whether this is measured in terms of real-time or combinatoric efficiency (Rosenbloom et al., 1985; Schoppers, 1987; Darwiche, 2003). With respect to combinatoric efficiency, it has, for example, been shown that Bayesian networks for probabilistic parsing are inherently exponential whereas asymmetric variants, such as sum-product networks, can be solved in polynomial time (Poon & Domingos, 2011).

With respect to generality, as was discussed in section 7.3, the focus here is on the generality of an entire system that is built from some type of technology rather than on the generality of the individual content elements that may be supported. Thus, the fact that individual symmetric structures are typically more general than individual asymmetric structures is not the issue here, although it does show up under efficiency. With respect to problem generality in particular, asymmetric structures can directly encode some types of knowledge—such as noninvertible/nonreversible computations—that cannot be so encoded in corresponding symmetric structures. In the reverse direction, if we consider discriminative versus generative learning, the latter can directly generate full instances whereas the former cannot.

Thus, while there may be many problems solvable via either pole, there can be gaps in either that don't exist for the other.

With respect to improvability, a distinction arises that is akin to that for problem generality—asymmetric structures naturally advantage learning inherently asymmetric information, whereas symmetric structures advantage learning to generate full instances.

With respect to operability, asymmetric more directly supports effectivity, as seen in variants 14–15 (action-centered versus non-action-centered; doing versus thinking). It also more directly supports reactivity, focusing as it does on directly computing the results necessary for appropriate behavior rather than on the larger space of possibilities that may be generated symmetrically. Variant 4 (reaction versus reflection) makes this explicit, with the inclusion of reaction as asymmetric, but other variants implicitly support this as well. In contrast, symmetric more directly supports safety, as it encourages robust thinking about the consequences of actions that are being considered.

Table 8.3 summarizes these trade-offs. A personality trait of *A* indicates an advantage for asymmetric and *S* for symmetric. Under efficiency, a subscript of "1" indicates an advantage when a solution is sought for an individual problem, whereas an "n" indicates an advantage for multiple related problems. For problem generality and learnability, a subscript of "a" indicates a necessarily asymmetric computation, whereas a "g" indicates the ability to generate instances of the problem.

According to this analysis, whereas usability isn't particularly relevant for this dichotomy, aspects of accuracy, efficiency, generality, improvability, and operability may be used either to select an appropriate pole for an application or to argue more broadly for spanning the (a)symmetric dichotomy to leverage both poles as necessary across the range of problems to be faced. In brief, the analysis shows that asymmetric yields precision, temporal efficiency for single problems, problem generality and learnability for inherently asymmetric computations, and reactivity, whereas symmetric yields correctness and robustness, spatial and temporal efficiency for multiple related problems, and problem generality and learnability for generating instances.

As was foreshadowed in chapter 6, an alternative way to view these results is as a 1D map in the shape of table 8.1, but that directly lists potential criterial advantages for each location instead of technologies. This

Table 8.3
Criterial trade-offs for the (a)symmetric dichotomy

Criteria	Advantage
Accuracy (A)	
Correctness (AC)	*S*
Precision (AP)	*A*
Robustness (AR)	*S*
Efficiency (E)	
Time (ET)	A_1/S_n
Space (ES)	S_n
Generality (G)	
Problem (GP)	A_a/S_g
Improvability (I)	
Learnability (IL)	A_a/S_g
Operability (O)	
Effectivity (OEf)	*A*
Reactivity (OR)	*A*
Safety (OSa)	*S*

might be thought of as a third type of map, in addition to architectural and disciplinary maps, but we will see these maps used both for disciplines and architectures, so it is simplest at this point to just consider them as variants of each of these types of maps.

Table 8.4 is such a map, and one in which the shape of the CA language can begin to be seen more clearly. It reuses the abbreviations from the first column of table 8.3 for criterial advantages and the subscripts from the second column to narrow the application of these advantages, such as ET_1 to designate a temporal efficiency advantage for individual problems. Two additional extensions here are: (1) Using parentheses to group subcriteria where possible, such as A(C, R) to represent the combination of AC and AR; and (2) abbreviating a complete set of parentheticals—at least complete in terms of the criteria considered here—by omitting them, such as reducing A(P, C, R) to A.

Both perspectives on (a)symmetric criteria reveal a within-accuracy trade-off of precision versus correctness and robustness; a within temporal-efficiency trade-off over single problems versus multiple related problems, with spatial

Table 8.4
1D map of criterial advantages partitioned according to the (a)symmetric dichotomy

Asymmetric (A)	Symmetric (S)
AP, ET_1, GP_a, IL_a, O(Ef, R)	A(C, R), E_n, GP_g, IL_g, OSa

efficiency also included in the latter; within problem-generality and learnability[1] trade-offs over the ability to perform/learn necessarily asymmetric computations versus the ability to (learn to) generate problem instances; and a within operability trade-off of effectivity and reactivity versus safety.

In this case, there is little additional insight gained by including the second perspective found in table 8.4, other than providing an encoded summary of what has already been described, but when interactions among dichotomies come to the fore in chapter 11 this perspective comes much more into its own.

8.4 Spanning

Both AI and cognitive science have explored a variety of technologies on both sides of this dichotomy, attesting to the individual importance of each. But, beyond this, a good argument can also be made that integrated systems ultimately need both asymmetric and symmetric capabilities—whether cast as procedural and declarative memory, shallow and deep reasoning, control and model learning, or other pairs—so that they can play on both sides of these trade-offs. This appears to be true both when building integrated systems for A(G)I applications and when building integrated models of cognition that are to match what is known about the human mind. A critical question then becomes how to support such combinations.

The addition approach simply combines multiple modules, with at least one supporting asymmetric structures and one supporting symmetric structures. This can be seen, for example, in approaches that provide distinct procedural and declarative memories (Anderson, 2007; Laird, 2012; Laird et al., 2017), or distinct deep and shallow reasoning mechanisms (Fink et al., 1985; Rosenbloom et al., 1985). It can also be seen in models of analogy, whether symbolic or neural, that use a symmetric module up front that is akin to a declarative memory for retrieving analogs followed by an asymmetric module that extracts useful results for the target from the retrieved analog (Falkenhainer et al., 1989; Gentner et al., 2001; Mikolov et al., 2013; Ustun et al., 2014).

The Clarion cognitive architecture nominally includes a 2×2 arrangement of action-centered versus non-action-centered subsystems—which, as noted in section 8.1, correspond respectively to procedural and declarative memories—and implicit versus explicit representations, appearing to triple-dip on the (a)symmetric dichotomy. However, as will be seen in sections 9.1 and 10.1, the first of these three distinctions in Clarion is actually more appropriately thought of as (un)controlled and the third as (meta)data. It is only its procedural versus declarative nature that fits truly under (a)symmetric. Still, its 2×2 arrangement provides a classic example of spanning via addition.

To communicate among the modules yielded by addition, one typical approach is to include in the resulting integrated system a short-term declarative working memory (Jones & Ritter, 2003; Chai et al., 2018) or blackboard (Erman et al., 1980) that the (now long-term) asymmetric and symmetric modules can read from and write to. Such short-term memories were hypothesized early on in support of integrated systems possessing just procedural long-term memories (e.g., Laird et al., 1987), but here they also support communication across distinct modules (Laird et al., 2017). The notion of a global workspace (Baars, 2002) also fits comfortably here, although it does add the further assumption that consciousness is directly associated with presence in the workspace.

When the focus is on combining asymmetric and symmetric learning via addition, a working memory is typically not needed for communication between them, as the results of such learning go directly into their respective long-term memories, which then themselves communicate via working memory. Many cognitive architectures do in fact use different learning mechanisms for each long-term memory (Anderson et al., 2004; Laird, 2012; Sun, 2017).

The first variation on reduction natively supports only an asymmetric or a symmetric capability but is then able to implement or simulate the other in some manner. For example, although the ACT-R cognitive architecture includes both procedural and declarative memories, the tradition includes work on simulating procedural memory via rulelike structures in declarative memory, in conjunction with an interpreter for these structures in procedural memory (Anderson, 1982). Although, this approach does require a minimal amount of domain-independent procedural structure for the interpreter, it does then allow domain-dependent rules to be represented in declarative memory.

In the reverse direction, early versions of the Soar cognitive architecture had only one long-term memory, for asymmetric rules (Laird et al., 1987). However, it did also have the kind of declarative working memory just mentioned into which structures instantiated by rule actions could be generated. It thus could use the conditions of rules to specify access paths to declarative structures that were encoded in the rules' actions, with this content then capable of participating in fully symmetric processing when in working memory. What it missed was the symmetric flexibility of access available in a true declarative memory, instead requiring a distinct asymmetric rule for each access path that was to be available.

As with this pair of examples, such simulations may require at least a stub of the pole to be simulated, whether a procedural interpreter of declarative knowledge or a declarative working memory; however, even in these cases, how the other long-term memory stores and accesses its knowledge is simulated. Other simple examples include using symmetric Horn clauses to simulate asymmetric rules (Kowalski, 1988); symmetric generative learning methods for asymmetric classification (Ng & Jordan, 2001); heteroassociative neural networks to simulate autoassociative neural networks via autoencoders built from sequences of heteroassociative encoder and decoder networks for which the encoder inputs are the same as the decoder outputs (Kingma & Welling, 2019); and asymmetric reinforcement learning to simulate symmetric search, at least for situations that have been previously experienced.

Such approaches can also be found in learning, typically by having a single learning mechanism of one type or the other while handling the differential aspects of acquiring asymmetric versus symmetric structures by learning from distinct types of knowledge. For example, early versions of Soar that had only a rule-based procedural memory also had only a single learning mechanism—chunking (Laird et al., 1986a)—that can be viewed as a form of asymmetric rule composition. To acquire symmetric declarative knowledge, albeit simulated within procedural memory, additional procedural structures were added that had the side effect of yielding a declarative learning capability known as data chunking (Rosenbloom et al., 1987).

In contrast, the Sigma cognitive architecture has only a symmetric Hebbian-like learning mechanism for parameter tuning (Rosenbloom et al., 2016a) that was originally developed for Bayesian networks (Russell et al., 1995), but it also supports asymmetric backpropagation by adding knowledge structures that communicate error rates backward, with the same symmetric learning mechanism—based on gradient descent—then locally leveraging both the forward and backward information for appropriate learning (Rosenbloom et al., 2017).

It should be noted, however, that while each of these two approaches enable forms of both asymmetric and symmetric learning, neither has proven adequate for supporting all of the forms of asymmetric and symmetric learning required for integrated systems. With respect to Soar, the response to this in the late version was to shift to the additive approach, with new learning mechanisms for symmetric facts and episodes in declarative semantic and episodic memories, and for asymmetric reinforcement learning. Although Sigma's single symmetric learning mechanism has been shown to support learning for both declarative and procedural memory, as well as reinforcement learning, it does not yield the kind of structure learning provided by chunking.

One final comment is needed on this first variation of reduction, by simulation, before moving on to the second variation. When such reduction is possible it may enhance in subtle manners the criterial advantages listed in section 8.3 for the pole doing the simulating, although possibly at the cost of other advantages. For example, ACT-R's simulation of procedural memory by interpreting rulelike structures in declarative memory enables the latter to extend to the problem generality of the former while likely worsening real-time efficiency on single problems.

The second variation on reduction delves more deeply into the pair of poles simultaneously, seeking a single technology that enables both directly. The examples I have been able to identify with respect to (a)symmetric involve adding an optional tweak to a technology originally aimed at only one of the poles, with this tweak then enabling the now-extended technology to directly support the other pole as well. This does bear some relationship to addition, in including a switch, but rather than there being a coarse-grained switch between two distinct modules, there is only one module that includes within it a fine-grained switch for the optional tweak.

Table 8.5
0D map of criterial advantages for the maximum across the (a)symmetric dichotomy

A, E(T, S_n), GP, IL, O

For example, it is possible to directly support symmetric structures, such as probabilistic graphical models, while enabling asymmetric structures by selectively turning off influences in particular directions along the models' links (Rosenbloom et al., 2016b). This enables using the same extended graphical models to support not only a declarative memory but also a procedural memory capable of representing such things as rules, sum-product networks, and heteroassociative neural networks, while also representing memory structures that blend characteristics of both types of memories (Rosenbloom, 2010a). In the reverse direction, in Soar and Sigma an asymmetric deliberation capability is extended to enable symmetric reflection via a metacognitive functionality that can detect, represent in working memory, and resolve problems in operator selection and application (Rosenbloom et al., 1988; Rosenbloom et al., 2016a).

With a sufficiently spanned (a)symmetric dichotomy it is possible in theory to achieve the maximum over the criterial advantages across the (a)symmetric dichotomy. Table 8.5 shows how spanning this one dichotomy could yield a combination of accuracy, temporal efficiency, spatial efficiency over multiple related problems, problem generality, learnability, and operability. It is important to note though that such a combination is truly achievable only if the best of each pole is available wherever it is needed. Thus, as mentioned in the previous section, this effectively makes these expressions upper bounds on what is possible. When the best of both poles is not achievable as needed, only approximations to this upper bound may be feasible. This same qualifier also applies with respect to what is achievable when spanning the two remaining dichotomies.

8.5 Summary

In summary, (a)symmetric turns out to be a major dichotomy that underlies a host of distinctions that have been articulated across AI, and even more broadly across computer and cognitive science. Included in this are variants that cover both task performance and learning. The dichotomy is further grounded in criterial trade-offs involving accuracy, efficiency, generality, improvability, and operability. Under the assumption that integrated systems need to support both asymmetric and symmetric capabilities, ideas for how to do this have been explored, along with the potential this raises for maximizing over the combination of each's criterial advantages.

9 (Un)Controlled

The core dichotomy here centers on the computational question of whether processing is *controlled*, in the sense that decisions are required before changes occur, versus *uncontrolled*, where changes may occur without prior decisions. As mentioned in chapter 6, this is a conflation of the earlier dichotomies of (non)monotonicity and (non)combinatory, although via a path that went through intermediaries of (in)consistent and (non)combinatoric.

The distinction between monotonicity and nonmonotonicity ultimately derives from logic, where it refers to inferences that remain true once they have been made versus ones that are defeasible, and thus retractable or modifiable even after having been made; and from mathematics, where it refers to functions that either never increase or never decrease. The distinction between consistent and inconsistent instead concerns whether the structures of interest represent a single consistent situation versus multiple inconsistent situations.

(Non)monotonicity in logic is like (in)consistency in considering whether what is explicitly known remains consistent with what was known in the past. In mathematics the connection relates to the (in)consistency of direction in changes to functional values. Given this reframing in terms of (in)consistent, it is then a relatively short step to (un)controlled, as the processing of consistent structures can typically be uncontrolled due to consistency guarantees among the potential changes and what already exists, whereas inconsistent structures typically imply control is required to choose among inconsistent alternatives. The overlap isn't perfect, as there can be other pressures to choose between controlled and uncontrolled—such as the possibility discussed next of needing to control combinatorics even when there is consistency—but it is close enough to enable many of

Table 9.1
1D map of technologies partitioned according to the (un)controlled dichotomy

Uncontrolled (U)	Controlled (C)
Arithmetic Circuits and Sum-Product Networks; Assumption-Based TMS; Bayesian Learning; Causal Association; Conjunctive Reasoning; Constraint Propagation; Dynamical Systems; Fast Thinking (and Learning); Hidden Markov Models; Implicit Processing; Knowledge (and Acquisition); Knowledge Search; Memory (and Learning); Monotonic Functions; Monotonic Logic; Neural Networks (and Learning); Non-Action-Centered; Parallel Processing; Perception; Policies; Probabilistic Graphical Models and Statistical Relational AI; Reaction; Reasoning; Rules (and Composition); (Self-)Attention Networks; Shallow Reasoning; Skills (and Acquisition); Vector Symbolic Architectures	Action-Centered; Action Selection and Execution; Causal Counterfactuals; Causal Intervention; Constraint Conditioning; Decision Trees (and Learning); Deliberation; Deep Reasoning; Disjunctive Reasoning; Doing; Explicit Processing; Justification-Based TMS; Markov Decision Processes; Nonmonotonic Functions; Nonmonotonic Logic; Operators; Planning; Problem Space Search; Programs; Reflection; Reinforcement Learning; Serial Processing; Slow Thinking (and Learning); Symbolic Analogy

the variants that were considered earlier to now be considered under (un) controlled.

The (non)combinatory dichotomy focused on whether representation was relational versus propositional, and whether processing was lifted—reflecting directly whatever first-order structure existed—or grounded to the point of being limited to a propositionalized processing of everything captured by the first-order structure. Whereas there was the potential for combinatorics in combinatory structures—where processing is (at least) exponential in the number of features—it was not the direct focus until the shift. As with (in)consistent, it is a short step from (non)combinatoric to (un)controlled, as combinatoric typically demands control, whereas non-combinatoric may often get away without it.

Table 9.1 displays a range of AI technologies across a 1D map that distinguishes uncontrolled versus controlled. As with table 8.1, brief explanations of each of these technologies can be found in appendix A.

9.1 Variants

Table 9.2 (adapted and expanded from Rosenbloom, 2019a) identifies many of the prior variants across AI and related fields that map onto (un)controlled.

Table 9.2
Identified variants of the (un)controlled dichotomy

	Uncontrolled (U)	Controlled (C)
1	Monotonic Logic	Nonmonotonic Logic
2	Monotonic Functions	Nonmonotonic Functions
3	Local Optimum	Global Optimum
4	Automatic Processing	Controlled Processing
5	Implicit Processing	Explicit Processing
6	Unconscious Processing	Conscious Processing
7	Biological Band	Cognitive Band
8	Fast (System 1) Thinking	Slow (System 2) Thinking
9	Expert	Novice
10	Fast Learning	Slow Learning
11	Memory	Action Selection and Execution
12	Rules	Operators
13	Non-Action-Centered	Action-Centered
14	Thinking	Doing
15	Reaction	Deliberation
16	Reaction	Reflection
17	Parallel Processing	Serial Processing
18	Assumption-Based TMS	Justification-Based TMS
19	Knowledge Search	Problem Space Search
20	Neural Network	Problem Space Search
21	Backpropagation	Reinforcement Learning
22	Bayesian Learning	Decision Tree Learning
23	Shallow Reasoning	Deep Reasoning
24	Conjunctive (And) Reasoning	Disjunctive (Or) Reasoning
25	Policies	Planning
26	Constraint Propagation	Constraint Conditioning
27	Proto-Affect	Primitive Emotions
28	Proto-Affect	Cognitively Elaborated Emotions
29	Causal Association	Causal Intervention
30	Causal Association	Causal Counterfactuals
31	Instances	Analogies
32	Vector Symbolic Architectures	Analogies

Variants 1–2 cover the two canonical forms of (non)monotonicity. In variant 1 (monotonic versus nonmonotonic logic), inference in monotonic logic naturally proceeds in an uncontrolled fashion given that any available inference can be made at any time without interfering with any other available inferences. It can be controlled, but this would be done to shift the resulting system into a different region of the trade-off space, where combinatoric efficiency is gained, but perhaps to the detriment of other criteria. In contrast, inference in nonmonotonic logic inherently involves conflicts that must be resolved and is thus inherently controlled.

The connection from functional monotonicity to (un)controlled in variant 2 (monotonic versus nonmonotonic functions) is a bit more tenuous. However, it can show up, for example, when applied to value functions in searches. With algorithms such as hill climbing and gradient ascent/descent, any action can be allowable that improves the value function, leading to a local optimum. To reach a global optimum, decisions must be made as to when to select an action that reduces the value function so that movement among hills becomes possible.

Variant 3 (local versus global maximum) is a specialization of variant 2 directly concerned with the difference between an uncontrolled search for a local optimum versus a controlled search for a global optimum. This can also be considered as a different perspective on variant 6 of (a)symmetric in table 8.2 (greedy versus exhaustive search).

Variants 4–8 all stem from cognitive science, reflecting multiple takes on a dual-system notion of cognition, in which one component operates quickly, automatically, and without awareness, whereas the other operates in a slower more controlled fashion that includes awareness (Newell, 1990; Schneider & Chien, 2003; Kahneman, 2011). The key to mapping these variants onto (un)controlled is the automatic-versus-controlled aspect they all have in common (Schneider & Shiffrin, 1977). Yet, the terminology used in these variants, and thus their emphases, does differ. Variant 4 (automatic versus controlled processing) directly considers the absence or existence of deliberate control. Variant 5 (implicit versus explicit processing) echoes the implicit versus explicit distinction found under (a)symmetric, but here concerned with awareness of processing rather than memory, assuming that awareness can be associated only with controlled processing. Variant 6 introduces the (un)conscious dichotomy, but only in the limited form that aligns with the awareness of processing found in variant 5.

Variant 7 (biological versus cognitive band) emphasizes the timescale of processing and that the fastest deliberate act—that is, where control becomes available temporally—occurs at the bottom of the cognitive band (chapter 2), giving rise to the ~50 ms cognitive cycle mentioned in chapter 1. Variant 8 (fast versus slow thinking) emphasizes in a more direct way the relative speeds of processing, but with many ancillary contrasts included in the full analysis (Kahneman, 2011), including automatic versus controlled processing. Some of these contrasts also crop up via other variants included here—such as implicit versus explicit processing, unconscious versus conscious processing, and parallel versus serial processing—whereas others may be worth further consideration.

Variants 9–10 are defined analogously to variant 8 but span both cognitive science and AI. Variant 9 (expert versus novice) characterizes fast uncontrolled behavior as what is exhibited in pattern-directed expert performance and slow controlled behavior as what is exhibited by novices who have not yet acquired the compiled expertise necessary to directly recognize solutions. Variant 10 concerns learning that is fast (and automatic) versus slow (and deliberate) (Laird & Mohan, 2018).

Variant 11 (memory versus action selection and execution) stays within cognitive science but shifts into more explicitly functional terminology; in particular, mapping memory onto uncontrolled processing while mapping action selection and execution onto controlled processing, much as was seen in table 6.1. Memory retrieval does often lead directly into decision-making, whether it concerns the best element to retrieve from declarative memory or the best action to be performed based on the results of procedural memory. Arguments have been made that decision-making is part of the function of the memories—for example, conceiving of action selection as the function of procedural memory—but in my view, memory retrieval is about more than just decision-making, and decision-making in general can be based on more than just what is immediately accessible from a single long-term memory. So, for our purposes here, these will be considered as distinct technologies. As will be discussed in section 9.4, the general idea of following memory retrieval by a decision then becomes a major approach to spanning (un)controlled, with memory providing uncontrolled processing and decision-making providing controlled processing.

Variant 12 (rules versus operators) is a specialization of variant 11. Rules provide one form of procedural memory, but there has been a long-term tension between the kinds of rules found, for example, in Soar versus ACT-R. Both sorts of rules are matched in parallel, but Soar's rules may also execute in parallel, whereas only one ACT-R rule can execute per cognitive cycle, thus requiring decisions. The resolution of this difference, at least from a Soar perspective, has been to map ACT-R's rules onto operators in Soar, with Soar's rules thus thought of as finer grained than ACT-R's rules; that is, multiple of the former are required for a single operator that would comprise one rule in ACT-R. Variant 12 adopts the Soar terminology, referring to the former as rules and the latter as operators.

Variant 13 (non-action-centered versus action-centered) stems from the Clarion cognitive architecture, as has already been discussed in section 8.1. In table 8.2, this variant aligned with the (a)symmetric distinction between declarative versus procedural memory. Here it aligns with whether processing amounts to uncontrolled memory access versus action selection, that is, with variants 11 and 12. Clarion also includes an implicit versus explicit distinction within both poles of this dichotomy, but because this is really a (meta)data variant concerning neural networks versus rules rather than an example of variant 5 here, it will be picked back up on in section 10.1.

As with the relationship between variants 15 and 14 in table 8.2, variant 14 here (thinking versus doing) generalizes variant 13 to the everyday context. However, the kind of thinking implicated in table 9.2 is limited to what occurs during uncontrolled memory access. While this might sound quite restrictive, in principle it can include reasoning with a full monotonic logic or a large language model (LLM). What is ruled out is engaging in either deliberation or reflection, as seen in the next two variants.

Variants 15–16 follow closely along the lines of variants 4–5 under (a) symmetric. In fact, variant 16 (reaction versus reflection) here is identical to variant 4 there, although here it emphasizes how reaction is uncontrolled, whereas reflection is controlled. In other words, the reaction versus reflection variant can be thought of as aligning these two underlying overall dichotomies to yield *AUX* processing versus *SCX* processing. On the other hand, variant 15 here switches things up versus variant 5 of (a)symmetric, contrasting reaction with deliberation rather than deliberation with reflection. Although deliberation is asymmetric like reaction, it is akin to reflection in being controlled. In general, and as will be discussed further in section 11.1, there is an orthogonality between these two dichotomies that engenders a 2×2 space of critical capabilities, and within which the alignment introduced above is merely the major diagonal (Rosenbloom, 2019a).

Variant 17 (parallel versus serial processing) has a cognitive science context plus a more directly functional one within AI and computer science, where uncontrolled implies the opportunity for parallelism, whereas controlled implies the centrality of sequential decision-making. It is clearly possible to perform what is nominally uncontrolled processing in a controlled sequential manner, although only at the cost of shifting its trade-offs to those found for controlled processing. It is also possible to leverage some amount of parallelism within controlled processing, such as with parallel game-tree search (Ferguson & Korf, 1988). However, in such a case there is at least still inherent seriality along the individual paths through the trees. These types of considerations will be seen again in section 9.4.

Variant 18 can be seen as a special case of variant 17 that concerns assumption-based truth maintenance systems (ATMS) (de Kleer, 1986) versus justification-based truth-maintenance systems (JTMS) (Doyle, 1979). Both focus on maintaining consistent sets of inferences, but the former maintains and processes all such sets in parallel whereas the latter considers only one such set at a time.

Variant 19 (knowledge versus problem space search) also stems from both cognitive science and AI and, in particular, integrated systems such as cognitive architectures. In many ways it is simply a variation on variants such as 4 (automatic versus controlled processing), 8–9 (fast versus slow thinking; expert versus novice), and 11 (memory versus action selection and execution), although it focuses on how, despite search being a central concept in both memory retrieval and action selection/execution, the two yield qualitatively different forms of search. The former is a fast, automatic search over memory structures, whereas the latter is a slower, controlled search over a problem space. Variant 20 (neural network versus problem space search) is a special case of variant 19 where the memory structures are neural networks. The notion of problem space search here may sometimes be cast as symbolic problem-solving, but both capture essentially the same idea.

Variant 21 (backpropagation versus reinforcement learning) mirrors variant 20, but with a learning focus. Backpropagation operates on uncontrolled neural networks whereas reinforcement learning operates across controlled sequences of actions, whether in the context of deliberation (such as sequential action execution in the world) or reflection (such as internal problem space search). As was seen in tables 8.1 and 8.2, both are asymmetric learning mechanisms, but as will be discussed further in section 9.2 on learning, they are distinguished here based on whether control is central to them.

Variant 22 (Bayesian versus decision tree learning) also focuses on learning, but here on Bayesian learning that acquires in an uncontrolled manner a posterior distribution given a set of examples versus decision tree learning that in a controlled manner sequentially extends a tree that performs classification. The former derives from probabilistic AI whereas the latter from symbolic AI. This variant will also show up again in section 9.2.

Variant 23 (shallow versus deep reasoning) revisits a distinction from AI and expert systems that was previously seen in the context of (a)symmetric. There, variant 3 contrasted shallow rule-based reasoning versus some form of deeper reasoning, such as logical or first-principles reasoning. Here, the contrast is instead between the automatic associational aspect of shallow reasoning versus the slower, more in-depth forms of deep reasoning that require deliberative control to keep them in reasonable check. In other words, the shallow versus deep variant—as with reaction versus reflection earlier—aligns these two overall dichotomies along the major diagonal of the 2×2 space of possible interactions.

Variants 24–26 stem from AI, but from multiple subareas within the field. Variants 24 (conjunctive versus disjunctive reasoning) and 25 (policies versus planning) are both from the subarea of planning, with the former arising from the realm of hierarchical plan instantiation (Lotem et al., 1999) and the latter from the realms of Markov decision processes and reinforcement learning versus the realm of planning (Ghallab et al., 2004). It should be noted that Markov decision processes are themselves a form of planning (Boutilier et al., 1999), although what they yield are policies rather than more conventional plans. In variant 24, *And* nodes delineate elements that all must be accomplished, but freely in any order, and which can thus be accomplished in an uncontrolled fashion. *Or* nodes delineate inconsistent alternatives out of which only one is to be accomplished.

In variant 25, policies are a form of procedural memory that are akin to reactions, but, as has been mentioned is often the case with procedural memories, accessing them is immediately followed by a decision. This combination of access and decision could therefore be considered one step of controlled deliberation rather than as uncontrolled reaction, but the contrast with planning is sharper and more informative if the focus remains on just the memory access piece here. Planning then involves deliberate search over problem spaces of action models or partial plans. This is thus another (un) controlled variant that aligns with the corresponding (a)symmetric variant.

Variant 26 (constraint propagation versus conditioning) is derived from the area of constraint satisfaction, where problems are specified in terms of constraints on possible solutions (Dechter, 2003). This same approach was discussed under (a)symmetric as a form of declarative programming, but here it is partitioned into uncontrolled and controlled subtechnologies. Constraint propagation takes the results of imposing constraints over some variables and propagates them to other constraints that share any of the same variables. A simple example is how the possible solutions to constraints in one area of a map-coloring problem influence the solutions available to constraints in neighboring locations. Constraint conditioning, in contrast, hypothesizes individual values for variables, with each choice of conditioning variables yielding a qualitatively different path than would be yielded by alternative choices. In analogy to variant 19 (knowledge versus problem space search), propagation thus becomes a form of uncontrolled knowledge search while conditioning becomes a controlled problem space search over the space of variable assignments.

Variants 27–30 stem from the same two trilevel hierarchies, concerning emotions (Ortony et al., 2005) and causality (Pearl & Mackenzie, 2018), which yielded variants 30–31 in table 8.2. Here, each pair of variants embodies the appropriate specializations of variants 15–16 (reaction versus deliberation; reaction versus reflection) to their respective domain. Variant 27 contrasts the reactive nature of proto-affect with the deliberative behavior associated with primitive emotions, whereas variant 28 contrasts proto-affect with the reflective behavior of cognitively elaborated emotions. Variant 29 contrasts the uncontrolled probabilistic reasoning underlying causal association with the controlled deliberation underlying causal intervention, whereas variant 30 contrasts causal association with the reflection underlying causal counterfactuals.

Variants 31 and 32 both concern analogies. Variant 31 contrasts instance-based approaches with general analogical processing (Falkenhainer et al., 1989; Aha et al., 1991; Gentner et al., 2001). The former match identical feature labels and can proceed in an uncontrolled manner. Both frames (Minsky, 1975; Fikes & Kehler, 1985) and cases (Kolodner, 1992; Leake, 1996) can also be viewed as forms of this. In contrast, general analogical processing must generate, evaluate, and select among alternative possible matches, thus requiring control. Variant 32 (vector symbolic architectures versus analogies) then concerns whether choices must be made during analogical processing versus a blending of all alternatives being carried through the process in an uncontrolled manner.

In the process of conflating the (in)consistent and (non)combinatoric dichotomies, four of the entries in table 9.2—variants 19–20 (knowledge search versus problem space search; neural network versus problem space search) and 25–26 (policies versus planning; constraint propagation versus conditioning)—were carried over from both dichotomies; twenty-one of the entries—variants 1–13, 15–17, 21, 23–24, and 29–30—were carried over solely from (in)consistent; and only two of them—variants 31–32 (instances versus analogies; vector symbolic architectures versus analogies)—were carried over solely from (non)combinatoric. Variant 14 (thinking versus doing) is new, having been added as an appropriate generalization of variant 13 (non-action versus action centered), as it also was added for (a)symmetric. Variants 18 (assumption-based versus justification-based TMS) and 22 (Bayesian versus decision tree learning) have also been added here, as they fit well with (un)controlled while not having been a good match to either of its predecessors. Variants 27–28 are also new here, having not been seriously considered under either of the previous formulations.

In the reverse direction, two variants from (in)consistent and eight variants from (non)combinatoric were dropped, as they did not map appropriately to (un)controlled. This does signal that some relevant distinctions have been lost via this conflation, particularly with respect to (non)combinatoric. Yet, the overall gain in simplicity combined with a sense that many of the dropped variants either showed up elsewhere—such as feedforward neural networks versus rules in (a)symmetric—or were not that critical at this time—such as Monte Carlo tree search versus alpha-beta search—led to a sense that this was worthwhile, at least at this stage of the overall exploration of dichotomic analysis.

9.2 Learning

Only three of the variants listed in table 9.2 explicitly concern learning: 10 (fast versus slow learning); 21 (backpropagation versus reinforcement learning); and 22 (Bayesian versus decision tree learning). Variant 10 explicitly addresses point (1)—of the processing that drives learning breaking down according to the dichotomy—that was introduced in chapter 6. It contrasts learning that proceeds via reactive architectural processes versus learning that requires deliberation.

Variant 21 contrasts backpropagation with reinforcement learning, rather than contrasting the former with Hebbian learning as was seen in section 8.1. Both backpropagation and Hebbian learning are uncontrolled along points (1)–(3), concerning the learning process and the input and output structures, and so fall on the same side of the present dichotomy. However, backpropagation does provide a contrast with reinforcement learning (RL) here. Although both are thoroughly asymmetric, operate in an uncontrolled manner, and generate uncontrolled memory structures, the input to backpropagation is uncontrolled memory structures, whereas the input to RL is a controlled sequence of actions and their resulting states. RL then operates by converting this controlled input into uncontrolled output—that is, into a policy. Variant 22 is simpler, contrasting Bayesian and decision tree learning, where the former is uncontrolled along points (1)–(3) and the latter is controlled along these same three points.

From the list of other learning mechanisms in table 8.2, those marked as asymmetric—such as macro/composition formation, skill learning, genetic programming, and control learning—tend toward controlled on input and either controlled or uncontrolled on output. Those listed as symmetric—such as fact storage, knowledge learning, inductive logic programming, and model learning—tend toward uncontrolled on input and output. However, there are exceptions.

Macro/composition formation is most naturally thought of as unitizing in a controlled manner sequences of controlled knowledge into new controlled structures. However, some technologies of this sort, such as chunking in Soar (Laird et al., 1986a), combine in an uncontrolled manner multiple rule traces and decisions extracted from controlled problem-solving episodes into new rules that run in parallel in an uncontrolled manner. In a sense it thus converts a combination of controlled and uncontrolled inputs—decisions and rules respectively—into uncontrolled outputs, providing with respect to point (4) a blending of what was seen between backpropagation and reinforcement learning. This compilation through practice of controlled processing into uncontrolled memory access turns out to be a key aspect of the development of expertise (Simon & Chase, 1973; Rosenbloom et al., 1985), and can now also be found in the world of generative AI (Yu et al., 2024).

Two other learning technologies are also worth mentioning in this context. Model learning typically yields symmetric outputs, whether uncontrolled—that is, as declarative memory structures over which general reasoning can occur—or as a state space over which controlled search can occur. Universal planning, as mentioned under (a)symmetric in section 8.1, is much like reinforcement learning even though it is not conventionally considered a form of learning. What it does is convert a controlled input in terms of a state space—although one that is symmetric rather than asymmetric since it is capable of full combinatoric search rather than just sequences of action executions—into a policy-like *AUX* memory structure.

9.3 Trade-Offs

The trade-offs that have been identified as being of interest with respect to (un)controlled concern accuracy, efficiency, generality, learnability, usability, and operability. Starting with accuracy, controlled will typically have an advantage with respect to correctness, as control lends itself more easily to establishing properties with certainty. This does not necessarily imply that correctness cannot be established for uncontrolled processing, such as for monotonic logic, but in such cases the ability to establish correctness arises from a source that is orthogonal to this dichotomy. There may also be trade-offs involving both accuracy and robustness, but if so, they are too subtle to be teased out here.

There is a clear split within temporal efficiency. Controlled dominates for combinatoric efficiency due to its potential to eliminate whole regions of computation that are either provably irrelevant or heuristically less likely to be relevant. This matters because without at least some combinatorics, it is impossible to perform at any reasonable level in domains that require search and planning before action, such as board games and logistics, particularly when there has been insufficient prior experience for techniques like reinforcement learning to condense combinatoric experience into a noncombinatoric policy. Uncontrolled dominates for real-time efficiency due to both the increased potential for parallelism and the reduction in overhead from eliminating any extra processing required for decision-making. Without such real-time efficiency, domains that exhibit rapidly evolving complexity, such as many real-world domains faced by both humans and robots, cannot adequately be approached.

The analysis of generality involves two subtle trade-offs, but ones that can be sufficiently well characterized as to be worth mentioning. The first such trade-off concerns problem generality, where controlled tends to dominate when the content available for use with the technology is relatively small, whereas uncontrolled tends to dominate when the available content is relatively large. In the former case, the dearth of content is compensated for by controlled combinatoric search over combinations of it, potentially widening in a dramatic way the set of problems that can be approached. This is what is seen, for example, with classical problem-solving and planning approaches, where a small set of operators can be combined combinatorically to yield solutions to a wide variety of problem.

In contrast, uncontrolled may more readily take advantage of an abundance of content from which to learn, to cover a wide range of possibilities without requiring combinatoric combinations of it. A similar phenomenon was identified, for example, by Berliner and Ebeling (1989) in the context of chess, where iso-performance curves reveal a trade-off in performance when independently varying the amount of knowledge and search—a form of variant 19—that is leveraged in choosing moves. A more direct example, however, is the trade-off between symbolic problem-solving with small amounts of knowledge versus neural networks learned from massive amounts of data, as is found in variant 20.

The second generality trade-off concerns entity generality, where controlled processing is characteristic of novices whereas uncontrolled processing is characteristic of experts. Both can in fact leverage controlled processing when there is sufficient time, but experts have acquired knowledge and skill that enables them to use fast, uncontrolled knowledge search over the vast amounts they have learned to perform quickly in many situations (de Groot, 1965; Simon & Chase, 1973), as well as to use this capability to influence whatever slow, controlled search may still be required in other situations (Bilalić et al., 2008). Novices have very little at their disposal other than some very basic knowledge and problem-space search. Thus, the essence of expert performance cannot be modeled without an uncontrolled capability, and the essence of novice performance cannot be modeled without a controlled capability (although without such a capability it is also impossible to model the full range of expert performance). A novice versus expert distinction has been included in table 9.2 as one of the dichotomies that maps onto (un)controlled; however, here it instead differentiates between how two classes of entities solve problems and, thus, which entities can be modeled.

With respect to learnability, uncontrolled technologies have generally proven easier to learn, possibly due to the lack of combinatorics and/or discontinuous sequential dependencies, both of which are common with controlled processing. Consider learning a neural network or even an individual rule in a parallel rule system versus learning a sequential program. Techniques such as reinforcement learning and chunking that acquire uncontrolled knowledge from controlled inputs may, however, provide an interesting hybrid when both types of processing are available. With such techniques, the controlled input may embody the issues that make learning more difficult while the uncontrolled output is still relatively easy to learn.

With respect to usability, uncontrolled processing dominates for ease of use. Assuming control is not actually necessary, it is much easier to not have to worry about it. This may explain part of the ease of use of neural networks versus more traditional symbolic AI systems. In contrast, controlled processing provides an advantage for programmability, understandability, and explainability. Programming is essentially about a human exerting control over how the computer works, making uncontrolled processing—such as parallel processing—typically more difficult to program unless the program fragments are sufficiently independent that no control is truly required. The abilities to understand and explain are both advantaged by processing that proceeds in a simple and clear manner with sequential dependencies, particularly causal ones, easily discernable.

With respect to operability, there are advantages for controlled with respect to both effectivity and safety, as control more directly supports the execution of appropriate actions that avoid unintended harmful consequences. In contrast, there is a large advantage for uncontrolled with respect to reactivity. This is clear explicitly from variants 15–16 (reaction versus deliberation; reaction versus reflection), both of which mention reaction, but also implicitly from a variety of other variants, such as 20 (neural network versus problem space search) and 25 (policies versus planning).

Table 9.3 summarizes the criteria for which uncontrolled and controlled have been identified as having potential advantages. A *U* denotes uncontrolled, and a *C* controlled. The subscript "l" denotes a large amount of content whereas "s" denotes a small amount of content. Similarly, the subscript "e" denotes an expert whereas "n" denotes a novice. These additional subscripts further extend the CA language.

Table 9.3
Criterial trade-offs for the (un)controlled dichotomy

Criteria	Advantage
Accuracy (A)	
Correctness (AC)	*C*
Efficiency (E)	
Time (ET)	
Combinatoric (ETC)	*C*
Real-Time (ETR)	*U*
Generality (G)	
Problem (GP)	U_l/C_s
Entity (GE)	U_e/C_n
Improvability (I)	
Learnability (IL)	*U*
Usability (U)	
Ease of Use (UEa)	*U*
Programmability (UP)	*C*
Understandability (UU)	*C*
Explainability (UEx)	*C*
Operability (O)	
Effectivity (OEf)	*C*
Reactivity (OR)	*U*
Safety (OSa)	*C*

As with table 8.3, this table includes aspects of accuracy, efficiency, generality, learnability, and operability, although not necessarily the same aspects. It also adds aspects of usability. In brief, the analysis shows that uncontrolled yields real-time efficiency, problem generality when a large amount of content is available, entity generality focused on experts, learnability, ease of use, and reactivity. In contrast, controlled yields correctness, combinatoric efficiency, problem generality when a small amount of content is available, entity generality focused on novices, programmability, understandability, explainability, effectivity, and safety. A 1D disciplinary map of these advantages, analogous to the one found in table 8.4 for (a)symmetric, and using the same notation, as slightly extended here, can be seen in table 9.4.

Both of these perspectives on (un)controlled criteria reveal a within-temporal-efficiency trade-off of real-time versus combinatoric efficiency; a

Table 9.4
1D map of criterial advantages partitioned according to the (un)controlled dichotomy

Uncontrolled (U)	Controlled (C)
ETR, G(P_l, E_e), IL, UEa, OR	AC, ETC, G(P_s, E_n), U(P, U, Ex), O(Ef, Sa)

within-problem-generality trade-off over large versus small bodies of content; a within-entity-generality trade-off of experts versus novices; a within-usability trade-off of ease of use versus programmability, understandability, and explainability; and a within-operability trade-off of reactivity versus effectivity and safety. While not a within-criterion trade-off, uncontrolled also advantages learnability here.

9.4 Spanning

The range of (un)controlled variants shown in table 9.2 reveals that, as with (a)symmetric, there has been extensive investigation into both sides of this dichotomy. The same three general approaches to spanning that dichotomy—addition and two variations on reduction—are also used here.

With respect to addition, many integrated systems already have both uncontrolled and controlled capabilities, following one form or another of the dual-system, or separate module, paradigm. This can, for example, be found in robotic, AGI, and cognitive architectures (Kortenkamp & Simmons, 2008; Goertzel, 2014; Sun, 2016). It can also be found in neural-network approaches to game-playing; for example, AlphaZero used an uncontrolled neural network to evaluate states plus a controlled Monte Carlo tree search to explore across states and to support decision-making (Silver et al., 2018).

Universal AI (Hutter, 2005), which is unrelated to universal planning, provides a canonical example from the learning side of AGI, based on a diagonalization of variants 21 (backpropagation versus reinforcement learning) and 22 (Bayesian versus decision tree learning) that combines uncontrolled Solomonoff induction (Solomonoff, 1964), a form of Bayesian learning that guarantees asymptotic accuracy but is incomputable, with controlled reinforcement learning. Deep reinforcement learning (Wang et al., 2020), which can be considered an approximation to universal AI that responds to its incomputability, is a similar combination of uncontrolled and controlled modules based purely on variant 21. Effectively there is a trade-off between universal AI and deep reinforcement learning of (asymptotic) accuracy for the former versus efficiency for the latter.

One classical example of simulation is the situation calculus (McCarthy, 1963), which leverages monotonic logic to simulate controlled processing by explicitly representing distinct states as *situation variables* within uncontrolled monotonic logic. It thus spatializes in an uncontrolled manner what would otherwise be a controlled temporal process. Nonmonotonic logic that bakes assumption processing, in a controlled manner, into the definition of the logic became mainstream in AI only later (McCarthy, 1980; Reiter, 1980).

Similarly, hidden Markov models (Baum & Petrie, 1966) are controlled models simulated by uncontrolled probabilistic graphical models (Koller & Friedman, 2009), and recurrent neural networks (Elman, 1990) are controlled models simulated by uncontrolled feedforward networks (Fine, 1999). Both attention and self-attention can also be thought of in this way, but instead of simulating control going forward, they simulate control over what is attended to in the past. In table 9.1 these kinds of simulated technologies show up as uncontrolled, based on how they are implemented, but they could instead have been listed as controlled, based on what is simulated, or they could have instead been explicitly considered as some form of hybrid.

In the reverse direction, leveraging controlled processing to control what would otherwise be inherently combinatoric uncontrolled processing amounts to simulating uncontrolled processing via controlled processing. One classical example here also involves monotonic logic, but in this case using heuristic search to control theorem proving (Slagle & Bursky, 1968; Bledsoe, 1971).

It is important to note that simulation in either of these directions can blur the trade-offs discussed in section 9.3. For example, situation calculus loses the ability to achieve the combinatoric efficiency that should be possible with native controlled processing, whereas processing uncontrolled knowledge in a controlled manner loses the real-time efficiency attainable with true uncontrolled processing.

The typical approach for developing a native capability that supports both uncontrolled and controlled processing is to add some kind of decision mechanism to what is otherwise an uncontrolled memory retrieval capability. When the decision mechanism is used, processing is controlled, and when it is not the processing is uncontrolled. What distinguishes this from simple addition is that the uncontrolled memory capability is common

Table 9.5
0D map of criterial advantages for the maximum across the (un)controlled dichotomy

AC, ET, G, IL, U, O

to both functionalities rather than there being two completely distinct modules. For example, Bayesian networks can be augmented with decision (and utility) nodes to yield influence diagrams (Howard & Matheson, 1984), neural networks can follow uncontrolled generation of distributions with choices based on these distributions, and the cognitive cycle at the heart of the common model of cognition adds a decision mechanism that is applied once uncontrolled procedural-memory retrieval has been completed (section 1.2.3). In such cases it is possible to consider the overall combination as yielding a deliberation mechanism for individual decisions whereas reaction results when no decisions are required. It is also possible to consider the reactive mechanism as being nested within a larger-scale deliberative mechanism, although thinking about it as a smaller grained additive approach is also not far from the mark.

As with table 8.5 for (a)symmetric, appropriately spanning the (un)controlled dichotomy can potentially yield the maximum of the criterial advantages across it. Table 9.5 shows how spanning this dichotomy all on its own—for example, by including a common model-like cognitive cycle that includes access to procedural memory followed by a decision about an action to be performed—can yield a combination of correctness, temporal efficiency, generality, learnability, usability, and operability.

9.5 Summary

The (un)controlled dichotomy is broadly relevant across AI and related fields in many of its variant forms. It involves criterial trade-offs rather widely, among accuracy, efficiency, generality, improvability, usability, and operability. The need to span the trade-off is akin to the need for spanning (a)symmetric, with the generic approaches to spanning also being analogous. There are also learning mechanisms along both sides of the dichotomy.

10 (Meta)Data

The original form of the (meta)data dichotomy, as (sub)symbolic, was one of the three dichotomies proposed to underlie the coarser traditional notion of symbolic/probabilistic versus neural/ML approaches to architecting minds (Rosenbloom et al., 2019a). Although the terminology can be confusing, this traditional notion can itself be thought of as a coarser (sub)symbolic—or (non)symbolic—dichotomy, in contrast with the more refined version that acted as just one out of three dimensions of it. The more refined version focused specifically on associating symbolic data with quantitative metadata, as found in both the common model of cognition and Sigma, and as further elaborated in chapter 3. What is mostly changed in moving from this refined notion of (sub)symbolic to (meta)data is foregrounding the more computational distinction between symbolic data and quantitative metadata.

As defined here, the (meta)data dichotomy does not, however, completely distinguish between whether there is data versus metadata, as both are assumed to be present in general, but instead, whether the meaning of a data/metadata dyad derives primarily from its data versus its metadata. The data pole in its purest form does omit the hybrid option for its symbols/data, implying that the meaning of structures is derived solely from the meaning of its data and how this data is composed into graphs. But some minimal metadata may also be included at this pole, such as 0/1 for false/true (as might be used in logic) or present/absent (as might be used in working memories of rule systems). Anything more than this would encroach into the notion of spanning this dichotomy (section 10.4).

At the other end, the metadata pole must obviously include some data, as metadata is always associated with data. However, the meaning of structures here stems primarily from the pattern of metadata values over the structures, with the structures themselves typically just vectors, or more generally tensors, of placeholders. Distributed representations in hidden layers of feedforward neural networks provide the classical example here, where the data—that is, the nodes in these layers—have little to no meaningful grounding, essentially acting simply as placeholders, and where composition is limited to forming vectors within layers and links across layers (section 3.4). In contrast, technologies that combine meaningful metadata and meaningful data, such as is found in local connectionist networks (Shastri, 1988), probabilistic graphical models (Koller & Friedman, 2009), and graph neural networks (Scarselli et al., 2009; Zhou et al., 2019) are considered to be outside of the scope of the metadata pole. Instead, these are all parts of the rich space of (meta)data hybrids that will be discussed in section 10.4.

Table 10.1 displays a range of AI technologies across a 1D map as they divide across data versus metadata. As with tables 8.1 and 9.1, brief explanations of each of these technologies can be found in appendix A. This map heavily depends on splitting various technologies into "numeric" versus "symbolic" variants, with "numeric" variants assuming that numbers are encoded as quantitative metadata rather than as data that externally designates numbers (section 3.3), as opposed to "symbolic" variants that omit or only minimally leverage metadata. For example, a 64-bit computer word

Table 10.1
1D map of technologies partitioned according to the (meta)data dichotomy

Metadata (M)	Data (D)
Dynamical Systems; Implicit Memory and Processing; Machine Learning; Motor Control; Neural Networks (and Learning); Numeric Computation; Numeric (Soft) Constraint Propagation; Numeric Functions; Numeric Induction; Numeric Memory (and Learning); Numeric Operators; Numeric Reaction; Numeric Reasoning; Numeric Shallow Reasoning; Numeric Skills (and Acquisition); Parameters (and Learning); Perception; Peripheral Processing; Probability and Statistics; (Self-)Attention Networks; Sensory-Motor Behavior; Tensor Processing; Utilities	Central Processing; Explicit Memory and Processing; Goals; Graph Processing; Knowledge (and Acquisition); Logic; Rules (and Composition); Search; Structures (and Learning); Symbolic Analogy; Symbolic Computation; Symbolic (Hard) Constraint Satisfaction; Symbolic Functions; Symbolic Memory; Symbolic Operators; Symbolic Skills (and Acquisition); Truth Maintenance Systems

amounts to a structure in which each of the 64 bits is an individual symbol whose associated 0/1 value amounts to its metadata. The word as a whole can externally designate a quantity, but its metadata is directly processable as numbers without needing to consult this designation.

10.1 Variants

The set of prior dichotomies that have been identified as mapping onto (meta)data can be found in table 10.2. Variant 1 (probability and statistics versus logic) embodies the classic distinction between observation-driven approaches to inference versus more formal ones, with the former yielding probabilities as metadata on alternatives and the latter representing alternatives as structural data. Although in general probabilities may be associated with meaningful data, yielding a hybrid such as probabilistic logic (Nilsson, 1986), the pure metadata version views alternatives simply as placeholders. Variant 2 then is a slight modification of this, where probabilities are generalized to (metadata) parameters and logical statements as (data) structures. In principle, this variant might provide a more appropriate label for the core dichotomy here, as it is also computational while assuming a bit less about the relationship between the two poles. I chose to go with (meta)data partly because it fit the (inverter)term form introduced in section 6.1, partly because of the nature of the hybrid symbol systems hypothesis introduced in chapter 3, and partly out of loyalty to the common model.

Variant 3 (utilities versus goals) specializes variant 2 to an aspect of problem-solving where what is to be achieved can be expressed as parameters that represent utilities or structures that represent goals. Variants 4–5 can then be seen as more abstract versions of the first three. Variant 4 echoes the original label for this dichotomy, focusing on the sub/nonsymbolic nature of metadata and the symbolic nature of data. Variant 5 relabels these poles in terms of whether the processing should be considered cognitive. Variant 6 is then a more specific variant of these two, focusing on neural networks as a canonical embodiment of subsymbolic noncognitive processing versus rules as a canonical embodiment of symbolic cognitive processing (Sloman, 1996; Sun, 2016). This provides the distinction between the two that was not found in table 6.6, which crossed (a)symmetric and (un)controlled.

Variant 7 focuses on the distinction between implicit and explicit memory and processing. It aligns closely with the previous three, while harking

Table 10.2
Identified variants of the (meta)data dichotomy

	Metadata (M)	Data (D)
1	Probability and Statistics	Logic
2	Parameters	Structures
3	Utilities	Goals
4	Sub/Nonsymbolic	Symbolic
5	Noncognitive	Cognitive
6	Neural Network	Rule
7	Implicit Memory and Processing	Explicit Memory and Processing
8	Unconscious	Conscious
9	Pre-attentive	Attentive
10	Peripheral	Central
11	Sensory-Motor Behavior	Knowledge Representation and Reasoning
12	Modular	Nonmodular
13	Skill	Knowledge
14	Machine Learning	Knowledge Representation and Reasoning
15	Skill Acquisition	Knowledge Acquisition
16	Parameter Learning	Structure Learning
17	Distillation	Chunking
18	Integer and Floating Point	Bit, Byte, and Word
19	Tensor Processing	Graph Processing
20	Regular Structures	Irregular Structures
21	Parallel Processing	Serial Processing
22	Soft Constraint Propagation	Hard Constraint Propagation
23	Quantitative	Qualitative
24	Continuous	Discrete
25	Differentiable	Nondifferentiable

back to variant 13 in table 8.2 and variant 5 in table 9.2. Here, though, the focus is on how data is examinable in memory or at least supports examinable processing, and thus is explicit, versus how metadata is inherently unexaminable and thus implicit with respect to both memory and processing. This is a usage found, for example, in the Clarion cognitive architecture (Sun, 2016).

The presence of a form of this variant across all three core dichotomies provides three complementary perspectives that help dissect and flesh out what has traditionally been a rather fuzzy distinction. More on this alignment can be seen in chapter 11, which focuses more strongly on interactions among the dichotomies, but it is worth saying a bit about it here first, focused specifically on aligning the (meta)data perspective on it with each of the two previous perspectives.

Aligning the (meta)data and (a)symmetric perspectives yields explicit memory for symmetric data (e.g., logic), and implicit memory for asymmetric metadata (e.g., heteroassociative neural networks). Misaligning the two perspectives—as in asymmetric data (e.g., rules) or symmetric metadata (e.g., autoassociative neural networks)—yields no clear result a priori. The common model, for example, considers a rule-based procedural memory to be implicit, whereas Clarion considers such a memory to be explicit.

A similar alignment of the (meta)data and (un)controlled versions yields explicit processing for controlled data (e.g., deliberation) and implicit processing for uncontrolled metadata (e.g., neural networks). Once again, a misalignment of these two perspectives—as in uncontrolled data (e.g., symbolic memory retrieval and reasoning) or controlled metadata—which, as will be seen more concretely in section 11.3, actually turns out to be difficult to conceptualize—yields no clear result a priori. The common model, for example, considers a symbolic long-term declarative memory to be implicit, whereas monotonic logic would typically be considered explicit.

Variant 8 (unconscious versus conscious) tracks variant 7, much as it did the corresponding variant in table 9.2, with a focus on (un)awareness, but here due to metadata versus data. This yields an alignment of uncontrolled metadata for unconscious and controlled data for conscious, although with a caveat that in architectures with both long-term and working memories, presence in working memory may also be considered necessary for consciousness.

Variant 9–13 can all be seen as alternative formulations of variant 5 (noncognitive versus cognitive), with many of them perhaps more familiar in cognitive science than AI. Variant 9 (pre-attentive versus attentive)—as was mentioned but not included under (a)symmetric in section 8.1—and variant 10 (peripheral versus central) leverage temporal and spatial notions, respectively, with the first separating out noncognitive processing that occurs prior to cognitive processing and the second separating out noncognitive processing that occurs outside of cognitive processing. Variant 10, for example, can be seen to reflect the traditional distinction between peripheral sensory and motor processing versus central cognitive processing. The former is traditionally assumed to be subsymbolic (metadata) and the latter symbolic (data). Mental imagery (Pearson et al., 2015) is often conceived of as the interface between peripheral and central processes (Laird, 2012) and may thus serve as an intermediate point along this spatial variant; although it may actually amount to composing quantitative data rather than being quantitative metadata (section 3.3), as occurs in Sigma via continuous region tensors (Rosenbloom, 2011a; Rosenbloom, 2012c).

Variant 11 is a more AI interpretation of variant 10, where sensory-motor behavior is based on continuous control over numeric metadata whereas knowledge representation and reasoning typically assume data in the form of something like symbolic logic. Variant 12, which is grounded in philosophy of mind (Fodor, 1983), is a slight shift from variants 10–11. It considers peripheral sensory-motor processing as being modularized according to individual modalities versus a nonmodal, and thus nonmodular, central capability for knowledge representation and reasoning.

Variant 13 (skill versus knowledge) focuses more on the distinctive nature of the processing rather than on its temporal or spatial location. It has already been seen as variant 16 in table 8.2, where skill was asymmetric and knowledge symmetric. Here, skill is grounded in continuous metadata and knowledge in symbolic data, rather than skill being grounded in procedural memory versus knowledge in declarative memory. Both are aspects of how the traditional skill versus knowledge dichotomy has appeared, but which may get differentially highlighted—or even seen as distinct concepts—depending on the circumstances. This is thus another locus where the alignment of these dichotomies yields a canonical interpretation, of procedural metadata versus declarative data, although the distinct aspects highlighted by the two dichotomies should not be lost.

Despite skill providing a canonical example of metadata in this variant, a more nuanced understanding of it requires distinguishing between this kind of continuous skill and the cognitive or symbolic skill that is traditionally based on rules in the procedural memories of cognitive architectures. Table 10.1 reflects this nuance, in distinguishing numeric versus symbolic skills rather than assuming all skills must be based on metadata. The comparable distinction in that table between numeric and symbolic procedural memory also reflects this nuance.

Variant 14 shifts the focus to one previously seen as variant 26 in table 8.2, concerning machine learning versus knowledge representation and reasoning. Here, machine learning is presumed to be central to metadata processing, whereas knowledge representation and reasoning are presumed to be central to data processing. This is, of course, somewhat of an apples-to-oranges comparison. Comparing machine learning to knowledge engineering would be more appropriate in some sense, as both concern how the content is acquired, but the comparison as expressed here focuses on where the key action is in metadata versus data approaches. With metadata, the interesting action is typically in the learning, with the actual use of the metadata involving, for example, simple calculations within fixed neural networks. In contrast, with data the interesting action is typically in what is done with the data once it is acquired, with its acquisition then considered as subservient to this.

Variants 15–17 shift more completely into learning. Variant 15 (skill versus knowledge acquisition) is a learning-oriented analog of variant 13 (skill versus knowledge), and thus also has direct ties back to the skill versus knowledge acquisition pair seen as variant 18 in table 8.2. Variant 16 (parameter versus structure learning) is a learning-oriented analog of variant 2 (parameters versus structures). It can also be seen as a version of variant 15 but expressed more in a direct (meta)data format, where skill learning is primarily the tuning of parameters that are encoded as quantitative metadata, and structure learning is the acquisition of new forms of symbolic data and of compound expressions over it. Variant 17 (distillation versus chunking) can be seen as a version of variant 16, where System 2 processing is converted into either parameters or structures that become available for System 1 processing. Distillation, as the term is used in general in neural networks, is broader than this (e.g., Hinton et al., 2015) but here the focus is on this kind of use.

Variants 18–21 are most directly due to computer science and AI, with accompanying computational and functional concerns. Variant 18 (integer and floating point versus bit, byte, and word) concerns the support of numeric versus symbolic computing by computer hardware. Symbolic computing is based on the core representation of bits, bytes, and words designating meaningful notions, with metadata of 1/0 denoting T/F or present/absent for individual bits, and aspects of composition and designation—both declarative and procedural—across words playing key roles. Numeric computing is based on metadata of 1/0 actually denoting 1/0 and quantitative composition based on base-2 positions within words. The result is that such words directly designate quantities to be processed.

Variant 19 (tensor versus graph processing) concerns how the data to which the metadata is attached is typically structured. When the focus is on metadata, structures such as vectors or larger-dimensioned tensors are the norm. When the focus is on data, graph structures are typical. Variant 20 abstracts this to regular structures for metadata versus irregular structures for data, with variant 21 noting that the former can typically be processed in parallel while the latter more often requires serial processing, as was previously seen as variant 18 in table 9.2. It also directly reflects the use of graphics processing units (GPUs) and tensor processing units (TPUs) for metadata and central processing units (CPUs) for data.

Variant 22 contrasts soft versus hard constraint propagation. There are ways of viewing these two poles as simple variations of each other, with both maintaining meaningful data and there being just a difference in how hard the constraints are. The contrast intended here though is between such technologies as Boltzmann machines (Ackley et al., 1985), where the hidden units are simply placeholders, and more standard symbolic constraint satisfaction (Dechter, 2003).

Variants 23–25 are all general variations on the quantitative nature of metadata versus the symbolic nature of data. Variant 23 (quantitative versus qualitative) is just this, but with an explicit acknowledgment of the qualitative nature of symbols. As was mentioned in section 3.3 and elsewhere, data can be used to represent numbers, but this does not obviate its inherently qualitative nature. The possibility of numeric data does, however, explain why the dichotomy of quantitative versus qualitative physics (Forbus, 2019) is missing from table 10.2. For it, both poles are based on data, with the latter simply a quantized abstraction of the former. So, the contrast here does not map onto (meta)data even though it distinguishes a form of numeric versus symbolic representation and reasoning.

Variant 24 (continuous versus discrete) goes beyond variant 23 in assuming that metadata is not only quantitative but continuous, with data being discrete. As has been seen, this doesn't eliminate the possibility of using integral values, such as 0 or 1, for metadata but those are just special cases of something that is inherently continuous.

Variant 25 (differentiable versus nondifferentiable) is a further specialization of variant 24, to where the metadata is not only continuous but also differentiable. This distinction comes to the fore when considering neural network learning algorithms such as backpropagation that require differentiable functions to be applicable.

10.2 Learning

Variant 14 in table 10.2 contrasts machine learning with knowledge representation and reasoning, implying the centrality of the former for overall metadata processing. However, learning is not just relevant to metadata, so variants 15–17 explicitly contrast forms of learning for metadata versus data. As mentioned in section 10.1, variant 15 (skill versus knowledge acquisition) is a reappearance of variant 18 from table 8.2. It points out the need for mechanisms capable of performing both quantitative skill/metadata acquisition and symbolic knowledge/data acquisition. Variant 16 (parameter versus structure learning) similarly points out the need to have mechanisms capable of quantitative parameter acquisition/tuning and symbolic structure learning. Variant 17 (distillation versus chunking) points to the need to be able to compile System 2 processing, whether based on parameters or symbolic structures, into System 1 processing of the same sort.

Pooling together variants 15 and 16, canonical forms of quantitative skill/parameter acquisition/tuning include Bayesian learning (Bissiri et al., 2016), neural-network learning (Hebb, 1949; Rumelhart et al., 1986), and reinforcement learning (Sutton & Barto, 2018). These are all focused on improving accuracy, although reinforcement learning may also yield a form of efficiency when it induces a shorter path. They are all also grounded in metadata in their processing, input, and output, addressing points (1)–(3), and thus also point (4), from chapter 6. Model-based reinforcement learning (Ray & Tadepalli, 2011), however, extends the input to include not only metadata but also data, in terms of a model of the environment that can be used for prediction.

Canonical forms of symbolic knowledge/data/structure acquisition include traditional symbolic approaches to inductive learning, whether in the form of decision tree learning (Quinlan, 1986), rule learning (Cohen & Singer, 1999), inductive logic programming (Muggleton & de Raedt, 1994), or genetic algorithms and programming (Holland, 1992; Koza, 1992); as well as approaches to macro/composition formation (Korf, 1985; Alphonse, 2004; Laird et al., 2017). These two forms of symbolic learning—inductive and macro—have been previously differentiated as knowledge-level versus symbol-level learning (Dietterich, 1986), with the former improving accuracy—although it can also improve problem generality if a preciseness close to 0 is identified with an inability to work on a problem—and the latter improving efficiency. Metadata learning that impacts accuracy would also be considered as a form of knowledge-level learning.

As with metadata learning, the approaches to data learning discussed here are all grounded in data, thus addressing points (1)–(4). There are examples, however, where metadata is also used, such as in the learning of decision trees, where an information-gain metric may be used in deciding incrementally on which feature to split at any node in the tree (Quinlan, 1986).

10.3 Trade-Offs

The trade-offs of interest for (meta)data concern accuracy, efficiency, generality, improvability, usability, and operability. With respect to accuracy, a trade-off can be seen akin to that for (a)symmetric in table 8.3, but here with data having an advantage for correctness and metadata having an advantage for precision. The former is due to a difference in how easy it is to prove properties of data versus metadata, whereas the latter is due to the difference in the graduated nature of numeric metadata versus the all-or-none nature of symbolic data. As opposed to what was seen for robustness in table 8.3, there does not appear to be a significant advantage here with respect to robustness for symbolic data versus numeric metadata.

With respect to efficiency, historically there was a significant overall advantage for data when focusing on real-time efficiency. Digital computers are most naturally symbolic, with the simplest operations involving logical operations on bits, bytes, and words, whereas numeric processing requires more complex sequences of operations. In ENIAC, for example, a floating-point operation took 250 times as long as a simple symbolic operation ("FLOPS," 2024). However, with the advent of parallel and specialized hardware of various sorts, the picture gets murky enough that no clear overall advantage remains apparent here.

Data does however retain a real-time efficiency advantage for symbolic/discrete/qualitative problems, whereas metadata has a comparable advantage for numeric/continuous/quantitative problems. This within-criterion trade-off, and in particular the slowness of simulating numeric processing on top of symbolic cognitive architectures, and how its processing in this fashion intrudes upon and distracts from problem-oriented sequential deliberation, is what led traditional symbolic architectures such as Soar to become hybrid by including the processing of quantitative metadata in the architecture itself. Slowness and intrusiveness are appropriate for deliberate processing of numeric data, such as in the conscious performance of arithmetic calculations; however, the processing of quantitative metadata is properly a background activity. This need for hybrid architectural processing was made explicit in the common model of cognition, where the combination of (symbolic) data plus (quantitative) metadata was developed. It is also reflected, in chapter 3, in both the hybrid symbol systems and hybrid cognitive architecture hypotheses.

Space efficiency favors symbolic data due to its ability to be variablized, as is seen, for example, in both rules and first-order logic, so that single statements can cover a wide variety of instances. Nearest neighbor algorithms (Taunk et al., 2019) that may be applicable to numeric problems may be able to partially compensate for this advantage, by enabling individual examples to stand in for whole regions around them, but do not handle the full scope of compression possible with variables.

Problem generality provides a within-criterion trade-off akin to the one found for (un)controlled in table 9.3. Here, the combinatorics available with symbolic data provides an advantage when a small amount of content is available, whereas with numeric metadata there can be an advantage when there is a large amount of content on which learning can operate. The same kinship with (un)controlled also exists for the trade-off between learnability and usability, with metadata advantaged for the former—as seen with deep neural networks of all sorts—and data for the latter, if, again ease of use is ignored. Data has an advantage with respect to safety, where it can be seen to derive from a combination of understandability and explainability.

Metadata's learnability advantage could be considered more of a trade-off if there were a sufficient advantage for data of learning simple facts, which it seems there ought to be, but the situation is murky enough with respect to metadata fact learning so as not to require introducing this additional complexity at this point. The ease-of-use advantage for metadata stems from the ability to create packages for learning and inference in neural networks that a wide range of people can use fairly easily in comparison to the more difficult challenge of making symbolic AI systems available for general use.

Table 10.3 summarizes the potential advantages for metadata and data, respectively, across the criteria. An *M* denotes metadata and a *D* data. The subscripts "c" and "d" denote continuous and discrete problems, respectively, further extending the CA language. This may more appropriately be considered as a distinction between qualitative and quantitative problems—to include integer arithmetic in the latter rather than the former—but since these both start with "q," "c" versus "d" becomes a simpler way to abbreviate this distinction. As in table 9.3, the subscript "l" denotes a large amount of content whereas "s" denotes a small amount of content.

In summary, the analysis shows that metadata has an advantage for precision, real-time efficiency with continuous problems, problem generality with a large amount of content, learnability, and ease of use. Data has an advantage for correctness, real-time efficiency with discrete problems, space efficiency, problem generality with a small amount of content, programmability, understandability, explainability, and safety. A 1D disciplinary map of these advantages, analogous to those seen in tables 8.4 and 9.4 can be found in table 10.4.

Both perspectives here reveal an accuracy trade-off between precision and correctness, a real-time efficiency trade-off as a function of whether the focus is on continuous problems versus discrete problems, an advantage for data with respect to spatial efficiency and safety, a problem generality trade-off that depends on whether there is a large or small amount of content available, an advantage for metadata with respect to learnability, and a usability trade-off between ease of use and programmability, understandability, and explainability.

10.4 Spanning

Approaches that combine significant forms of both data and metadata processing are typically referred to as *hybrid* systems. Casting this dichotomy

Table 10.3
Criterial trade-offs for the (meta)data dichotomy

Criteria	Advantage
Accuracy (A)	
Correctness (AC)	*D*
Precision (AP)	*M*
Efficiency (E)	
Time (ET)	
Real-time (ETR)	M_c/D_d
Space (ES)	*D*
Generality (G)	
Problem (GP)	M_l/D_s
Improvability (I)	
Learnability (IL)	*M*
Usability (U)	
Ease of Use (UEa)	*M*
Programmability (UP)	*D*
Understandability (UU)	*D*
Explainability (UEx)	*D*
Operability (O)	
Safety (OSa)	*D*

Table 10.4
1D map of criterial advantages partitioned according to the (meta)data dichotomy

Metadata (M)	Data (D)
AP, ETR_c, GP_l, IL, UEa	AC, E(TR_d, S), GP_s, U(P, U, Ex), OSa

in terms of data versus metadata, rather than as qualitative versus quantitative, directly implies one approach to creating such hybrids, by blending varying amounts of meaning from both data and metadata. Nonetheless, as with the previous two dichotomies, it still makes sense to start with the additive approach of combining multiple modules before moving on to the two forms of reduction: implementing/simulating one on top of the other or, as with the association of quantitative metadata with data, leveraging a deeper native generalization of both.

The approach of implementing distinct modules for data versus metadata processing is a common one, particularly when metadata processing takes the form of inference in neural networks. The result is typically distinct symbolic and neural modules (Jilk et al., 2008; Sun, 2016). At a hardware level, it is also possible to consider the inclusion of both CPUs and GPUs/TPUs as an additive approach to spanning (meta)data.

Shifting now to reduction, it isn't truly possible to implement metadata processing via data processing, or vice versa, due to their playing qualitatively different roles. However, simulation in both directions is possible. Simulating data processing via metadata processing is effectively what is imagined when symbolic processing is said to emerge from neural processing (McClelland, 2010). Embedding is an alternative approach in which symbols and relations are explicitly encoded as metadata patterns (Jones & Mewhort, 2007; Mikolov et al., 2013). Tensor product representations (Smolensky & Legendre, 2006) that implement logical operations on combinations of vectors also fit this pattern. Vector symbolic architectures, or hyperdimensional computing, more generally can be seen to include these last two classes of approaches and many more (Kleyko et al., 2022). All of these approaches support symbolic processing in a manner that is slower than a native implementation of it, but Newell's layering of timescale bands implies that cognitive processing should proceed more slowly than biological processing, so this could be appropriate.

In the reverse direction, as was mentioned in section 8.3, the slowness of simulating metadata processing on top of data processing has led some traditional symbolic cognitive architectures to shift toward hybrid technologies that combine both in some essential manner. This slowness isn't an issue when considering the deliberative processing of quantitative data, such as when humans perform mental arithmetic, but it is critical when considering the variety of forms of metadata processing that must seamlessly integrate with nearly all forms of data processing.

A wide variety of technologies are based on some form of a native generalization that combines meaningful data and metadata. One general class of such technologies begins with meaningful data but then associates with it probabilities, utilities, or heuristic values. This includes, for example, probabilistic logics and probabilistic graphical models (Nilsson, 1986; Pearl, 1988; Darwiche, 2002; Koller & Friedman, 2009; Poon & Domingos, 2011;

Table 10.5
0D map of criterial advantages for the maximum across the (meta)data dichotomy

A(P, C), E(TR, S), GP, IL, U, OSa

de Raedt et al., 2016), as well as heuristic and probabilistic search (Hart et al., 1968; Newell & Simon, 1976; Coulom, 2006) and decision-theoretic planning (Boutilier et al., 1999).

A second general class of such technologies begins with meaningful metadata but then extends its data to include some significant amount of structure. This includes convolutional neural networks (LeCun et al., 1989; Waibel et al., 1989), recurrent neural networks (Elman, 1990), recursive neural networks (Goller & Kuchler, 1996; Frasconi et al., 1998), local connectionist networks (McClelland & Rumelhart, 1981; Feldman & Ballard, 1982), tensor product representations (Smolensky & Legendre, 2006), graph neural networks (Scarselli et al., 2009; Zhou et al., 2019), and transformers (Vaswani et al., 2017).

The hybrid approach to symbol systems proposed in chapter 3, which flexibly blends data and metadata, can be seen as an attempt at starting from a more neutral perspective to enable subsuming both of these classes of technologies under a single generic umbrella. Yet, whatever perspective is taken, it should be clear that hybrid (meta)data technologies are often at the leading edge of current research. The criterial advantages potentially available from such hybrids can be seen in table 10.5. These include precision, correctness, real-time temporal efficiency, spatial efficiency, problem generality, learnability, usability, and safety.

10.5 Summary

The (meta)data dichotomy gets at a core aspect of subsymbolic versus symbolic processing, concerning whether the processing is focused not simply on numbers versus symbols—as, for example, in the contrast between quantitative and qualitative reasoning (Forbus, 2019)—but on whether it concerns (quantitative) metadata versus (symbolic) data. Approaches to spanning this dichotomy have been discussed, with a considerable focus on those that enable both data and metadata to be derived as special cases of a single more general approach.

This dichotomy maps onto a number of preexisting ones that have played important roles in AI and cognitive science while also largely partitioning learning into parameter tuning versus structure acquisition (versus both). Although some may still argue for the use of just one of metadata versus data, it should be clear from the discussion here that even in the most extreme cases, the deprecated other is (almost) always there to at least some extent, whether as Boolean values for metadata or placeholders/units for data.

With the hybrid nature of the common model of cognition and the advent of statistical-relational AI it is also clear that significant segments of the community that were traditionally focused on symbolic processing have moved to hybrid technologies that span the dichotomy in some form. Likewise, with the surge in work on structured neural networks and the addition to standard neural networks of functionalities like probabilistic search, as is found in game-playing programs, we are seeing some recognition of this on the neural side as well. Chapter 3 attempted to add more of a theoretical basis behind this approach, with the introduction of the hybrid symbol systems hypothesis.

11 Interactions

The full power of dichotomic analysis becomes apparent once interactions among the dichotomies are considered. Sections 11.1–11.3 explore the three binary interactions afforded by the dichotomies discussed in chapters 8–10. Each yields 2D maps of AI technologies, both disciplinary and architectural. Section 11.4 then explores the full trinary interaction among them, yielding 3D maps. In addition to maps of AI technologies, these analyses also include multidimensional maps of interactions among evaluation criteria that yield insights into the more complex trade-offs that occur when multiple dichotomies are considered in combination.

11.1 (A)Symmetric × (Un)Controlled

An abstract map representing the cross product between (a)symmetric and (un)controlled can be seen in table 11.1. It is much like the map in table 6.1, with the same dimensions and the same use of deliberation and reflection in the controlled row. For the uncontrolled row, table 6.1 leveraged the procedural memory versus declarative memory variant of (a)symmetry whereas here, with a focus more on AI, reaction rather than procedural memory is the label used for *AUX* processing, and the generic term of reasoning—whether based on logic, probabilities, or other forms of content—is used instead of declarative memory for *SUX* processing.

Each adjacent pair of locations in this map, whether horizontal or vertical, yields a distance of 1 between its abstract technologies. So, an entire distance-1 loop is implicitly defined in which reaction is 1 from reasoning, which is 1 from reflection, which is 1 from deliberation, which is 1 from reaction. There is then a distance of 2 between the abstract technologies

Table 11.1
Abstract, AI-oriented 2D map of the cross product between (a)symmetric and (un)controlled

	Asymmetric (A)	Symmetric (S)
Uncontrolled (U)	Reaction	Reasoning
Controlled (C)	Deliberation	Reflection

along each of the two diagonals. In other words, both of these diagonals distinguish locations that are opposites with respect to this map.

Rosenbloom (2019a) mapped a trilevel control hierarchy and the two trilevel content hierarchies that appeared in chapter 9—for affect and causation—onto an earlier version of this map by bending them into L-shaped paths, each step of which moved a distance of 1 and which all together covered three out of the four locations. The control trichotomy, for example, went from procedural memory (i.e., reaction or *AUX*) down to deliberation (i.e., *ACX*) and then right to reflection (i.e., *SCX*), omitting reasoning (i.e., *SUX*). This earlier work also yielded architectural maps for the common model of cognition (see table 6.4) and the three cognitive architectures that initially inspired it—ACT-R, Soar (see table 6.2), and Sigma (see table 6.3). Although these three architectures all have capabilities that fit each of the four locations, the common model itself is missing a reflection component, as no consensus has yet been developed on this topic.

Architectural maps were also included in the earlier work for the two content hierarchies and for AlphaZero (see table 6.5), which, like the control hierarchy, turned out to be missing a reasoning component. Although not included in the earlier work, both universal AI (Hutter, 2005) and deep reinforcement learning turn out to only span the asymmetric column, albeit with quite different instantiations for *AUX*, while omitting the symmetric column due to a focus on direct/shallow performance rather than on any form of deeper thought. In a crude anthropomorphic sense, the distinction between the asymmetric and symmetric columns can be thought of as that between doing and thinking (as seen in variants found in both tables 8.2 and 9.2). The distinction between the rows can be thought of as System 1 versus System 2.

The combination of this pair of dichotomies thus seems to tap into essential aspects of the structuring of integrated systems, yielding architectural maps that can characterize existing systems while ultimately implying the hypothesis that any attempts to build complete AGI or cognitive architectures, or to model human cognition, will need to be complete with respect to this map. Table 11.1 can thus be viewed as an abstract architectural map that generalizes across the more concrete ones that fit particular architectures. It can, however, also be viewed as an abstract disciplinary map in which each location generalizes over all of the different technologies that might be used there.

Fleshing out this abstract disciplinary map with a variety of more concrete technologies yields the more detailed disciplinary map in table 11.2. Multidimensional disciplinary maps such as this one can be created as cross products of lower dimensional maps, with the one here derived from the 1D maps found in tables 8.1 and 9.1, although this may at times require either considering technologies that are only implicit in one of the source maps or splitting the technologies in one source map according to the dichotomy that underlies the other. Such maps provide a straightforward way of categorizing the included technologies according to the underlying 2×2 framework.

As with the earlier 1D disciplinary maps, all technologies situated at a single location in table 11.2—such as rules and heteroassociative neural networks (and backpropagation) (*AUX*)—are identical with respect to the map. Likewise, technologies in adjacent locations are separated by a distance of 1; for example, moving horizontally from rules (*AUX*) to monotonic logic (*SUX*) involves just a change in (a)symmetry, whereas a vertical shift from knowledge search (*SUX*) to problem space search (*SCX*) involves just a change in (un)controlled. Locations along a diagonal are instead opposites that are separated by a distance of 2—for example, from policies (*AUX*) to planning (*SCX*) along the main diagonal and from operators (*ACX*) to dynamical systems (*SUX*) along the minor diagonal. In general, when considering direction in a map along the minor diagonal, the assumption will be, as here, that movement is from the bottom left to the top right. Movement along the major diagonal follows the natural direction—at least for readers of English and comparable languages—from the top left to the bottom right.

Digging deeper into this cross product, it turns out that five of the variants in tables 8.2 and 9.2 are labeled identically between the two dichotomies: 3|23 (shallow versus deep reasoning); 4|16 (reaction versus reflection);

Table 11.2
A 2D disciplinary map that adds more concrete interpretations of the cross product between (a)symmetric and (un)controlled to the abstract entries in table 11.1

	Asymmetric (A)	Symmetric (S)
Uncontrolled (U)	Arithmetic Circuits and Sum-Product Networks; Attention Networks; Causal Association; Heteroassociative Neural Networks (and Backpropagation); Hidden Markov Models; Implicit Memory; Monotonic Functions; Policies; Procedural Memory (and Learning); Reaction; Rules (and Composition); Shallow Reasoning; Skills (and Acquisition)	Assumption-Based TMS; Autoassociative Neural Networks (and Hebbian Learning); Constraint Propagation; Declarative Memory (and Learning); Dynamical Systems; Explicit Memory; Knowledge (and Acquisition); Knowledge Search; Monotonic Logic; Non-Action-Centered; Probabilistic Graphical Models and Statistical Relational AI; Reasoning; Self-Attention Networks; Thinking
Controlled (C)	Action-Centered; Action Selection and Execution; Causal Intervention; Decision Trees (and Learning); Deliberation; Doing; Nonmonotonic Functions; Operators; Procedural Programs; Reinforcement Learning	Analogy; Causal Counterfactuals; Constraint Conditioning; Declarative Programs; Deep Reasoning; Explicit Processing; Justification-Based TMS; Nonmonotonic Logic; Planning; Problem Space Search; Reflection

13|5 (implicit versus explicit), assuming consolidation of memory and processing; 14|13 (action-centered versus non-action-centered), although the ordering here is switched between the two tables; and 15|14 (doing versus thinking), also with the order switched. This can be viewed as a diagonal alignment between the two core dichotomies in table 11.2 for these five variants, along the major diagonal—*AUX* to *SCX*—for the first three pairs, and the minor diagonal—*ACX* to *SUX*—for the remaining two pairs.

Another way to view these alignments is as revealing how these five dichotomies each amounts to a *coarse*, or *compound*, dichotomy that can usefully be decomposed into multiple more refined, or basic, ones. An earlier example of this approach appeared in Rosenbloom, Joshi, and Ustun (2019a), where the coarse symbolic/statistical versus neural/ML distinction was reinterpreted in terms of the major diagonal among three more basic dichotomies. Here, the major diagonal amounts to a 2D projection of that 3D diagonal. An updated 3D view will be seen in section 11.4.

Table 11.3
Interaction of criteria across 2D map of the cross product between (a)symmetric and (un)controlled

	Asymmetric (A)	Symmetric (S)
Uncontrolled (U)	AP, $ET(C_1, R_{(1^2, n)})$, $G(P_{(as, al^2, gl)}, E_e)$, $IL_{(a^2, g)}$, UEa, $O(Ef, R^2)$	A(C, R), $E(T(C_n, R_{(1, n^2)}), S_n)$, $G(P_{(al, gs, gl^2)}, E_e)$, $IL_{(a, g^2)}$, UEa, O(R, Sa)
Controlled (C)	A(P, C), $ET(C_{(1^2, n)}, R_1)$, $G(P_{(as^2, al, gs)}, E_n)$, IL_a, U(P, U, Ex), $O(Ef^2, R, Sa)$	$A(C^2, R)$, $E(T(C_{(1, n^2)}, R_n), S_n)$, $G(P_{(as, gs^2, gl)}, E_n)$, IL_g, U(P, U, Ex), $O(Ef, Sa^2)$

Table 11.3 replaces the technological contents of tables 11.1 and 11.2 with the cross product of the criteria from the two dichotomies. Each location here thus includes all of the criteria from either dichotomy on which it is advantaged. This 2D disciplinary map uses the CA language already introduced, but with three additional extensions. First, when there is a double advantage for a criterion, due to both dichotomies agreeing on the advantage, the criterion is shown as squared, such as for reactivity in *AUX* here. Second, when the dichotomies yield alternative narrowings of the same type—as for real-time efficiency in *AUX* here—they appear as a comma-separated parenthetical subscript, denoting a disjunctive combination of the two. Third, when the dichotomies yield narrowings of different types—as for problem generality in *AUX* here—they are simply abutted, denoting a conjunctive combination of the two. As witnessed by problem generality for AUX here, the second and third extensions can be combined to denote disjunctive combinations of conjunctive pairs, effectively yielding a disjunctive normal form.

There is a lot of detail in this map, but shifts between adjacent locations do simply yield the same criterial trade-offs found along the corresponding individual dimensions. More complex trade-offs are found along the diagonals as they contrast opposites with respect to the map. For example, along the main diagonal the following trade-offs can be seen:

- Precision for robustness and a double dose of correctness.
- A double dose of real-time efficiency on single problems for a double dose of combinatoric efficiency on multiple problems.
- A double dose of problem generality on inherently asymmetric problems over large amounts of content for the same on generating instances over small amounts of content.

- Expert behavior for novice behavior.
- A double dose of learnability over asymmetric structures plus ease of use for programmability, understandability, and explainability.
- A double dose of reactivity for a double dose of safety.

This is roughly what one might expect for a projection along these two dimensions of a coarse symbolic versus neural dichotomy. However, viewing it in this somewhat more rigorous manner makes these trade-offs both more explicit and more precise. It is only somewhat more rigorous because of the subjectivity involved in the original ascription of advantages to the individual dichotomies. Given the abstract characterization of the 2D interaction in table 11.1, this set of trade-offs can more generally be viewed as what happens with a shift from reaction to reflection.

Based on the abstract characterization in table 11.1, the minor diagonal can be interpreted generally as a shift from deliberation to reasoning. The following trade-offs can be seen here:

- Precision for robustness.
- Enhanced efficiency on single problems for enhanced efficiency on multiple problems.
- A double dose of problem generality on inherently asymmetric problems over small amounts of content for a double dose of the same on generating instances over large amounts of content.
- Novice behavior for expert behavior.
- Programmability, understandability, explainability, plus a double dose of effectivity for a double dose of learnability with respect to instance generation and ease of use.

The one trade-off included here that may appear odd concerns novice versus expert behavior, which is due to the (un)controlled dichotomy. This same trade-off appeared along the major diagonal, where it does seem appropriate in terms of the standard caricature of novice versus expert behavior, but here the difference is murkier. One way to rationalize this might be to hypothesize that novice versus expert should also have been included as a variant of (a)symmetric, essentially aligning novices with thinkers and experts with doers along this dimension. That would shift the trade-off along the major diagonal to involve double doses of novice versus expert while neutralizing it along the minor diagonal. On its own, the idea

Table 11.4
0D map of criterial advantages for the maximum across the (a)symmetric and (un)controlled dichotomies

A(P, C^2, R), E(T^2, S_n), G(P^2, E), IL^2, U, O^2

of including this variant in table 8.2 did not appear to be sufficiently compelling, but this interaction effect may imply that it should be considered.

Table 11.4 shows what may be possible when (un)controlled and (a symmetric are jointly spanned. Here, as with the earlier spanning maps in tables 8.5, 9.5, and 10.5, the combination is best modeled as occurring via a maximum, rather than as a conjunction or disjunction, with the combination of a single dose and a double dose thus being a double dose. This 0D map reveals the potential for precision, robustness, spatial efficiency over multiple related problems, entity generality, and usability; plus, double doses of correctness, temporal efficiency, problem generality, learnability, and operability. Ideally, architectures such as Soar and Sigma that are complete in some sense with respect to the map in table 11.1 could achieve such a combination, although how well these architectures actually fill each location may remain debatable.

Another way to understand table 11.3 is to narrow the focus to just those criterial advantages that are common to the two dichotomies at each location; that is, those that yield double doses. Table 11.5 shows that this narrowing yields:

- Reaction (*AUX*) particularly advantages real-time efficiency on single problems, problem generality on inherently asymmetric problems over large amounts of data, learnability on asymmetric structures, and reactivity.
- Reasoning (*SUX*) particularly advantages real-time efficiency on multiple related problems, problem generality on generating instances over large amounts of data, and learnability on generating instances.
- Deliberation (*ACX*) particularly advantages combinatoric efficiency on single problems, problem generality on inherently asymmetric problems over small amounts of data, and effectivity.
- Reflection (*SCX*) particularly advantages correctness, combinatoric efficiency over multiple related problems, problem generality on generating instances over small amounts of data, and safety.

Table 11.5
Interactions between (a)symmetric and (un)controlled that yield double doses of criterial advantages

	Asymmetric (A)	Symmetric (S)
Uncontrolled (U)	ETR_1^2, GP_{al}^2, IL_a^2, OR^2	ETR_n^2, GP_{gl}^2, IL_g^2
Controlled (C)	ETC_1^2, GP_{as}^2, OEf^2	AC^2, ETC_n^2, GP_{gs}^2, OSa^2

Table 11.6
0D map of double-dose criterial advantages for the maximum across the (a)symmetric and (un)controlled dichotomies

AC^2, ET^2, GP^2, IL^2, O^2

These all appear appropriate as caricatures of these locations within the overall space of AI technologies. If novice versus expert were added as a variant of (a)symmetric, as discussed earlier, this perspective would include it along the main diagonal for entity generality while being silent about it along the minor diagonal.

A double-dose map that spans both dichotomies can be seen in table 11.6. It can be derived either from extracting the double doses from table 11.4 or maximizing across the four locations in table 11.5. The result shows particular criterial advantages for correctness, real-time efficiency, problem generality, learnability, and operability.

11.2 (A)Symmetric × (Meta)Data

Table 11.7 shows a 2D architectural map based on the cross product between (a)symmetric and (meta)data. The entries here, rather than being abstract as in table 11.1, are examples, or perhaps even prototypes, of what might appear in an architecture that is complete with respect to this map. This map in fact turns out to be a rather good fit to core parts of the Clarion cognitive architecture (Sun, 2016)—with asymmetric being action-centered, symmetric being non-action-centered, metadata being implicit and data being explicit—although Clarion uses a form of logical rules for *SXD*, rather than a simple factual knowledge base.

A more fleshed-out 2D disciplinary map of technologies can be seen in table 11.8. As with table 11.2, this amounts to a cross product of the

Table 11.7
A prototypical 2D architectural map for the cross product between (a)symmetric and (meta)data

	Asymmetric (A)	Symmetric (S)
Metadata (M)	Feedforward Neural Networks	Autoassociative Neural Networks
Data (D)	Rules	Facts

Table 11.8
A 2D disciplinary map of the cross product between (a)symmetric and (meta)data

	Asymmetric (A)	Symmetric (S)
Metadata (M)	Attention Networks; Heteroassociative Neural Networks (and Backpropagation); Implicit Action-Centered; Motor Control; Numeric Action Selection and Execution; Numeric Computation; Numeric Deliberation; Numeric Doing; Numeric Functions; Numeric Implicit Memory and Processing; Numeric Policies; Numeric Procedural Memory (and Learning); Numeric Procedural Programs; Numeric Reaction; Numeric Shallow Reasoning; Numeric Skills (and Acquisition); Sensory-Motor Behavior	Autoassociative Neural Networks (and Hebbian Learning); Dynamical Systems; Numeric (Soft) Constraint Propagation; Numeric Declarative Memory (and Learning); Implicit Non-Action-Centered; Numeric Declarative Programs; Numeric Implicit Memory and Processing; Numeric Reasoning; Numeric Thinking; Parameters (and Learning); Probability and Statistics; Self-Attention Networks; Utilities
Data (D)	Causal Association; Causal Intervention; Decision Trees (and Learning); Explicit Action-Centered; Operators; Rules (and Composition); Symbolic Computation; Symbolic Deliberation; Symbolic Doing; Symbolic Functions; Symbolic Implicit Memory & Processing; Symbolic Induction; Symbolic Policies; Symbolic Procedural Memory (and Learning); Symbolic Procedural Programs; Symbolic Reaction; Symbolic Shallow Reasoning; Symbolic Skills (and Acquisition)	Causal Counterfactuals; Explicit Memory and Processing; Explicit Non-Action-Centered; Goals; Knowledge (and Acquisition); Logic; Planning; Reflection; Search; Symbolic (Hard) Constraint Satisfaction; Symbolic Declarative Memory (and Learning); Symbolic Declarative Programs; Symbolic Deep Reasoning; Symbolic Thinking; Truth Maintenance Systems

technologies seen in tables 8.1 and 10.1, with the same kinds of adjustments as were made earlier, rather than an exhaustive catalogue of such technologies.

The major diagonal here—from *AXM* to *SXD*—can be interpreted as a projection of a coarse subsymbolic/neural versus symbolic/logic dichotomy. The minor diagonal—from *AXD* to *SXM*—is less compelling but can be thought of as cognitive skill—as epitomized by rules—versus subcognitive reasoning, as epitomized by the kind of soft constraint satisfaction yielded by autoassociative neural networks such as Boltzmann machines.

Technologies that are immediate neighbors, and thus at a distance of 1, can be seen to share traits along one dichotomy while contrasting along the other. For example, heteroassociative and autoassociative neural networks—*AXM* and *SXM*, respectively—are both based on metadata but contrast with respect to (a)symmetric. In the vertical direction, decision trees (*AXD*) share their nature as classifiers with heteroassociative neural networks (*AXM*) but differ with respect to (meta)data. This map thus summarizes the combination of the two dichotomies, while explicitly showing that it isn't all just subsymbolic/neural versus symbolic/logical by laying out how a range of example technologies array themselves across the two dimensions. The opposite locations along the minor diagonal can in general be seen as providing mixtures of what might be thought of as pure neural and logical traits.

These minor-diagonal locations also provide reasonable paths for gradually moving between the two opposites along the major axis. For example, we can follow the (meta)data variant of (heteroassociative) neural networks versus rules to move vertically between related forms of procedural knowledge, from *AXM* to *AXD*, and then follow the (a)symmetric variant of rules versus logic to move horizontally between related forms of data processing, from *AXD* to *SXD*. The resulting L-shaped path provides a two-step transition between the opposites along the major diagonal. Moving along the other path between the major-axis opposites, we can first use the (a)symmetric variant of heteroassociative versus autoassociative neural networks to move horizontally, from *AXM* to *SXM*. Viewing the latter as forms of soft constraint satisfaction we can then move vertically, from *SXM* to *SXD*, by shifting to hard constraint satisfaction.

The contrast along the major diagonal also helps clarify some of the confusion that can exist concerning skill versus knowledge and implicit versus explicit memory. The epitome of skill is asymmetric metadata processing

Table 11.9
Interaction of criteria across 2D map of the cross product between (a)symmetric and (meta)data

	Asymmetric (A)	Symmetric (S)
Metadata (M)	AP^2, $ET(C_1, R_{(1c^2, 1d, nc)})$, $GP_{(as, al^2, gl)}$, $IL_{(a^2, g)}$, UEa, O(Ef, R)	A, $E(T(C_n, R_{(1c, nc^2, nd)}), S_n)$, $GP_{(al, gs, gl^2)}$, $IL_{(a, g^2)}$, UEa, OSa
Data (D)	A(P, C), $E(T(C_1, R_{(1c, 1d^2, nd)}), S)$, $GP_{(as^2, al, gs)}$, U(P, U, Ex), O	$A(C^2, R)$, $E(T(C_n, R_{(1d, nc, nd^2)}), S_{(1, n^2)})$, $GP_{(as, gs^2, gl)}$, U(P, U, Ex), OSa^2

(*AXM*), but each aspect individually—*A* and *M*—may lead to a technology being considered skill based. Likewise, each is individually sufficient to yield a form of implicit memory that cannot be examined, leading to both locations along the minor diagonal also being labeled as implicit even though each has a dimension along which it in isolation would be considered explicit (*S* and *D*). The epitome of knowledge is symmetric data (*SXD*), although the term may also be used for forms of asymmetric data, such as rules. Both symmetry and data appear necessary for explicitness in memory, so it is only listed for *SXD*.

As table 11.3 did for the interaction between (a)symmetric and (un)controlled, table 11.9 maps the evaluation criteria for the interaction between (a)symmetric and (meta)data. Along the major diagonal the following tradeoffs can be seen:

- A double dose of precision for robustness and a double dose of correctness.
- Combinatoric efficiency on single problems for the same on multiple related problems.
- A double dose of real-time efficiency on single continuous problems for a double dose of the same on multiple related discrete problems and varying amounts of spatial efficiency.
- A double dose of problem generality on inherently asymmetric problems over large amounts of data for a double dose of the same on generating instances over small amounts of data.
- A double dose of learnability on asymmetric structures, learnability on generating instances, and ease of use for programmability, understandability, and explainability.
- Effectivity and reactivity for a double dose of safety.

Table 11.10
0D map of criterial advantages for the maximum across the (a)symmetric and (meta) data dichotomies

$A(P^2, C^2, R), E(T(C, R^2), S_{(1, n^2)}), GP^2, IL^2, U, O(Ef, R, Sa^2)$

Given the less compelling nature of the contrast along the minor diagonal here, identifying its trade-offs is left as an exercise for the interested reader.

As with table 11.4, table 11.10 shows the criterial advantages yielded when (a)symmetric and (meta)data are jointly spanned. It reveals the potential for robustness, combinatorial efficiency, spatial efficiency on single problems, usability, effectivity, and reactivity, plus, double doses of precision, correctness, real-time efficiency, spatial efficiency over multiple related problems, problem generality, learnability, and safety. Ideally then, this should be what is possible with Clarion.

As with tables 11.5–11.6, tables 11.11–11.12 narrow the focus here to those criterial advantages that receive double doses from the interaction between (a)symmetric and (meta)data. Table 11.11 reveals the following double doses:

- *AXM* particularly advantages precision, real-time efficiency on single continuous problems, problem generality on inherently asymmetric problems over large amounts of data, and learnability on asymmetric structures.
- *SXM* particularly advantages real-time efficiency on multiple related continuous problems, problem generality on generating instances over large amounts of data, and learning with respect to generating instances.
- *AXD* particularly advantages real-time efficiency on single discrete problems, and problem generality on inherently asymmetric problems over small amounts of data.
- *SXD* particularly advantages correctness, real-time efficiency on multiple related discrete problems, spatial efficiency on multiple related problems, problem generality on generating instances over small amounts of data, and safety.

Table 11.12 shows that, when (a)symmetric and (meta)data are jointly spanned, double-dose criterial advantages can exist for precision, correctness, real-time efficiency, spatial efficiency over multiple related problems, problem generality, learnability, and safety.

Table 11.11
Interactions between (a)symmetric and (meta)data that yield double doses of criterial advantages

	Asymmetric (A)	Symmetric (S)
Metadata (M)	AP^2, $ETR_{1c}{}^2$, $GP_{al}{}^2$, $IL_a{}^2$	$ETR_{nc}{}^2$, $GP_{gl}{}^2$, $IL_g{}^2$
Data (D)	$ETR_{1d}{}^2$, $GP_{as}{}^2$	AC^2, $E(TR_{nd}, S_n)^2$, $GP_{gs}{}^2$, OSa^2

Table 11.12
0D map of double-dose criterial advantages for the maximum across the (a)symmetric and (meta)data dichotomies

$A(P, C)^2$, $ET(R, S_n)^2$, GP^2, IL^2, OSa^2

Table 11.13
1D map of double-dose criterial advantages across (a)symmetric for hybrid (meta)data technologies

	Asymmetric (A)	Symmetric (S)
Hybrid (H)	AP^2, $ETR_1{}^2$, $GP_a{}^2$, $IL_a{}^2$	AC^2, $E(TR_n, S_n)^2$, $GP_g{}^2$, $IL_g{}^2$, OSa^2

As mentioned in section 10.4, significant effort across the field(s) has been put into developing hybrid (meta)data approaches. These technologies are also often at the leading edge of current research. As shown in table 11.13, the double-dose criterial advantages for hybrid (a)symmetric approaches can be determined by maximizing across the two rows in table 11.11. The trade-offs here are:

- Precision for correctness.
- Real-time efficiency on single problems versus the same plus spatial efficiency on multiple related problems.
- Problem generality and learnability on inherently asymmetric problems versus the same on generating instances.
- Nothing versus safety.

Three technological contrasts of note that correspond to the structure of this map are: (1) arithmetic circuits and sum-product networks versus probabilistic graphical models and statistical relational AI; (2) structured heteroassociative neural networks versus structured autoassociative networks; and (3) attention versus self-attention networks. In the first case, both alternatives perform probabilistic computation over meaningful nodes. In the second case, both perform neural-like computations over networks embodying meaningful structure. In the third case, both identify relevant prior units in temporal neural networks. What distinguishes the two alternatives in each pair is (a)symmetric.

Table 11.13 does, however, also provide a prime example of how maximizing across criterial advantages may prove overly optimistic for insufficiently complete hybrids. Consider, for example, large language models, which are hybrid to some extent (with structured graphs) and also symmetric to some extent (with self-attention), but certainly do—at least at present—provide neither correctness nor safety, although they do strongly support problem generality on generating instances.

11.3 (Un)Controlled × (Meta)Data

Table 11.14 shows a 2D disciplinary map based on the cross product between (un)controlled and (meta)data. Here the rows maintain the same interpretation as in table 11.8 while the columns correspond to the rows in table 11.2. The top-right location (*XCM*) turns out to be sparsely populated here and may actually seem a bit oxymoronic. It is possible to argue that, given control needs to select something, there must be data involved to be selected, and thus, at least a hybrid is required rather than pure metadata. However, as with the notion of metadata itself requiring data to which it is associated, it is possible to separate out the data that is being selected from what determines its selection. Thus, the assumption here will be that, although *XCM* selects among alternative data, this selection is based on content that is primarily metadata. Still, this complexity has led me to be quite cautious about what to include at this location in the map—excluding from consideration, for example, any forms of reflection based purely on metadata—implying that there truly may be much more than is shown here.

The major diagonal in table 11.14 can be viewed as distinguishing varieties of numeric reaction and reasoning (*XUM*) from varieties of symbolic deliberation and reflection (*XCD*), with the minor diagonal contrasting varieties of symbolic reaction and reasoning (*XUD*) with varieties of numeric deliberation (*XCM*). Table 11.15 maps the criteria for the interaction between (un)controlled and (meta)data.

Table 11.14
A 2D disciplinary map of the cross product between (un)controlled and (meta)data

	Uncontrolled (U)	Controlled (C)
Metadata (M)	Dynamical Systems; Neural Networks (and Learning); Numeric (Soft) Constraint Propagation; Numeric Fast Thinking; Numeric Memory (and Learning); Numeric Monotonic Functions; Numeric Policies; Numeric Reaction; Numeric Shallow Reasoning; Numeric Skills (and Acquisition); Peripheral Processing; Probability and Statistics; (Self-)Attention Networks; Uncontrolled Tensor Processing	Numeric Action Selection and Execution; Numeric Deliberation; Numeric Nonmonotonic Functions; Numeric Operators
Data (D)	Assumption-Based TMS; Knowledge (and Acquisition); Knowledge Search; Monotonic Logic; Rules (and Composition); Symbolic (Hard) Constraint Propagation; Symbolic Induction; Symbolic Memory; Symbolic Reaction; Symbolic Shallow Reasoning; Symbolic Skills (and Acquisition)	Constraint Conditioning; Decision Trees (and Learning); Deep Reasoning; Explicit Memory and Processing; Justification-Based TMS; Nonmonotonic Logic; Planning; Problem Space Search; Symbolic Analogy; Symbolic Deliberation; Symbolic Operators; Symbolic Procedural Programs; Symbolic Reflection

Table 11.15
Interaction of criteria across 2D map of the cross product between (un)controlled and (meta)data

	Uncontrolled (U)	Controlled (C)
Metadata (M)	AP, $ETR_{(c^2, d)}$, $G(P_1^2, E_e)$, IL^2, UEa^2, OR	A(P, C), $ET(C, R_c)$, $G(P, E_n)$, IL, U, O(Ef, Sa)
Data (D)	AC, $E(TR_{(c, d^2)}, S)$, $G(P, E_e)$, IL, U, O(R, Sa)	AC^2, $E(T(C, R_d), S)$, $G(P_s^2, E_n)$, $U(P, U, Ex)^2$, $O(Ef, Sa^2)$

Along the major diagonal here the following trade-offs can be seen:

- Precision for a double dose of correctness.
- A double dose of real-time efficiency on continuous problems for combinatoric and spatial efficiency.

- A double dose of problem generality on large datasets for the same on small datasets.
- Expert behavior for novice behavior.
- Double doses of learnability and ease of use for double doses of programmability, understandability, and explainability.
- Reactivity for effectivity and a double dose of safety.

Systems such as AlphaZero that combine some form of neural networks (*XUM*) for positional evaluation and search (*XCD*) for action selection likely succeed so well at gaming because they can span the trade-off space across this major diagonal.

Perhaps the most interesting pattern witnessed by the minor diagonal is that, in contrast to the major diagonal where there are numerous double doses, there is but one here. Instead, the minor diagonal tends to blur the distinctions highlighted for the major diagonal by combining complementary single doses from the two dimensions for things like problem generality and usability. Whether technologies exist along the minor diagonal that can actually benefit from such complementary strengths is an independent question, as is whether a broader number of single doses is preferable in at least some situations to a smaller number of double doses.

Consider rules (*XUD*) on the minor diagonal, for example. Part of their appeal through the years has been what they provide with respect to problem generality, learnability, usability, and operability. However, the advantages now accrued by neural networks (*XUM*) with respect to problem generality—particularly given large bodies of content—learnability, and ease of use go a long way toward explaining the paradigm shift that has been seen in the field. Indeed, their ease of use has even enabled high school students to publish research based on them (Lee, 2025). Yet, the trade-offs between these two approaches of precision for correctness, and ease of use for programmability, understandability, explainability, and safety have continued to raise important issues for neural networks.

Table 11.16 shows the criterial advantages yielded when (un)controlled and (meta)data are fully spanned. It reveals the potential for precision, combinatoric and spatial efficiency, entity generality, effectivity, and reactivity; plus, double doses of correctness, real-time efficiency, problem generality, learnability, usability, and safety. While AlphaZero has no presence along the minor diagonal here, it does play in this overall space due to its strengths along the major diagonal.

Table 11.16
0D map of criterial advantages for the maximum across the (un)controlled and (meta) data dichotomies

A(P, C^2), E(T(C, R^2), S), G(P^2, E), IL^2, U^2, O(Ef, R, Sa^2)

Table 11.17
Interactions between (un)controlled and (meta)data that yield double doses of criterial advantages

	Uncontrolled (U)	Controlled (C)
Metadata (M)	ETR_c^2, GP_l^2, IL^2, UEa^2	
Data (D)	ETR_d^2	AC^2, GP_s^2, U(P, U, Ex)^2, OSa^2

Table 11.18
0D map of double-dose criterial advantages for the maximum across the (un)controlled and (meta)data dichotomies

AC^2, ETR^2, GP^2, IL^2, U^2, OSa^2

Tables 11.17–11.18 narrow the focus here to those criterial advantages that receive double doses from the interaction between (un)controlled and (meta)data. Table 11.17 reveals the overall pattern already mentioned—whereas there are a significant number of double doses along the major diagonal, there is only one along the minor diagonal. Moreover, *XUM* shows no double doses at all, perhaps justifying the relative lack of technologies at this location. From this table, it should also be clear that AlphaZero, in ignoring the minor diagonal, may be missing only a double dose of real-time efficiency for discrete problems, as might be provided by rules (*XUD*), at least with respect to this 2D map. However, considering the bigger 3D picture that is the focus of section 11.4, it is also noticeably missing any form of reasoning (*SUX*).

To wrap up here, table 11.18 shows that, when (un)controlled and (meta) data are jointly spanned, double-dose criterial advantages can exist for correctness, real-time efficiency, problem generality, learnability, usability, and safety.

11.4 (A)Symmetric × (Un)Controlled × (Meta)Data

Table 11.19 shows the result of combining the 2D disciplinary maps in tables 11.2, 11.8, and 11.14 into a single 3D map, formatted as separate (un) controlled planes, each of which crosses (a)symmetric with (meta)data. This disciplinary map is analogous to the one developed by Rosenbloom, Joshi, and Ustun (2019a), albeit much more fleshed out while still remaining only illustrative rather than exhaustive. It also has (un)controlled substituted for (non)combinatory and (meta)data substituted for (sub)symbolic. The switch from (non)combinatory to (un)controlled has left a relative dearth of alternatives for symmetric controlled metadata (*SCM*), although there is a rich space of hybrid technologies (*SCX*), such as probabilistic search, and forms of heuristic search that use numeric evaluations, rewards, and utilities.

As with the earlier map, feedforward and recurrent—that is, heteroassociative—neural networks (*AUM*) end up at one extreme along the major diagonal. However, the shift from (non)combinatory to (un)controlled implies that the other extreme is not combinatory approaches such as first-order/monotonic logic and statistical relational AI but controlled approaches such as nonmonotonic logic and problem space search. This does make the contrast even more striking while, as discussed shortly, turning out to correspond to the two most critical technologies it was found necessary to combine in achieving champion-level performance in AlphaZero, if straight symbolic search is replaced with a hybrid in the form of probabilistic search.

Still, as with the prior analysis, the overall story is not limited to the major diagonal, with other significant approaches found by stepping away from the major diagonal in various directions. In general, adjacent cells in this map have many properties in common due to sharing the same pole along two out of the three dichotomies, with distances and thus dissimilarities increasing as fewer poles are shared. Yet, this does imply the existence of gradual paths that connect any two locations, including those along the major diagonal. For example, there is a gradual path from heteroassociative neural networks, through rules, to monotonic logic, and finally to nonmonotonic logic. There is also a distinct path between the same two locations that goes from heteroassociative neural networks, through autoassociative neural networks and numeric (soft) constraint propagation, to symbolic (hard) constraint propagation, and finally to constraint conditioning.

Table 11.19
A 3D disciplinary map of the cross product among (a)symmetric, (un)controlled, and (meta)data shown as distinct (un)controlled planes. Each plane comprises a 2D map of (a)symmetric by (meta)data

Uncontrolled (U)	Asymmetric (A)	Symmetric (S)
Metadata (M)	Attention Networks; Implicit Action-Centered; Fast Numeric Doing; Heteroassociative Neural Networks (and Backpropagation); Numeric Monotonic Functions; Numeric Policies; Numeric Procedural Memory (and Learning); Numeric Reaction; Numeric Shallow Reasoning; Numeric Skills (and Acquisition)	Autoassociative Neural Networks (and Hebbian Learning); Dynamical Systems; Fast Numeric Thinking; Numeric (Soft) Constraint Propagation; Implicit Non-Action-Centered; Numeric Declarative Memory (and Learning); Numeric Reasoning; Parameters (and Learning); Probability and Statistics; Self-Attention Networks
Data (D)	Causal Association; Fast Symbolic Doing; Rules (and Composition); Symbolic Policies; Symbolic Procedural Memory (and Learning); Symbolic Reaction; Symbolic Shallow Reasoning; Symbolic Skills (and Acquisition)	Assumption-Based TMS; Explicit Non-Action-Centered; Fast Symbolic Thinking; Knowledge (and Acquisition); Knowledge Search; Monotonic Logic; Symbolic (Hard) Constraint Propagation; Symbolic Declarative Memory (and Learning); Symbolic Reasoning

Controlled (C)	Asymmetric (A)	Symmetric (S)
Metadata (M)	Numeric Action Selection and Execution; Numeric Deliberation; Numeric Nonmonotonic Functions; Numeric Operators; Numeric Procedural Programs; Slow Numeric Doing	Numeric Declarative Programs; Slow Numeric Thinking
Data (D)	Causal Intervention; Decision Trees (and Learning); Explicit Action-Centered; Slow Symbolic Doing; Symbolic Deliberation; Symbolic Operators; Symbolic Procedural Programs	Causal Counterfactuals; Constraint Conditioning; Deep Reasoning; Explicit Memory and Processing; Justification-Based TMS; Nonmonotonic Logic; Planning; Problem Space Search; Reflection; Slow Symbolic Thinking; Symbolic Analogy; Symbolic Declarative Programs

This 3D structure also implies that an architectural map for an integrated system that is complete with respect to it can, at least in principle, be built by additively including any one technology from each location. Such an approach could work if the technologies are sufficiently mutually compatible or if they are loosely enough coupled through something like a working memory.

Alternatively, reductive approaches could be considered that span multiple locations to potentially yield simpler and more elegant systems (section 7.4). This latter approach has, for example, been the focus with Sigma's desideratum of functional elegance (section 1.2.2). Such approaches can yield hybrids along one or more of the three dimensions of this table that amount to valuable technologies such as vector symbolic architectures and statistical relational AI that are not found in the table itself.

Table 11.20 shows the individual criterial advantages for the locations in table 11.19. In its entirety, this map exhibits significant amounts of both complexity and subtlety. One way to simplify all of this is to consider the ideal profile that results from spanning the entire map—that is, the potential criterial advantages for systems that are complete with respect to this map—yielding the 0D map in table 11.21. This map includes triple doses

Table 11.20
Interaction of criteria across 3D map of the cross product among (a)symmetric, (un) controlled, and (meta)data

Uncontrolled (U)	Asymmetric (A)	Symmetric (S)
Metadata (M)	AP^2, $ET(C_1, R_{(1c^3, 1d^2, nc^2, nd)})$, $G(P_{(as, al^3, gl^2)}, E_e)$, $IL_{(a^3, g^2)}$, UEa^2, $O(Ef, R^2)$	A, $E(T(C_n, R_{(1c^2, 1d, nc^3, nd^2)}), S_n)$, $G(P_{(al^2, gs, gl^3)}, E_e)$, $IL_{(a^2, g^3)}$, UEa^2, O(R, Sa)
Data (D)	A(P, C), $E(T(C_1, R_{(1c^2, 1d^3, nc, nd^2)}), S)$, $G(P_{(a^2, g)}, E_e)$, $IL_{(a^2, g)}$, U, O(Ef, R^2, Sa)	$A(C^2, R)$, $E(T(C_n, R_{(1c, 1d^2, nc^2, nd^3)}), S_{(1, n^2)})$, $G(P_{(a, g^2)}, E_e)$, $IL_{(a, g^2)}$, U, $O(R, Sa^2)$

Controlled (C)	Asymmetric (A)	Symmetric (S)
Metadata (M)	$A(P^2, C)$, $ET(C_{(1^2, n)}, R_{(1c^2, 1d, nc)})$, $G(P_{(a^2, g)}, E_n)$, $IL_{(a^2, g)}$, U, $O(Ef^2, R, Sa)$	$A(P, C^2, R)$, $ET(C_{(1, n^2)}, R_{(1c, nc^2, nd)})$, $G(P_{(a, g^2)}, E_n)$, $IL_{(a, g^2)}$, U, $O(Ef, Sa^2)$
Data (D)	$A(P, C^2)$, $E(T(C_{(1^2, n)}, R_{(1c, 1d^2, nd)}), S)$, $G(P_{(as^3, al, gs^2)}, E_n)$, IL_a, $U(P, U, Ex)^2$, $O(Ef^2, R, Sa^2)$	$A(C^3, R)$, $E(T(C_{(1, n^2)}, R_{(1d, nc, nd^2)}), S_{(1, n^2)})$, $G(P_{(as^2, gs^3, gl)}, E_n)$, IL_g, $U(P, U, Ex)^2$, $O(Ef, Sa^3)$

Table 11.21
0D map of criterial advantages for the maximum across the (a)symmetric (un)controlled, and (meta)data dichotomies

$A(P^2, C^3, R), E(T(C^2, R^3), S_{(1, n^2)}), G(P^3, E), IL^3, U^2, O(Ef^2, R^2, Sa^3)$

of correctness, real-time efficiency, problem generality, learnability, and safety. It also includes double doses of precision, combinatoric efficiency, spatial efficiency over multiple related problems, usability, effectivity, and reactivity. There is also some potential for robustness, spatial efficiency over single problems, and entity generality.

Given these three maps, several more focused analyses can help exemplify how full-scale dichotomic analysis might be used. Section 11.4.1 analyzes in detail the main diagonal of table 11.20—from *AUM* to *SCD*. Section 11.4.2 explores what can be said about the similarities and differences between the common model of cognition and AlphaZero beyond what was apparent from their 2D architectural maps in section 6.3. Section 11.4.3 examines Sigma, one cognitive architecture that is complete with respect to these maps, in a manner that complements what was seen in section 4.3.3. Section 11.4.4 similarly complements the analysis of large language models found in section 4.3.2 by exploring them from a dichotomic perspective. Section 11.4.5 follows the pattern established in sections 11.1–11.3 of concluding with analyses of double-dose criterial advantages, by analyzing the triple-dose criterial advantages made possible by the interactions among all three dichotomies.

11.4.1 The Major Diagonal (from *AUM* to *SCD*)

The beginning of the major diagonal (*AUM*) can be canonically exemplified by heteroassociative neural networks (and backpropagation). With respect to accuracy, this location exhibits a double dose of precision from the combination of asymmetry and metadata while lacking any advantage with respect to either correctness or robustness. With respect to efficiency, it highlights real-time efficiency for single continuous problems while providing some amount of benefit for various other forms. With respect to generality, it highlights inherently asymmetric problems over large bodies of content, while also yielding a double dose for generating instances over large bodies of content and some advantage for expert behavior. It does, however, omit any advantage for generating instances over small bodies of content. With respect to learnability, it highlights inherently asymmetric problems while still yielding a double dose for generating instances. It also shows double doses of ease of use and reactivity while ignoring programmability, understandability, explainability, and safety.

Much of this would seem to match expectations for such technologies, except perhaps for suggesting double doses of problem generality and learnability for generating instances. Instance generation is primarily a function of symmetric technologies, as can be seen with the triple doses for *SUM*, but surprisingly double doses are still seen for asymmetric technologies here. This stems from uncontrolled × metadata yielding double doses for both problem generality on large datasets and learnability (table 11.15). When combined with asymmetric, these split into double doses for both inherently asymmetric problems and instance generation, yielding what is seen here for *AUM*. Although this does retain the expected relationship of symmetric technologies being advantaged over asymmetric technologies for these criteria, it would seem to overvalue what *AUM* provides, revealing that, given how these expressions are computed, the upper bounds they provide can be quite loose at times.

The end of the major diagonal (*SCD*) can be canonically exemplified by traditional problem space search and planning technologies. Unsurprisingly, this location yields the opposite advantages in many cases. With respect to accuracy, it exhibits a triple dose of correctness and a single dose of robustness (from symmetric), but without the presence of numbers precision is meaningless here. It achieves efficiency, but more focused on multiple related discrete problems. With respect to generality, it highlights generating instances over small bodies of content. It only yields a single dose of learnability, and even there only for generating instances. It also shows double doses of programmability, understandability, and explainability, plus a triple dose of safety, while ignoring ease of use and reactivity.

Comparing the relative advantages of these two opposites does support an argument that spanning them is likely to be necessary for achieving human-level intelligence, and particularly ones that are actually usable by humans.

11.4.2 The Common Model of Cognition and AlphaZero

In section 6.3, 2D architectural maps of (a)symmetric × (un)controlled were introduced for the current version of the common model of cognition and for AlphaZero, both of which omitted one of the four locations in the abstract map shown in table 11.1—reflection (*SCX*) for the former and reasoning (*SUX*) for the latter. A deeper understanding and comparison of these two integrated systems begins here by expanding these earlier maps to 3D while updating their terminology and refining the assignment of technologies to locations, yielding the maps in tables 11.22–11.23.

Table 11.22 shows essentially the same pattern for the common model as did the original 2D map, due to the technologies that it embraces being hybrid, even if a more nuanced analysis must acknowledge that much more of the action stems from data than from metadata. Table 11.23 shows the equivalent map for AlphaZero. It introduces an additional distinction based on the assumption that AlphaZero's reactivity is due to metadata (*AUM*) rather than to a hybrid technology (*AUX*), thus introducing an additional source of incompleteness in its map, for *AUD*.

Table 11.22
A 3D architectural map of the common model of cognition

Uncontrolled (U)	Asymmetric (A)	Symmetric (S)
Metadata (M) *Data (D)*	Procedural Memory (and Composition)	Declarative Memory (and Fact Learning)

Controlled (C)	Asymmetric (A)	Symmetric (S)
Metadata (M) *Data (D)*	Deliberation (and Reinforcement Learning)	

Table 11.23
A 3D architectural map of *AlphaZero*

Uncontrolled (U)	Asymmetric (A)	Symmetric (S)
Metadata (M)	Heteroassociative Neural Networks (and Backpropagation)	
Data (D)		

Controlled (C)	Asymmetric (A)	Symmetric (S)
Metadata (M) *Data (D)*	Deliberation	Monte Carlo Tree Search

The next step in the analysis is to explore in an abstract manner the relative criterial advantages yielded by the combination of locations included in each. This is pursued by generating 0D criterial maps from the two 3D architectural maps and then comparing each with the ideal map in table 11.21 to reveal what potential criterial advantages are lost by incompleteness in the respective architectural maps.

The criterial advantages for the common model can be seen in table 11.24. From this map it can be seen that its lack of reflection reduces its potential for correctness, combinatoric efficiency on multiple related problems, problem generality on generating instances over small bodies of content, and safety. If we were also to consider its relative emphasis on data over metadata, further deficits would start to appear as well. Table 11.25, for example, shows additional reductions of potential criterial advantages with respect to precision, real-time efficiency, problem generality, learnability, and ease of use when no metadata is included. The true story likely lies somewhere between these two maps.

The criterial advantages for AlphaZero can be seen in table 11.26. Its deficits with respect to the ideal are more minimal, concentrated as they are in the details of real-time efficiency, problem generality, and learnability. On the face of it, this may seem surprising—particularly with respect to problem generality—as AlphaZero was built narrowly to learn to play games rather than to cover the full range of problems requiring intelligence. Part of this may be due to how powerful the major diagonal is all by itself, as is

Table 11.24
0D map of criterial advantages for the common model of cognition

$A(P^2, C^2, R), E(T(C_{(1^2, n)}, R^3), S_{(1, n^2)}), G(P_{(a^3, gs^2, gl^3)}, E), IL^3, U^2, O^2$

Table 11.25
0D map of data-only criterial advantages for the common model of cognition

$A(P, C^2, R), E(T(C_{(1^2, n)}, R_{(1c^2, 1d^3, n^2)}), S_{(1, n^2)}), G(P_{(as^3, al^2, g^2)}, E), IL^2, U(Ea, P^2, U^2, Ex^2), O^2$

Table 11.26
0D map of criterial advantages for *AlphaZero*

$A(P^2, C^3, R), E(T(C^2, R_{(1c^3, 1d^2, n^2)}), S_{(1, n^2)}), G(P_{(a^3, gs^3, gl^2)}, E), IL_{(a^3, g^2)}, U^2, O(Ef^2, R^2, Sa^3)$

Table 11.27

0D map of criterial advantages for *AlphaZero* without reflection

$A(P, C)^2, E(T(C_{(1^2, n)}, R_{(1c^3, 1d^2, nc^2, nd)}), S), G(P_{(a^3, g^2)}, E), IL_{(a^3, g^2)}, U^2, O^2$

leveraged by AlphaZero but not by the current common model. Part of it may also be due to giving too much credit to AlphaZero for its coverage of reflection (*SCX*)—Monte Carlo tree search is appropriate for game-playing but likely does not support the wider range of reflective abilities required of general intelligence.

One way of getting at this latter effect is to consider what the criterial advantages for AlphaZero would look like without reflection, as in table 11.27, and to assume that this is a more accurate characterization of it outside the scope of applicability of Monte Carlo tree search. This version reduces the advantages with respect to correctness, robustness, and safety, as well as some of the details involved in temporal efficiency and problem generality. As with the common model, the true story is likely somewhere in between these two criterial maps.

If this map is compared with the data-only analysis of the common model, a trade-off can be seen of extra doses of precision, learnability on asymmetric structures, and ease of use for a dose of robustness, plus some variations in the details of real-time efficiency and problem generality. An additional intriguing consequence of the overall comparison of the common model and AlphaZero is that it may not be necessary for an architecture to be complete with respect to the 3D maps to approach the potential of what can be yielded by completeness. Still, it must be recalled that when talking about potential we are talking about an upper bound on what can result rather than necessarily what would actually result.

11.4.3 Sigma

One cognitive architecture that is complete with respect to the full 3D maps is Sigma. The architectural map in table 11.28 shows how this plays out, including not only the core technologies implemented in its architecture, which are shown in bold, but also other technologies that Sigma's functional elegance has been shown to subsume through combinations of architectural mechanisms and bits of knowledge and/or skills. Many of these additional technologies can be found in the fleshed out disciplinary map in

Table 11.28
A 3D architectural map of Sigma plus other technologies it has been shown to subsume (core technologies in bold)

Uncontrolled (U)	Asymmetric (A)	Symmetric (S)
Metadata (M)	**Tensor-Based Extended Factor Graphs (and Gradient-Descent Learning)**	
Data (D)	Constraint Propagation; Declarative Memory (and Learning); Functions; Heteroassociative Neural Networks (and Backpropagation); Hidden Markov Models; Implicit Memory & Processing; Induction; Knowledge (and Acquisition); Knowledge Search; Parameters (and Learning); Policies; Probabilistic Graphical Models & Statistical-Relational AI; Probability and Statistics; Procedural Memory; Perception; Reaction; Reasoning; Rules; Shallow Reasoning; Skills (and Acquisition); Sum-Product Networks; Vector Symbolic Architectures	

Controlled (C)	Asymmetric (A)	Symmetric (S)
Metadata (M)		
Data (D)	**Tensor-Based Decision Procedure** Action Selection and Execution; Deliberation; Doing; Operators; Reinforcement Learning	**Tensor-Based Impasses and Subgoaling** Deep Reasoning; Explicit Memory and Processing; Planning; Problem Space Search; Reflection

table 11.19, but there are also some not found there either because they are hybrid or because of their having an ambiguous status along one of more of the dichotomies. Overall, this map can be viewed as an updated, extended, and simplified version of the 3D analysis of Sigma in Rosenbloom (2019a).

By including the subsumed technologies, the map in table 11.28 can help in assessing for Sigma what was difficult in the analyses of the common model and AlphaZero—that is, how well it supports the full functionality required at each location it covers. While this remains qualitative, it provides a complementary perspective on how well Sigma approaches what is needed for general intelligence from that found in section 4.3.3's analysis of how well it approximates full-spectrum intelligence.

Because Sigma is complete with respect to the 3D map, its 0D map of potential criterial advantages is just the one seen in table 11.21. Yet, it is important to reemphasize here that potential criterial advantages don't always reflect actual criterial advantages. For example, Sigma's uncontrolled technologies (*XUX*) do not at present yield anywhere near the real-time efficiency found in dedicated neural networks, such as are found in AlphaZero or large language models, despite the various efforts—some of which are mentioned in section 15.4—to improve its efficiency.

Although simplicity has not been a criterion considered in this overall dichotomic analysis, by making explicit how Sigma approaches (functional) elegance, table 11.28 can assist in evaluating this key aspect of simplicity with respect to it. One piece of this is how a single combination of native technologies spans both (a)symmetric and (meta)data in handling all uncontrolled processing (*XUX*). A second piece, and one shared with Soar, is the nesting that occurs in how deliberation (*ACX*) builds upon uncontrolled memory processing, in directly using what is retrieved in this manner for decision-making rather than postulating a separate memory module in support of it; and in how reflection (*SCX*) builds upon deliberation by reusing this cognitive decision-making process for metacognitive decisions. A third piece is how each native technology in the map supports many different technologies that might otherwise require separate modules for their implementation. An entry for elegance (SE) could thus be added to the overall expression for Sigma's criterial advantages found in table 11.21, but it is left implicit here due to the lack of a direct focus on simplicity in the overall analysis.

11.4.4 Large Language Models

Large language models (LLMs) were not the center of the AI universe in 2019, when the earlier analyses of the common model, AlphaZero, and Sigma were done, but it would be remiss at this point not to attempt an analysis of them in terms of a 3D architectural map of a transformer-based LLM (Vaswani et al., 2017). As with the architectural map of Sigma, this should complement the assessment of how well such an approach approximates full-spectrum intelligence that was seen in section 4.3.2. Table 11.29 is such an architectural map, at least to the best of my present understanding of LLMs. It does go beyond the kinds of minimal LLMs considered in the earlier analysis to include in *ACX* how reinforcement learning with human feedback (Ziegler et al., 2019) is used to tune these models.

Table 11.29
A 3D architectural map of an LLM

Uncontrolled (U)	Asymmetric (A)	Symmetric (S)
Metadata (M)	Structured Heteroassociative Neural Networks (and Backpropagation)	Embedding and Self-Attention (and Learning)
Data (D)		

Controlled (C)	Asymmetric (A)	Symmetric (S)
Metadata (M)	Softmax-Based Decisions (and Reinforcement Learning)	
Data (D)		

Perhaps the biggest open question about this map is whether something should be listed for reflection (*SCX*). Although there is no conventional mechanism for it in LLMs, there are at least three ways such a capability may be exhibited in some form. First, as hinted in section 8.4, reinforcement learning can capture aspects of what problem space search and planning yield for what has previously been experienced, although without any ability to reflect on more novel problems that arise during performance. Second, sequences of situations and actions learned as content by LLMs may also yield the ability to simulate forward search in limited ways without direct architectural support for such a capability. Third, self-attention may yield something akin to reflection through its ability to learn about dependencies among tokens in a sequence. None of these possibilities support what would conventionally be considered a general reflection, search, or planning capability, so that location has been left empty in table 11.28. Yet this ultimately may prove to be an underestimate of what LLMs provide here.

Without reflection, the overall shape of the LLM architectural map is much closer in form to that of the common model—in fact, it is identical to it, with three sets of hybrid technologies and the same omission of reflection—than it is to AlphaZero. LLMs are considered as hybrid rather than just as metadata because of the essential roles played by both structured graphs and word-based tokens; although, as with AlphaZero, the hybrids here are much more skewed to metadata than to data. The overall result is that the 0D map of criterial advantages for LLMs of this sort is just what was seen in table 11.24 for the common model.

Table 11.30

0D map of metadata-only criterial advantages for LLMs

$A(P^2, C, R), E(T(C_{(1^2, n)}, R_{(c^3, d^2)}), S_n), G(P_{(as^2, gs, 1^3)}, E), IL^3, U(Ea^2, P, U, Ex), O(Ef^2, R^2, Sa)$

As with the common model, but in the opposite direction, it is possible to consider the potential criterial advantages if LLMs were purely metadata. Table 11.30 shows that this yields reduced correctness, real-time efficiency on discrete problems, space efficiency, problem generality with small amounts of content, programmability, understandability, explainability, and safety; although, once again, the true story likely resides somewhere in between these two maps.

Comparing tables 11.25 and 11.30 clarifies the trade-off between the common model and LLMs, of (some amount of) correctness, space efficiency, programmability, understandability, explainability, safety versus precision, learnability, and ease of use. The correctness versus precision contrast is notable, as it implies that LLMs have the potential to be more precise about some things while still being less correct overall.

The notion of an LLM as an amplified but unconstrained subconscious (section 4.3.2.3) is reflected in table 11.29 by the heavy emphasis on uncontrolled technologies, the massive amount of content enabled therein by the learnability advantage of uncontrolled metadata (*XUM*), and the lack of reflection (*SCX*) to rein this in. Given their similar overall shapes, should the same appellation apply to the common model? It is indeed similarly largely subconscious, but its emphasis on data over metadata removes the amplification that LLMs achieve by learning over large amounts of content while at the same time providing more potential for constraint via increases in correctness and safety.

11.4.5 Triple-Dose Criterial Advantages

Sections 11.1–11.3, covering the interactions among pairs of dichotomies, each concluded with analyses of the double-dose criterial advantages yielded by their 2D interactions. Here, the equivalent is to consider the triple-dose criterial advantages in table 11.31. This map implies that triple-dose advantages are relatively few and follow particular patterns. Uncontrolled (*XUX*) broadly provides triple advantages for real-time efficiency, but with

Table 11.31
Interactions among (a)symmetric, (un)controlled, and (meta)data that yield triple doses of criterial advantages

Uncontrolled (U)	Asymmetric (A)	Symmetric (S)
Metadata (M)	ETR_{1c}^3, GP_{al}^3, IL_a^3	ETR_{nc}^3, GP_{gl}^3, IL_g^3
Data (D)	ETR_{1d}^3	ETR_{nd}^3

Controlled (C)	Asymmetric (A)	Symmetric (S)
Metadata (M)		
Data (D)	GP_{as}^3	AC^3, GP_{gs}^3, OSa^3

a different narrowing in each location. Uncontrolled metadata (*XUM*) similarly provides triple advantages for learnability, and the 2D diagonal from uncontrolled metadata (*XUM*) to controlled data (*XCD*) provides triple advantages for problem generality, both also with a different narrowing in each location. The one other triple advantage is safety, for symmetric controlled data (*SCD*).

Focusing on just the major diagonal here—from *AUM* to *SCD*—we can see a trade-off of triple doses of real-time efficiency on single continuous problems, problem generality on inherently asymmetric problems over large amounts of data, and learnability on asymmetric structures for triple doses of correctness, problem generality on generating instances over small amounts of data, and safety. If double doses from table 11.20 are also considered while abstracting out the complex patterns for efficiency and generality, a trade-off can also be seen of precision, learnability, ease of use, and reactivity for programmability, understandability, and explainability. This provides a canonical perspective on an extreme form of the traditional, coarse-grained neural-versus-symbolic trade off in the form, for example, of feedforward neural networks versus planning.

A rather different but no less interesting pattern is the complete lack of triple-dose advantages for controlled metadata (*XCM*) in this map. Could it be related to the relative dearth of entries for symmetric controlled metadata (*SCM*) in table 11.19, or do both perhaps stem from some ignorance or unconscious bias on my part in the pursuit of this analysis? I don't have an answer to this question at this time.

Table 11.32
0D map of triple-dose criterial advantages for the maximum across the (a)symmetric, (un)controlled, and (meta)data dichotomies

AC^3, ETR^3, GP^3, IL^3, OSa^3

To close, table 11.32 show the potential triple-dose advantages when all three dichotomies are jointly spanned. This comprises correctness, real-time efficiency, problem generality, learnability, and safety. These are important criteria on their own, but the full picture does become even more compelling when the entire set of advantages shown earlier in table 11.21 is considered.

11.5 Summary

Whereas chapters 8–10 explored the individual (a)symmetric, (un)controlled, and (meta)data dichotomies, this chapter has considered the interactions that can occur among these three dichotomies. In doing so, it has yielded a range of 2D and 3D maps of AI technologies. Disciplinary maps provide ways of structuring the overall space of technologies underlying AI and related fields plus means for understanding their similarities and differences. The 3D disciplinary map in table 11.19 brings a form of order to the world of AI technologies while simultaneously emphasizing the conceptual breadth they span. Immediate neighbors share many properties, as they agree along two of the three dichotomies, with kinship further waning with additional distance. Architectural maps provide ways of understanding integrated systems that combine technologies from across multiple locations in disciplinary maps. This has been exemplified through maps for the common model of cognition, AlphaZero, Sigma, and transformer-based LLMs.

Adding in the consideration of criterial advantages yields multidimensional maps that help understand both the trade-offs across locations in technology-based maps and the potential yielded by spanning multiple such locations. Summary 0D maps of criterial advantages build on this to provide compact expressions that provide upper bounds on what complex architectural maps might yield. These maps of criterial advantage can in turn be used for understanding, evaluating, comparing, and contrasting such approaches, complementing the kinds of analyses in section 4.3 that considered how well they approximated a multifaceted definition of intelligence.

The full 3D structure may be considered implicitly as a call for integrated systems that span all eight of its locations; that is, that are complete with respect to it. However, an analysis in terms of criterial advantages reveals that while this may be true with respect to fully maximizing the potential of an integrated system, it is possible that some locations may be omitted with little loss of overall potential.

12 Conclusion to Part II

The focus of this part has been on deepening our understanding of AI and related technologies, and of integrated systems that are based on these technologies. The approach has been to begin with a quartet of dichotomies that earlier proved useful in understanding either the pervasive but rather coarse distinction between symbolic and subsymbolic technologies for intelligence or the overall structure of integrated systems built from combinations of such technologies. Based on further experience with and analysis of these four dichotomies, they have been reformulated here into the three core dichotomies summarized in table 12.1.

These core dichotomies have been explored as individuals, in pairs, and altogether as a triplet. The overall result is a range of 0D, 1D, 2D, and 3D maps, yielding a more systematic understanding of the space of technologies via disciplinary maps, and the structure of integrated systems that combine multiple of them via architectural maps. Similarities and differences among the technologies are easily identified with respect to these maps in terms of whether they share dichotomic poles, forming the basis for an L1 distance metric that simply counts the number of unshared poles between technologies.

Table 12.1
Summary descriptions of the three core dichotomies used in this analysis

Dichotomy	Description
(A)Symmetric	Does uniform processing occur in a single direction or omnidirectionally?
(Un)Controlled	Do decisions precede actions or do they execute under their own impetus?
(Meta)Data	Does processing focus on symbolic data or numeric metadata?

Building further on this approach, it might be possible to develop a complete *Atlas of AI*. There have long been atlases of the earth. Atlases of the brain are also now being developed (e.g., Sunkin et al., 2013; Mai et al., 2015). There is even a recent proposal for an across-species atlas of intelligences, although its ultimate form now seems more likely to be encyclopedic rather than dimensional (Boyle et al., 2022). From its title, the *Atlas of AI* (Crawford, 2021) may appear to move in this direction for AI, but as its subtitle—*Power, Politics, and the Planetary Costs of Artificial Intelligence*—highlights, its focus is actually quite different.

As shown in chapters 8–10, each of the three core dichotomies can serve as a cluster label that subsumes many additional variants already familiar across AI and related disciplines. Additional dichotomies from these disciplines may either map onto one or more of these three, as further variants, or form the bases for new clusters. Similarly, additional technologies may slot into locations in the dichotomic maps presented here, or reflect ways of spanning one or more of these dichotomies, or even challenge the proposed structure in new ways.

A complementary extension would involve digging more deeply into the existing variants within each cluster to understand what if anything further differentiates them—that is, understanding what aspects of them aren't fully accounted for by the cluster's centroid. At times such differences may simply implicate another of the three core dichotomies, as the difference between the (a)symmetric variants of reaction versus reflection and deliberation versus reflection in table 8.2 turns out to depend simply on the (un)controlled distinction between reaction and deliberation. At other times the difference may merely be one of level of abstraction, as in the contrast between the (a)symmetric variants of rules versus logic and rule-based reasoning versus first-principles reasoning, where the key difference is the abstraction of logic in the first pair to first-principles reasoning in the second. A third form of differentiation that has clearly shown up is whether the dichotomies relate to performance versus learning, as in the (a)symmetric variants of skill versus knowledge and skill acquisition versus knowledge acquisition—perhaps indicating the utility of a fourth cluster focused on this distinction. Mapping all of this out could provide additional structure while identifying aspects that may be missing from the current analysis.

Once criteria that are relevant to evaluating AI technologies and integrated systems, such as those listed in table 7.1, are added to the overall analysis, it becomes possible to explicitly consider trade-offs that would otherwise be implicit in these maps, whether within individual locations, across multiple locations, or between maps. Such criteria provide a concrete means of understanding the trade-offs embodied by each of the individual dichotomies as well as those arising from interactions among them. They also provide a means of summarizing the potential advantages of integrated systems composed of technologies that span multiple locations and of comparing and contrasting such systems in terms of their potential criterial advantages. Beyond analyzing existing systems, there is potential for new systems to be designed in a top-down manner, choosing combinations of technologies from locations in these maps that yield desired combinations of criterial advantages, although this remains very much a speculative possibility rather than a demonstrated reality at this point.

Although not all of the criteria listed in table 7.1 have been used in analyzing trade-offs here, there is utility in summarizing them all in one coherent list. These criteria provide an understanding of both why different subcommunities have adhered so strongly to one or the other pole of various dichotomies, as well as why supporting both poles in integrated systems may often make the most sense. It is quite likely that there are additional criteria of note that are missing from the list provided here, but this effort to identify criteria does at a minimum continue the progress begun via previous lists.

In understanding how to apply the various criteria as part of a dichotomic analysis, the criterial advantage (CA) language has been defined that enables populating maps with concise expressions for potential criterial advantages. The language is based on a set of constant symbols, each of which provides an abbreviation for a criterial advantage, such as AP for precision (as part of accuracy). Subscripts denote the narrowing of applicability of such advantages, such as ET_1 for restricting temporal efficiency to when a solution is sought for an individual problem, with multiple conjoined subscripts denoting a combination of such constraints. Commas in subscripts indicate alternative narrowings. Superscripts denote the number of times a criterial advantage appears in the cross product of multiple dichotomies. Parentheses are used to group related elements that are to be treated in a comparable manner.

One of the more interesting questions raised by this overall analysis is whether it is necessary for an integrated system to be complete with respect to the 3D map in table 11.19—or whatever its equivalent might be in a more extended dichotomic analysis—for it to be capable of human-level cognition. The 0D map of potential criterial advantages from such an approach, as seen in table 11.21, might argue in favor of completeness, and the analysis of Sigma has shown that it is possible for an integrated system to be complete in this manner, even if it does not necessarily capture every aspect implicated at these locations nor leverage all of the criterial advantages potentially available at them. However, the relatively minimal reduction in potential criterial advantages identified in less complete systems such as AlphaZero does imply that the answer to this question is not that simple.

As a final note, it is worth recalling that this whole analysis has focused on the basic processing required for intelligent behavior, emphasizing such capabilities as memory, representation and reasoning, learning, and search, while largely ignoring other key functionalities that need to be part of any complete picture, such as sensory, motor, and natural language processing. In a modular approach to intelligence, where each of these functionalities is implemented semi-independently via its own unique technologies, much more may be required to extend the dichotomic analysis to include them. However, a more elegant approach to intelligence might find a strong overlap with the dichotomies and technologies already explored here.

III Integrations: My Life as an Architectural Explorer

13 Introduction to Part III

Throughout my life I have been a seeker, but of *intellectual insight* rather than spiritual enlightenment.[1] By trade, I am an academic computer and cognitive scientist, with the core of my search having been the *nature of mind*, specifically, in a manner that spans both artificial and natural variants, as studied in artificial intelligence (AI) and cognitive science.[2] My first experiences with "AI" occurred in 1960, with two putative question-answering machines. One was a Hasbro *Think-A-Tron*—"the machine that thinks like a man"—which I received for Channukah that year (figure 13.1[3]). The other was a fictional UNIVAC—"an electric machine that can answer anything"—but which had a human with a typewriter inside to actually compose the answers (Parkinson, 1960). My first experience with cognitive science, at least by that name, was attending the inaugural conference on this topic at the University of California San Diego (UCSD) in 1979. More on my time at UCSD can be found below and in chapter 15.[4]

Later the search for insight went meta, into the nature of the *disciplines that encompass (the study of) mind*—from AI and cognitive science most nearly, moving outward to computing science[5] within which both AI and the computational aspects of cognitive science find a home, and ultimately to the full scope of science and engineering and, in fact, all intellectual endeavors. I have approached these various searches more as an *explorer* than as a traditional theorist or experimentalist, with the aim of capturing the resulting insights as immutable aspects of integrated *architectures* (chapter 2). For example, the lion's share of my work has revolved around *cognitive architectures* (section 1.2) that are to capture what is discovered about the nature of mind.

This memoir predominantly looks back at my life that was, but it also includes what is and helps to lay the groundwork for what may still come.

Figure 13.1
Hasbro Think-A-Tron.

In form, it is a *personal intellectual memoir* (Garnham, 2005). It is a memoir rather than an autobiography in focusing on an overarching theme rather than on all of my life. The theme in this case is bipartite:

1. *Substantive*: Mind and its encompassing disciplines.
2. *Methodological*: Seeking insight via architectural exploration.

It is illustrative to understand this theme in terms of figure 1.5, which lays out my career less thematically but in somewhat more detail, in terms of five levels of subject matter plus three forms of analysis. The top three levels of subject matter cover minds—the focus of the substantive subtheme—plus two key encompassing disciplines: computing (under the assumption that minds are a special case of computing) and science (under the assumption that computing is part of science). This thus matches the substantive subtheme quite well. The two bottom layers narrow the methodological approach for the study of minds to cognitive architectures—the exploratory focus throughout most of my career—and the common model of cognition, a community consensus over such architectures.

The methodological subtheme—of seeking insight via architectural exploration—is discussed more generally in chapter 14. In the original versions of this memoir there was a footnote about this subtheme to the effect that "this is a discussion that might more typically be found in an academic paper in philosophy of science than a memoir, but most of it has been worked out in the process of writing this memoir, it helps to understand the later material, and it provides a prime example of how I think at this point in my life, so it is included here." Some of the material was in fact later revised and published at an artificial general intelligence (AGI) conference (Rosenbloom, 2022a). That material plus more of the original content of chapter 14 can now be found where it more appropriately belongs in the context of this book—in chapters 1 and 2. Yet, relating this methodological subtheme to figure 1.5 requires the further understanding of how the reflective method can be seen as a variant of architectural exploration, where the architectures of interest are the essences discussed in chapter 1 and the methodology guides exploration of how these essences are developed and applied in the context of their respective topics.

This memoir is intellectual in that the main story line concerns my research career rather than emotionally evocative episodes. It also includes speculations in philosophy of (computer and cognitive) science that proved essential in developing and applying the notion of architectural exploration, plus references (although not always as would be called for in a technical work). It is personal in being about my life, larded with informality and citations to websites, and strewn with personal reflections (von Goethe, 1892; Voegelin, 1989), confessions (Augustine, 400; Rousseau, 1782), meditations (Aurelius, 180), maxims (Gracián, 1647; La Rochefoucauld, 1665; von Goethe, 1892), apologies (Cibber, 1740), aphorisms, and anecdotes.[6,7,8] Maxims play an especially important role here, with a numbered list of them included in appendix B, and cited throughout this memoir (for example, maxim 13, which has applied here, of writing what you want to say before considering where to publish it).

It thus takes on a hybrid form that combines both the professional and the personal. The main text block focuses on the overall theme, but it is accompanied by endnotes,[9] asides, figures, and an *interregnum*[10] that cover both personal matters and key aspects of my career that don't fit the overall theme. This additional material was originally intended to enrich, explain, and motivate the main story line, while remaining peripheral rather than central to it. However, with it included, the possibility arises of alternative paths through the memoir, focusing on just the overarching theme, my career as a whole, my personal life, or the full story. In early versions of this memoir, aspects of the story outside of the main theme were highlighted differentially in some manner. To simplify things typographically, this highlighting is omitted here, but the same choice of paths is still available conceptually, just by ignoring whatever material does not appear to be on the path to be followed.

To start with a bit of background, the most essential personal aspects of my life can be found in asides 13.1 and 13.2, covering my family and nature. The basics of my professional background comprise my education (but see aside 13.3 for more on the rationales for these educational choices) and subsequent academic positions. The former included a BS in mathematical sciences—a mixture of mathematics, computer science, operations research, and statistics—at Stanford University (1976),[11] an MS (1978) and PhD (1983) in computer science (CS) at Carnegie Mellon University (CMU),[12] and a year (1978–1979) as a nondegree visiting graduate student in psychology at the University of California, San Diego (UCSD)[13] that was sandwiched between the MS and any serious work toward the PhD at CMU. As a result, I was formally trained on both computing and the mind but remain little more than a fitfully educated amateur in philosophy of science despite it having become increasingly important in my later work.

Aside 13.1: My family.

Pride of place goes to my wife, Sharon Elaine Lee, who has attracted, advised, stood with, intrigued, and loved me for over fifty years. We met in the fall of 1972, during freshman orientation weekend at Stanford, at a pickup volleyball game outside of my freshman dorm (Branner Hall). Elaine—she went by her middle name at the time, Pride of place goes to my wife, Sharon Elaine Lee, who has attracted, advised, stood with, intrigued, and loved me for over fifty years. We met in the fall of 1972, during freshman orientation weekend at Stanford, at a pickup volleyball game outside of my freshman dorm (Branner Hall). Elaine—she went by her middle name at the time, although she is known as Sharon in a number of other contexts so I will use both here just as I have throughout our life together—had walked over from her own dorm to visit Sue Rosse, a high school friend of hers who was living in the room next to mine. Elaine and I remained together across years of schooling and intense early professional activities, albeit with of course some bumps such as three years physically apart during graduate/medical school, before marrying in 1986, and having one son, Michael Lee Rosenbloom.

The rest of my immediate family consists of two sisters—Kate Pohl and Anne Schardt. Kate is the oldest and Anne the youngest, but all three of us were born within around five years. My father, Arnold Rosenbloom, died from lung cancer the year before Michael was born. He was a technological optimist throughout his life, but the silver bullet he anticipated did not arrive in time. My mother, Florence Simon Rosenbloom, died recently at age 96.

Aside 13.2: My nature as a human being.[14]

With respect to overall personality, my son, Michael, characterizes the two of us as high-verbal introverts. Myers–Briggs-style pop-psychology personality tests ("Myers–Briggs Type Indicator," 2024)[15] typically grade me out as Introverted, iNtuitive, Thinking, and Judging (INTJ). This type is known as the Architect, Scientist or Mastermind. The first label is particularly appropriate given the titles of both this book and this part. The second is hopefully appropriate as well, even with the methodological focus on exploration. I've particularly never thought of myself in terms of the third.

My family as I was growing up described itself as "terminally honest," possibly due to (a self-diagnosis as) existing in a mild/high-functional region of the autism spectrum (at least for most of us). Two trivial bits of this are that I rock incessantly when thinking hard about problems, often distracting those around me, and I can be overly sensitive to sensory—particularly auditory—stimulation, making noise-canceling headsets a true gift.[16] Notions of fairness also run deep in my family. We aren't really capable of manipulating others and feel betrayed when we are manipulated in turn. The Golden Rule is the core of my behavioral code, although ideally modulated with understanding how differences among people also matter in how to behave with respect to them.

Two additional traits are likely also due to this position on the spectrum. First, I have some difficulty in reading people, which likely isn't helped by my eyes tending to go up and to the left when thinking about how to respond to someone rather than looking directly at them. Second, my first instinct when talking to others is often to take whatever they say literally and provide what answers seem to be called for without first understanding if they are teasing, or if there is hidden meaning, or if the question is even one that I should be answering. All in all, it shouldn't be surprising that I have always highly valued both honesty and sincerity.

(continued)

Aside 13.2 (continued)

I can be insecure and self-conscious at times, with the former inducing a yen for recognition. I am also a perfectionist by nature, which indirectly makes me highly competitive, as losing is a failure to be as good as you ought to be. So, I hate losing, but from a fairness perspective, I also dislike making others lose. So, frankly, I don't respect my own competitiveness. I have partly overcome this by engaging less in outright competitions,[17] a mixed blessing. I have also tried to rise above this trait by modeling my behavior on a vague understanding of some of the better parts of the traditional notion of an English gentleman. So, basically, I have muddled through. However, the perfectionism also imbues me with a tendency to worry and a long-lasting sense of regret for when I have failed to satisfy my own standards. Still, I do what I can to attain peace, remain calm, and keep on an even keel. I eventually learned to be at least somewhat more accepting of my own limitations and failures.

I am an optimist and an idealist rather than a pessimist or a cynic, although hopefully both are tempered with a sufficient degree of realism; and generally open minded and cheerful. I am, I believe, also good at thinking clearly and logically; making connections across disparate content and finding order in chaos; listening, understanding other perspectives, and enabling others to contribute; and being able to combine the articulation of big visions with attending to the niggling details (maxim 4).[18] I'm impatient when papers or presentations provide too low a bit rate—that is, I get bored when I must attend to information that arrives too slowly—and am not terribly good at being patient with people who are not clear and logical thinkers. I am not naturally facile at social interaction and often most comfortable when thinking on my own in front of a computer screen. Much of my life is lived inside my own mind. I am also not as innately generous as I would like to think of myself being but try to overcome this by acting more as I think I ought, and thus also more like a traditional gentleman.

As implied by much of this, I can easily get lost in thought, making me inattentive to the outside world and easy to startle. In conversations, I will often listen and think more than talk, and keep thinking on earlier points while the conversation moves on, and then interrupt later with thoughts on the earlier topic. With also not being great at determining when it is appropriate to speak up in a multiparty dialogue, my conversation style can be awkward at times. As a child I was known in my family for a penchant for sitting quietly and observing what was going on around me, as opposed to my sisters who were more verbally or physically active.

Aside 13.3: Educational choices.

My educational choices were always tactical rather than strategic, based on the immediate circumstances rather than a long-term plan. I attended public schools for K–12 wherever my family lived. This included three elementary schools, two in Los Angeles (the San Fernando Valley to be precise), where I was born in late 1954, and one near the Manned—later Johnson—Spacecraft Center in Texas, where my family lived in the mid- to late 1960s due to my father's participation in the Gemini and Apollo programs (aside 13.4); one middle school in Texas; and two high schools, one in Texas and one back in Los Angeles. It was thus a fragmented early educational history, but it all went well enough other than providing a challenge in developing long-term friendships.

My choice of Stanford for undergraduate education was partly from hero worship of my father, who earned his MS in electrical engineering there; partly from having been awed by it during a family trip when I was young; partly because when I visited my sister Kate when she was a freshman there, it seemed fun;[19] and partly because it was far enough away from home without being too far off. I was already enrolled as an undergraduate at the University of California, Los Angeles (UCLA) as a senior in high school, through a program available at the time, but it was much too close to home to want to continue there for the full undergraduate experience.

Stanford allowed delaying choice of an undergraduate major until the end of sophomore year, a degree of flexibility I found essential due to an abundance of interests at this point and little initial insight into which would ultimately fit me best. My choice of mathematical sciences was due to ruling out other early interests (aside 13.5) while finding it sufficiently natural and agreeable to my way of thinking.

Oddly, I never really considered whether to continue for a PhD after my BS. Another aspect of hero worship of my father had me assuming at four years old that I would get a PhD, and for whatever reasons I never later questioned this early assumption. Fortunately, it worked out, although I was a number of years into graduate school until I fully understood what it meant to do research toward a PhD.

Still, I did need to decide on both a discipline and a university for graduate work. I was considering two disciplines within my major of mathematical sciences—computer science, because its focus on abstract mechanisms and algorithms fit well with how I think, and operations research, because I appeared to be a natural at linear programming. I eventually picked computer science both because I thought it would be more fun and because it enabled

(continued)

Aside 13.3 (continued)

me to pursue my long-term interest in understanding minds through trying to build them.

I had not heard of CMU until late in my undergraduate career at Stanford, when I took a graduate survey course on AI taught by Cordell Green. The course included a video of Allen Newell talking about CMU's work on speech understanding. When I visited CMU as part of my decision-making process for graduate school, I was greatly impressed both with Allen's vision for AI and the warm and communal nature of its Computer Science Department. It was small, including only a PhD program, and had a wonderful camaraderie among the students, and even with the faculty. So, although I had not heard of CMU until the year before, and it was in Pittsburgh of all places, I ended up choosing it over MIT and Stanford. Fortunately, Allen turned out to be a terrific advisor (aside 15.5), and CMU turned out to be a wonderful environment in which to pursue a CS PhD.

The year at UCSD was originally motivated by a need to be near Elaine, who was in medical school there at the time, but also by home sickness for the sun and open skies of California. I turn out to be one of those people whose mood is all too dependent on the weather and, in particular, on the presence or absence of sunshine. Having grown up largely in California, I hadn't realized this until spending a few years in Pittsburgh. Fortuitously, this year at UCSD working with Don Norman also ended up being truly valuable intellectually. At a minimum, it taught me how to think like a psychologist and to appreciate the literature in its own right, rather than always approaching it from an AI perspective. More about the impact of this year on me can be found in chapter 15.

Aside 13.4: Living near the Manned Spacecraft Center (MSC) in the mid- to late 1960s.

My family moved to Nassau Bay, Texas, across the street from MSC, in 1965, shortly before Gemini IV, and returned to LA in 1969, shortly before Apollo XI. On arriving in Nassau Bay, I recall a boat parade on the local bayou for the Gemini IV astronauts, Jim McDivitt and Ed White, after they returned to their homes in the area. We later learned to "crab" on this bayou, using chicken wings attached to coat hangers.

Astronauts there, at least at that time, were viewed much like movie stars in LA. I recall attending school with several of their kids, babysitting Rusty Schweikert's five children—he flew on Apollo IX and was a friend of my parents. I also recall, although the memory may be apocryphal, Buzz Aldrin—the lunar module pilot on Apollo XI—visiting with my Boy Scout troop to share with us the kinds of food they ate in space. The film "Apollo 10½: A Space Age Childhood" (2025) provides a taste of what it was like to be a kid in that area at that time.

My parents designed a house on a small lake—the eponymous Nassau Bay. We ended up owning a rowboat and a small sailboat, both simply stored on the grass at the end of our lot, and learned to water ski. It was all great fun, and we even managed to ignore that the lake was infested with water moccasins. However, I did get more up close and personal with one when it hid in the grass until run over with a lawn mower.

Moving to Texas from LA was rather a culture shock, fun at first in experiencing Texas-style western culture, from bull's horns for my wall when I first got there to eventually a live horse (Star) for us kids, although without sufficient training in working with horses. In the Scouts I even carved a rather ugly but ultimately winning Pinewood Derby car, an activity surprisingly but appropriately highlighted on "For All Mankind" (2024). However, I never learned to appreciate the climate (I hate humidity!) nor ever really came to feel like I belonged (figure 13.2[20]).

Figure 13.2
Me as a nerdy ninth grader winning the gold medal in math at the Clear Creek High School (Texas) science fair (1969).

(continued)

Aside 13.4 (continued)

I was quite happy to move back to LA (figure 13.3), and not just because we stayed that first summer in my aunt and uncle's Malibu beach house while our new home in LA was being renovated. We watched Apollo XI's progress from the beach house, and I recall my father being quite relieved when the part he had been responsible for was done. Figure 13.4[21] shows my continued evolution once back in LA.

Figure 13.3
Me as a happy tenth grader back in Los Angeles (1969 or 1970).

Figure 13.4
Me as a barber-averse twelfth grader (1972).

Aside 13.5: Early interests and phobias ↬ academia.

I took the math and physics honors sequences as a freshman at Stanford, and while I did fine grade-wise, I quickly learned that, as opposed to what I would later find in computer science (or even linear programming), neither was truly natural enough for me to consider as a career. I also ruled out other early interests, such as history, law, and economics—the first two due to a fear of writing, a phobia that in retrospect appears largely to have resulted from ignorance about how to write, as I did eventually overcome it in graduate school, and the last due to a lack of sufficient interest after the initial courses.

My early fear of speaking in public, likely due to self-consciousness and the spectrum, never went completely away,[22] although decades of practice did moderate it, as did getting into the flow of a talk.

I never planned to go academic, and with early phobias concerning both writing and speaking, and no initial clue about research, I was clearly not a natural at it from the start.[23] Yet, when I was finishing my PhD, Allen surprised me by suggesting I consider universities.[24] Once I explored the options, I did end up taking a chance and choosing an academic position. The squiggly arrow in the title of this aside reflects this oblique outcome.

Eventually I became semi-comfortable at teaching and leading groups, and quite comfortable with writing and research, enabling a challenging yet satisfying career in which I could largely determine my own path.

My major academic positions have been as a research computer scientist[25] at CMU (1983–1984) (figure 13.5);[26] returning to Stanford as an assistant professor of computer science and psychology (1984–1987); and then moving to what would become the Viterbi School of Engineering at the University of Southern California (USC), first as research staff (1987–1988), then as a research assistant professor of computer science (1988–1990), followed by a tenured (associate and then full) professor of computer science (1991–2022), before ultimately becoming a professor emeritus (2022). During my time at USC, I also had a home at its Information Sciences Institute (ISI) (1987–2007)—where I actually started at USC as research staff before joining the CS department—and its Institute for Creative Technologies (ICT) (2008–present).

The overall inspiration for this memoir was a pair of talks given near the end of their careers by two other CMU computer scientists, Allen Newell and Randy Pausch. Allen was my PhD advisor, mentor, second father figure,

Figure 13.5
Me as a research computer scientist at CMU (circa 1983) in my standard-issue cinder-block-walled office.

colleague, and friend. He often had more confidence in me through the early years than I had in myself. His talk on *Desires and Diversions* (Newell, 1991) discussed styles of research careers, with an emphasis on his own style—of focusing on a single *ultimate scientific question* (or *grand challenge*) throughout his career, which in his case was understanding the human mind—and the trajectory it yielded when supplemented by a number of significant diversions into other topics.[27] I have shown a grainy old video of this talk to groups of students a number of times over the years to help them understand his, and to some extent my, career path and to help them think prospectively about their own careers.

Randy and I overlapped as graduate students, although he was several years behind. His talk, officially titled *Really Achieving Your Childhood Dreams*, went viral as *The Last Lecture* (Pausch, 2007) before appearing in book form (Pausch, 2008). According to Wikipedia ("Randy Pausch," 2024) "The talk was modeled after an ongoing series of lectures where top academics are asked to think deeply about what matters to them, and then give a hypothetical 'final talk,' with a topic such as "what wisdom would you try to impart to the world if you knew it was your last chance?'"

This memoir, which was begun at a critical juncture in my life—on the cusp of retirement and battling cancer (see aside 19.2 for more on this topic)—has been an attempt at an extended written version of such a lecture,[28] with much of the wisdom I would try to impart listed as maxims in appendix B. In the process, it has provided an opportunity to both review and frame my life and work, with the intention of including both the good and the bad, and both strengths and weaknesses to the extent that they seem relevant and that I understand them. One key outcome has been a better understanding of both me and my overall research approach/trajectory. It was not, for example, until this effort that I was able to articulate, or even to truly understand, either part of the theme that has ruled my career.[29] When later I was asked to give a *Remarkable Trajectory Lecture* (Rosenbloom, 2022b)—USC Viterbi School of Engineering's own version of a "final talk" series—this memoir was the foundation for my presentation.

The story line moving forward from here focuses on the substantive subtheme, concerned with mind and its encompassing disciplines. Rather than following a chronological ordering, this material proceeds through four tiers of architectural exploration that together comprise my work on this subtheme (figure 13.6[30]). The innermost two tiers (1 and 2) focus on the mind. The outermost tier (4) focuses on computing, while also laying the groundwork to go significantly beyond this to span all intellectual endeavors. The intermediate tier (3) turns out to supplement both the inner tiers' focus on the mind and the outer tier's focus on disciplines, although with the latter narrowing to AI and cognitive science. Each tier is distinctively characterized by the nature of the architecture(s) involved.

This figure graphically captures an alternative perspective on the bipartite theme with which this memoir began. Thus, as with the theme itself, it

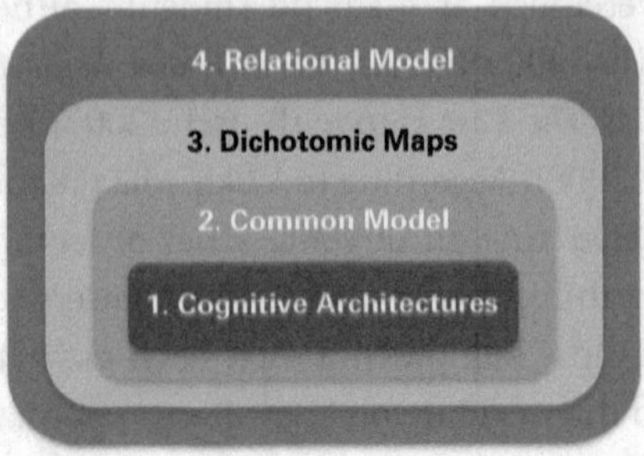

Figure 13.6
The four tiers of architectural exploration that comprise the substantive subtheme.

is helpful to relate it back to the newer and more comprehensive perspective presented in figure 1.5. Tiers 1 and 2 here, on cognitive architectures and the common model, map onto the bottom two subject matter levels, with this then broadening on up in figure 1.5 through the additional subject matter levels that focus on minds, computing, and science. Tier 3 then brackets the three most specific levels in figure 1.5, via dichotomic analysis (part II), whereas tier 4 brackets the two most general levels in figure 1.5 via relational analysis. Essential analysis, which does not appear in figure 13.6, turns out to bracket all five levels in figure 1.5, with its application to the three most specific levels found in part I and its application to the two most general levels, in defining terms such as computing, science and great scientific domains, to be found in chapter 18.[31]

With this mapping between the two figures—and thus, these two related perspectives on my career—in mind, the remainder of this memoir will continue to be structured around figure 13.6 rather than figure 1.5. Chapter 15 thus covers tier 1, concerning the grand challenge that has consumed most of my career (maxim 15), of developing cognitive architectures that aid in understanding the human mind and in understanding and building artificial minds. In particular, a sequence of three cognitive architectures—XAPS (Rosenbloom, 2006) → Soar (Laird et al., 1987) → Sigma (Rosenbloom et al., 2016a)—each of which perhaps might be better thought of as an *architectural lineage*,[32] will be examined, along with a bit of relevant work that occurred prior to this sequence. This work aligns naturally with Allen Newell's talk, given that much of the early segments were done in collaboration with him, and with a grand challenge akin to his ultimate scientific question.

Chapter 16 covers tier 2, concerning the common model of cognition.[33] It is quite short—barely more than a stub—due to shifting much of its content from earlier versions of this memoir to section 1.2.3. What is left here focuses on historical and social aspects of the work. Chapter 17 covers tier 3, concerning dichotomic maps, albeit also in a very condensed manner compared to prior versions of this memoir given the extensive coverage of this topic in part II. This is the newest body of work covered in this memoir, as the work leveraging essential analysis in part I largely occurred after this memoir was written, and there has not been a compelling need to extend the memoir further to include it.

Chapter 18 covers tier 4, concerning the relational model, which provides an architecture that is to span all of computing (Rosenbloom, 2012a) and potentially beyond that to all intellectual activity. Until writing this memoir and formulating its substantive subtheme, I had considered tier 4 as a diversion rather than as thematic; however, with the relationship of the relational model to both essential analysis and the reflective method, the overlap in how disciplinary maps in tier 3 and the relational model in tier 4 attempt to understand whole disciplines, and the architectural perspective that is common across all four tiers, my perspective on this has evolved.

Chapter 19 concludes with reflections on this memoir's theme. It also starts to consider what these four tiers and their interrelationships imply about my career so far and what it might make sense to continue pursuing into the future.

14 Methodology: Seeking Insight via Architectural Exploration

Methodologically, I am too often thrilled neither by new theorems and proofs, as would befit a theorist, nor by new experiments and their results, as would befit an experimentalist. Part of this is certainly that those that can be accomplished, which are largely on parts of architectures, so often do not get at the issues concerning architectures that I truly care about (see the discussion of criteria in section 2.6). However, part of my disinclination toward traditional theory may also be due to reaching an understanding as an undergraduate that, despite what might seem to be implied by figure 13.2, I wasn't going to be a mathematician (aside 13.5).[1] Another part may be due to too many attempts at formalization in AI that appear to have obscured rather than clarified the points being made. My disinclination toward traditional experiments may additionally be partly due to their inherent messiness.[2]

What turns out to really turn me on intellectually are new *insights*—*aha* moments that yield a new way of looking at things (maxim 1). This can be seen as part of what Ritz and Vaidyanathan (2025) recently termed the *beauty of understanding*. Their article was even titled "Aha=wow." Another recent article provides evidence for this as a third route to a good life, beyond the traditional two of pursuing *happiness* (*hedonics*) or *meaning* (*eudaimonics*), that instead seeks "*a psychologically rich life*—or a life characterized by a variety of interesting and perspective-changing experiences" (Oishi & Westgate, 2021). In other words, it proposes the rewards of being an *aha junkie*. Meaning still does matter to me, coming from pursuing, via architectural exploration (section 2.6)—and more recently the related reflective method (section 1.1)—impactful topics in which I have sufficient ability to make a difference (maxim 14). Traditional notions of happiness have played a much lesser role in my intellectual life, although they have not been totally absent.

Perhaps this makes me inherently prescientific, or nonscientific, or even a bit of a Kuhnian revolutionary (Kuhn, 1962) in always longing for a paradigm shift in my own thinking, but I get little thrill from either formalization or precision. As mentioned in chapter 13, my undergraduate degree was in mathematical sciences, and I have run a range of experiments over the decades, so it isn't that I don't understand these activities or can't be competent at them. But to me they are at most a means to an end rather than an end in themselves; and too often we seem to lose sight of the ends by overfocusing on such means, or metaphorically losing sight of the forest due to the trees. In combination with my natural inclinations, this has implied that although I deeply appreciate the role of formalizations and experiments in solidifying our understanding of the world, neither has ever been at the center of my own endeavors.

In search of insights, I have found myself repeatedly drawn to a research methodology that combines *exploring* complex phenomena with seeking to impose order on them by identifying a fixed conceptualization—an *architecture* (chapter 2)—that both embodies any core insights and authorizes a space of more flexible *variable content*[3] that flesh out its capabilities. As discussed in section 2.2, the combination of an architecture and a body of variable content yields a *system*, which can be semi-formalized as *system = architecture + variable content*. Such systems may be conceptual, mathematical, computational, physical, and so on. An *application* is then simply a system that yields value of some sort. If knowledge is power—in the everyday sense of the term rather than the technical sense introduced in chapter 5—and applications are payoff, we can think of architecture as *essence*, relating it back to the core of essential analysis in part I.

Power for power's sake has rarely appealed to me, even if it implies the ability to leverage that power for desirable aims. Although I wouldn't mind being wealthy—one obvious form of applications payoff—I've also rarely found it attractive as a goal. Developing applications that matter to funders, and beyond to society as a whole, is a much worthier goal, and often essential for acquiring and sustaining research funding, but one I have typically found to be a duty rather than an interest. Instead, getting at the essence of complex phenomena, whether for understanding them or producing them, always has thrilled me.

Thus, by nature, I am an architecture person who has too often given less than their proper due to the variable content or systems that are thus enabled (maxims 19 and 25). At various points in my career, I have put effort into variable content and systems, but they were often not the happiest of times, including one extended effort that likely contributed to a significant burnout in the late 1990s (see chapter 15 for more on this). Yet, underemphasizing these two aspects has likely also held me back in significant ways, such as by losing the architectural insight that can result from a thorough study of the variable content and applications it enables (or fails to enable),[4] reducing the availability of important forms of research funding, and denying the pathway of entrepreneurship that has been taken so productively by many of my colleagues. Such consequences are a major risk of focusing too exclusively on architectures.

Beyond simply being an architecture person, I have also always cared deeply about the simplicity, beauty, and elegance of the architectures I create,[5] even when I did not consciously realize this. These are notions more typical of physics than CS/AI or cognitive science and are even further from what is the norm in the biological sciences, including neuroscience, where the notion of evolution as a tinkerer (François, 1977) can imply an expectation for an architecture that might informally be characterized in CS terms as "hacks upon hacks." Even in physics the idea can be controversial (Hossenfelder, 2018), but if you don't search for beauty before settling for less you will likely never find it. And such a search should be less controversial when it is merely one path being followed within the overall community rather than the dominant culture that can make other paths appear less than respectable.

To a physicist, beauty may focus on the mathematical form of a theory. For me, the core question is whether it is possible to understand and/or produce a wide range of complex phenomena of interest via conceptually clean interactions among a small orthogonal set of fixed ideas or mechanisms plus appropriately limited variable content (maxim 20). In chapter 5, this was referred to as *exuberance*.

In my explorations into the Sigma cognitive architecture, I came to call this *functional elegance*. It is clearly related to Occam's razor—whether you spell his name in this manner or as the more erudite Ockham—but with a different emphasis. It also relates to Allen Newell's earlier maxim that one should *listen to the architecture*.[6] An important part of this was exploring what the existing architecture implies about a new capability before designing a new mechanism from scratch for it (maxims 16 and 21–22). In this way, it actively prioritizes mechanism reuse and adaptation,[7] and thus a reductive strategy rather than mechanism addition, as was highlighted in the discussion of deconstruction in section 5.3.3 and was further discussed in section 6.2 as a means of spanning the dichotomies introduced in part II. But functional elegance goes beyond just these incremental questions, of how an architecture can be extended to new capabilities, to concern the ultimate beauty of the resulting architecture as a whole (maxim 20), including its expression in terms of the most fundamental concepts or mechanisms that can be identified (maxims 22–23).

The extent to which I have achieved functional elegance across the various architectures I have developed is of course an independent question. The apparent lack of further avenues for such progress on the Soar cognitive architecture in the late 1990s was a second factor in the burnout just mentioned, and to my stepping away from not just Soar, but from cognitive architecture research as a whole for a decade (maxim 17). I began work on Sigma in the late aughts because I thought I saw a path for going way beyond what could be done in this manner with Soar. Whether or not Sigma is ultimately the best way to approach understanding and building minds—for example, the appeal to beauty could be misleading in designing cognitive architectures and/or Sigma may not end up be the most beautiful approach to doing so—Sigma has at least extended in fundamental ways our understanding of the functionally elegant approach to cognitive architectures.

Simplistically, exploration seeks insights and architectures capture them. A good example is the work discussed in section 15.1 on modeling the *power law of practice*. From exploring a regularity in human behavior an architectural learning mechanism was developed, called *chunking*, that could be used across a broad range of tasks. Yet, the full relationship between exploration and architecture is more multifarious than is indicated by this simple aphorism. Architectures can, for example, guide the exploration and interpretation processes that lead to new insights. They can also guide decisions about how and when the architecture itself needs to be modified.[8] As mentioned earlier, Allen Newell called this listening to the architecture. A more systematic exploration of this topic can be found in section 2.6, but a few additional examples may be helpful here.

- Linnaean taxonomy (Linnaeus, 1735) has aided in interpreting what is uncovered during exploration of the natural world. Although the notion of a taxonomy provides too weak an architecture to yield much guidance in exploration, once a new species is discovered the taxonomy can help interpret it in terms of the hierarchy of known species.
- The understanding of *data chunking* in Soar (section 15.2), as an interaction between search and chunking, arose from identifying major gaps in the architecture's ability to support declarative (fact) learning and memory. In this case, listening to the architecture led to an approach that surprisingly didn't require adding new architectural mechanisms. However, as will be discussed in aside 15.7 it was ultimately superseded in the late version of Soar by new declarative memories and learning mechanisms (section 1.2.1).
- A much less satisfactory example concerns how I struggled for years to implement a new architectural mechanism for chunking in Sigma, attempting to use the architecture as it has evolved over this time to guide how best to do this. Unfortunately, it is an example for which the guidance has never been quite sufficient, or, perhaps, simply an example of where I have been too tone deaf to hear it properly.[9]
- A much more positive example from Sigma is how two other capabilities that were already well studied in the field but that were new to Sigma—reinforcement learning (Rosenbloom, 2012b)[10] and episodic memory and learning (Rosenbloom, 2014)—were deconstructed in terms of combinations of preexisting mechanisms. As with data chunking, what were traditionally distinct modules were instead reinterpreted as combinations of more basic mechanisms. These novel reinterpretations yielded some of the strongest *aha* moments of my career, moments when it felt like a deeper understanding was achieved based on taking a new perspective.
- Another pair of positive examples from Sigma were how its basic architectural mechanisms were extended to handle feedforward neural networks, without adding a distinct neural module (sections 1.2.2 and 15.4), and to compute *surprise* as part of a theory of emotions via a side effect of gradient descent learning at factor nodes (Rosenbloom et al., 2015).
- A more recent push in Sigma to explore emotion, motivation, and personality exemplified how understanding a combination of the major gaps in an architecture's coverage—in this case concerning "personal"

aspects—and what topics it may be most ready to handle (or be extended to handle) can lead to appropriate selection of topics to explore and corresponding architectural modifications to implement.

- The *common model of cognition* (section 1.2.3 and chapter 16) guided new analyses of functional brain scans that have been used to compare it to more traditional approaches from neuroscience for modeling the high-level architecture of the human brain. Fortunately for the common model, it turned out to dominate the traditional approaches (Stocco et al., 2021).

Beyond simply adding mechanisms or capabilities to existing architectures, it is also possible to talk explicitly about exploring the space of architectures. This can amount to, for example, searching for a computer architecture that enables secure, energy-efficient, high-speed computation; or iteratively developing a cognitive architecture that yields a good model of the fixed structures and processes underlying the human mind. A lineage of architectures can thus be considered as a search within a relatively compact region of the space of all possible architectures, with each distinct lineage searching its own region of the overall space.

The cognitive architectures I have developed over the years have partially been intended as expressions of, and means for testing, hypotheses. Often these hypotheses start out rather vague,[11] such as some notion of the potential benefits deriving from the combination of: rules and neural-like activation in Xaps (Rosenbloom, 2006); rules, problem-solving, and learning-by-chunking in Soar (sections 1.2.1 and 15.2); and Soarlike cognitive architectures and probabilistic graphical models in Sigma (sections 1.2.2 and 15.4). Sometimes this develops into a somewhat more precise—and even at times grander—hypothesis, such as that chunking provides a general learning mechanism (Laird et al., 1986a) or that Soar is an architecture for general intelligence (Laird et al., 1987) that can support a *unified theory of cognition* (Newell, 1990). More often, crisper small hypotheses are spun off, such as that chunking in Soar could be used not just to speed up performance but also to learn new things (Rosenbloom et al., 1987), or that Sigma could straightforwardly be extended to incorporate neural networks (NNs; Rosenbloom et al., 2016b; Rosenbloom et al., 2017).

Each of the other architectures I've worked on—that is, ones that don't amount to traditional cognitive architectures—also embodies some form of grand yet vague hypothesis, as I've always found it more satisfying to work on entire forests rather than on individual trees. The common model of cognition is based on the hypothesis that there is useful commonality hiding behind the surface differences in many cognitive architectures. Dichotomic maps (part II and chapter 17) are based on the hypothesis that AI, cognitive science, and cognitive/A(G)I technologies and architectures can be understood in terms of the cross products among a small set of broadly applicable dichotomies. And the relational model (chapter 18) is based on the hypothesis that computing as a great scientific domain can be understood in terms of a small set of relationships with itself and the three other such domains—the physical, life, and social sciences. The same bigger-picture criteria apply here, but controlled experiments play an even smaller role. Instead, exploration is the dominant methodology.

To many of a more rigorous bent this likely seems like a fatal flaw (maxim 11), and driven by general community angst over this, there have been a number of workshops and papers over the years concerned with how to rigorously evaluate work on cognitive architectures. Yet none of them have truly succeeded. I have worried about this myself at various times over my career, with my latest thoughts on the topic being summarized in sections 2.6, 4.2–4.3, and 5.4.3, plus the material on evaluating and comparing integrated systems in part II; however, in understanding myself as an architectural explorer the point is to accept the resulting limitations and instead celebrate what can be yielded by such a methodology (maxim 12). This doesn't rule out such approaches as controlled experiments when doable and relevant, nor formalization—or semi-formalization, as found for example in the ME (chapter 18) and CA (part II) languages—when it appears helpful and I can understand how to do it, but it shifts the focus from such activities to exploration and the bigger-picture criteria relevant to it.

Still, it is also true that as a discipline matures, its architectures should as much as possible become more in the way of rigorous theories while the explorations themselves should morph increasingly into controlled experimentation. The relationship between architecture and exploration can be thought of as analogous to, and a precursor of, the more traditional relationship between theory and experiment. An architecture provides the framework that guides and summarizes exploration, while an exploration can help refine and flesh out the architecture. Early natural scientists, were, for example, accompanied in their explorations by simple architectures with associated variable content, such as maps and taxonomic hierarchies. While these explorations largely focused on further populating the architectures with variable content, such as new locations and species, they may also have helped to refine the architectures themselves.

My own focus, instead of being on the natural world, has been on the world of *information*, in terms of *computing* and its kissing cousin, *thought*. Simon (1969)[12] referred to these as *sciences of the artificial* that involve understanding what we create. But in my own attempts to understand the discipline of computing I ended up taking a more aggressive tack that acknowledges the existence of natural forms of computing (Denning, 2007) while arguing that all four great scientific domains—the physical, life, social, and computing sciences—inherently involve an inextricable intertwining of understanding and creating, and thus both natural and artificial phenomena, with computing only being somewhat out ahead of the other three in terms of its ability to create instances of the phenomena it studies (Rosenbloom, 2012a).

As discussed in section 2.6, exploration was the name of the game in the early years of AI, when there were many more unexplored areas than researchers to look into them and it was more fruitful to move on to new topics once initial explorations of existing ones yielded sufficient low-hanging fruit than to work over each existing topic very carefully first (aside 14.1). Now, however, exploration has fallen into relative disrepute as a method of investigating topics in AI—with a broader coverage of the space of topics presently under our belts, proofs and controlled experiments have become the coin of the realm.

Aside 14.1: Exploration and exploitation.

Reinforcement learning (RL) is an AI technology that learns to choose appropriate actions for tasks based on rewards received during experience in performing the tasks. One key aspect of it is the trade-off between exploration and exploitation. Exploration spreads out experience to uncover what might be found down (largely) unexplored paths whereas exploitation focuses on payoff from the most promising paths. In other words, exploration searches for better possibilities whereas exploitation seeks payoff from these better possibilities. Balancing the two appropriately can maximize total long-term reward. Science, like RL, cannot work effectively without exploration, even if the recognized payoff is mostly in the theory and experiments—or even the applications—that comprise exploitation.

Rosenbloom (2012a) argued that the existence of stronger research methods—in terms of the level of veracity they can guarantee—within a discipline tends to reduce the acceptance of weaker ones (maxim 11). This is completely appropriate when both strong and weak methods apply to the same problem, but it tends to carry over even to those areas of the discipline in which the stronger methods don't (yet) apply (maxim 12). Erecting new disciplinary boundaries that wall off the latter areas can sometimes protect them from thus being ignored due to no strong methods being available for them, but there is also then a concomitant risk of the walled off areas becoming insular and irrelevant.

One of the reasons I have enjoyed attending conferences on artificial general intelligence (AGI) over recent years is that the spirit of exploration is still very strong in this relatively new community that has returned to AI's original grand challenge of building systems capable of general human-level intelligence (or beyond). Much of the work that goes on in that community strays from what is considered academically respectable, but my mind is often more engaged by the thought-provoking work that is presented at such fora—and the potential for new insights often seems higher—than at many traditional AI or cognitive science conferences. It is worth noting that by relabeling themselves as AGI, rather than thinking of themselves as part of AI, they have achieved some of the necessary separation mentioned in the previous paragraph but also accompanied by the dangers inherent to such balkanization. I have dealt with this threat personally by spanning the divide—that is, by attending both AGI and more traditional academic conferences. I approach both with the goal of unearthing what insight might be available, while always trying to keep in mind how well the various possibilities are actually supported. It is now interesting to see how the overall topic of AGI is being approached anew from the perspective large language models (LLMs) (e.g., Bubeck et al., 2023; Agüera y Arcas & Norvig, 2023). Section 4.3.2 speculated further along this line, in analyzing the relationship between LLMs and intelligence in general.

My ongoing dedication to exploration is another reason why funding was more difficult during the latter part of my career (maxim 11), while making publication more difficult as well. Finding some useful experiments to add to papers when considering publication helps with this latter issue to some extent, but in general my later papers have appeared more in niche venues rather than in leading ones. Even leading venues that eagerly published papers that were largely exploratory in the early days of AI no longer consider such work appropriate (maxim 11). Yet, in understanding and accepting this self-architecture—as an architectural explorer—late in life, I freed myself to work on the problems I most care about via the methodology with which I am most comfortable (maxim 19).

15 Grand Challenge: Cognitive Architecture [Tier 1]

I have spent most of my career on the grand challenge of a unified understanding of thought in natural and artificial minds that enables modeling the former and building the latter (maxims 14–16); that is, on cognitive architectures (section 1.2) that span both A(G)I and cognitive science (maxims 5, 7, and 10). We now refer to such architectures as *humanlike*, as they span both human minds and artificial minds that are sufficiently similar to them (Laird et al., 2017). Similarity is a squishy concept here, but if you think of the space of all possible minds (chapter 4), humanlike minds should fall within a single compact region, or perhaps on a single hill if you consider hill climbing in the search for models of intelligent behavior.

The original goal for AI was to build complete artificial minds, but most of the field has focused on only individual pieces over the years, with the recent piece being almost exclusively machine learning (ML).[1] To reuse a metaphor, in searching for a unified understanding of thought, the focus is metaphorically on the forest of the overall mind rather than the trees of the individual capabilities into which it might be partitioned; the trees are typically attended to only in terms of how they can be combined, and how they can be developed in such a way as to be combinable, to yield a fully functioning forest. There are times when addressing the forest becomes fashionable (maxim 18), but most of the time those of us who view it as a mission rather than a fad are considered eccentric at best. We are also considered methodologically suspect, both because it is harder to prove theorems or run experiments on the whole forest (maxims 11–12) and because the focus on the forest yields neither as precise a picture of the individual trees nor narrow applications that are as rewarding.

Following in the footsteps of Allen Newell and Herb Simon, and the cognitive approach to AI more generally, I have also long believed that there must be at least one path toward unified AI that follows the blueprint laid out by human minds, at some appropriate level of abstraction, so that pursuing both fields in tandem is not only possible but enables pooling insights from both toward a single melded result (maxim 10). This is part of what has led to the focus on humanlike minds in the common model of cognition.

Working on a grand challenge that seeks to unify across both capabilities and disciplines requires delving into many subareas of each of the disciplines, inevitably yielding a broad background and perspective on them, while at the same time—and in multiple ways—making one feel like an outsider with respect to both disciplines (aside 15.1).

Aside 15.1: Being an outsider.

I have felt like an outsider throughout most of my career, and life. Several aspects of this derive from pursuit of this grand challenge. First, when pursuing any grand challenge, what matters is whatever is relevant to conquering it. Yet, much of what is necessary to succeed in such an effort can appear irrelevant to the majority who either don't pursue a grand challenge at all or who pursue a very different one. Second, seeking a unified understanding of thought in terms of cognitive architectures implies a focus on the minority approach of *general* rather than *narrow* AI (and cognitive science), and the associated approach of trading off local optimization of individual capabilities in favor of global optimization for general humanlike intelligence. Third, approaching natural and artificial minds as two sides of the same coin leads to push back from much of both fields with respect to not sufficiently optimizing the results for their respective (precision) goals of matching human behavior versus producing optimal computational behavior. Fourth, spanning AI and cognitive science, as with many other such interdisciplinary efforts, implies that you are neither quite fish nor fowl.

Another aspect of this outsider sense is due to splitting time and allegiance between an academic computer science department and an off-campus research institute. Situating my research at institutes that were not limited by traditional disciplinary bounds or notions of what basic research is facilitated many aspects of tackling the grand challenge. However, in the process, this split also tended to make me an outsider in both environments, particularly as a tenured faculty member making such a split.

Beyond these professional aspects, even early in life I often felt like an outsider. Part of this may just be the human condition, but it was enhanced by a range of factors, such as being a self-conscious introvert, on the spectrum, and a Jewish (even if barely so) Californian in southeast Texas during my early teens (see asides 13.2 and 13.4). Whatever the original causes, I ended up accepting it as integral to my identity, and even feeling an odd sense of pride in being different. This may have made me more willing, or perhaps even eager, to consider intellectual paths that were outside of the mainstream. It may have also made me feel like an outsider in personal and professional contexts that didn't necessarily require it, and thus unintentionally limited the level of interpersonal bonding that might otherwise have occurred.

I began my graduate career at Carnegie Mellon University (CMU) in 1976,[2] working on a project with Doug Lenat that eventually became Eurisko (Lenat, 1983), the predecessor to Cyc (Lenat & Guha, 1989). Eurisko would become an architecture for automated exploration—or *discovery*—within the space of concepts and heuristics. Although little progress was made in the year I was with the project,[3] it did provide an impetus to read Pólya's (1945) classic work on heuristics.[4] It was also my first involvement with something like a cognitive architecture. Eurisko was principally intended to be a computational architecture, although it could also be considered theoretical with respect to the abstract capability of exploration or discovery.

The next year I switched advisors to Allen Newell[5] and joined the Instructable Production System (IPS) project. The goal here was to develop an AI architecture based on *productions*—or *rules*—that combine context-sensitive conditions with actions to be executed in the specified contexts. Such systems can be very reactive, in terms of their ability to respond appropriately to whatever the current context is. They differ from traditional stimulus-response (S-R) systems primarily in including an internal *working memory* as part of the overall context.

The idea driving work on IPS was that you shouldn't have to program such a system, instead it should learn automatically how to perform tasks in a limited domain—in this case, job-shop scheduling—from instructions provided to it in English. Production systems seemed particularly appropriate for such a project because each production provides an independently learnable, local bit of knowledge. They turn out also to have roots in both theory of computation (Post, 1943) and the modeling of human cognition (Newell & Simon, 1972).

My contribution to IPS, other than participating in group meetings, was to help program a very small part of one of the early OPS—that is, the Official Production System of the IPS project—rule-based systems that provided the architecture for IPS. Rule-based systems thus became the first lineage of cognitive architectures—albeit[6] relatively minimal ones—with which I had any hands-on involvement.

The goal of the project unfortunately turned out to be terribly overambitious for the time, and ultimately premature by several decades. As Allen put it in *Desires and Diversions*, the IPS project was a classic failure, being described via a prospective paper (Rychener & Newell, 1978) and a retrospective paper (Rychener, 1983), but without any intervening papers containing actual results. The project did, however, develop rule systems, such as OPS5 and its underlying Rete algorithm (Forgy, 1982), that would be used as architectures for a number of expert systems during the 1980s boom. IPS would also articulate many of the questions concerning the relationship between knowledge and problem-solving that would later be addressed with the development of the Soar cognitive architecture. OPS5 also played a central role in the implementation of Soar's long-term procedural memory. Much later, goals similar to those of the IPS project were revisited in the form of Interactive Task Learning (Gluck & Laird, 2019). Large language models (e.g., Brown et al., 2020) can also be seen in a similar light.

Following these two largely unproductive research years,[7] I spent a year (1978–1979) as a visiting graduate student in psychology at the University of California, San Diego (UCSD), working with Don Norman. Although the initial inspiration for this move was largely personal (aside 13.3), it did enable me to truly learn about experimental and cognitive psychology—a long-term desire that previous attempts had failed to satisfy[8]—and to embed myself in both the topic and the mindset. Serendipitously, it also turned out to be the year that people like Jay McClelland, David Rumelhart, and Geoff Hinton[9] began the Parallel Distributed Processing (PDP) journey at UCSD that reinvigorated work on neural networks in cognitive science and AI. It was also the year that the first annual conference of the Cognitive Science Society was held at UCSD.

15.1 XAPS

While at UCSD, inspired by what was going on in the psychology department there that year and attempting to fuse this with what I had learned the previous year from the IPS project, I began exploring the first of three eXperimental Activation-based Production System (XAPS) architectures—the first lineage of cognitive architectures in which I took a leadership role.

The core idea behind XAPS was that rules would not only generate new symbol structures conditionally based on existing ones but would also guide the flow of neural-like activation associated with these symbol structures. In the terms of that era, these were local connectionist models (Shastri, 1988) rather than the distributed vector models that have since dominated the field. Still, they provided an initial experience with exploring what happens when an architecture includes symbols and associated quantitative values, an idea that reappears in subsequent explorations into cognitive architectures, and which becomes fully embraced in both Sigma and the common model of cognition.

One of the things I learned from this effort was that unconstrained activation processing is a morass within which it is sometimes possible to produce what appears to be magic, but which most often leads to an uncontrollable mess. Eliminating the morass requires an approach such as constraining the processing to match the semantics of probability theory, as in probabilistic graphical models (Koller & Friedman, 2009), or taming it by leveraging a learning algorithm that molds it to align with experience, as in the later development of the backpropagation algorithm (Rumelhart et al., 1986) that is central to deep learning. Ad hoc programming can work wonders with symbolic systems but won't do much of use in such activation-based systems beyond enabling a few toy examples.

When I returned to CMU in 1979 from my year at UCSD,[10] I continued with XAPS. John Laird—another student of Allen Newell's who had also participated in the IPS project—and I started to bat around ideas for how to create a general cognitive architecture (GCA).[11] But more pragmatically, by this point I had very little sense of a useful direction in which to head with XAPS. Allen helpfully reoriented me when he asked for assistance with a project concerned with the power law nature of human practice curves—simplistically, $T = BN^{-\alpha}$, where T is the time to perform a task, N is the number of times it has been performed, and α and B are constants. The general power law we worked with though was $T = A + B(N + E)^{-\alpha}$, where A is the asymptote of performance and E is the effective number of practice trials that occurred prior to the experiment being modeled.

The goal here was first to establish whether human practice did indeed show this regularity across all of the tasks for which we could either find or generate practice curves, and if so to then use this invariant from human behavior as inspiration for building an architectural learning mechanism that would inevitably yield such curves when applied to comparable tasks. This could be viewed as a classic instance of computational cognitive modeling, but it was also inherently architectural by focusing on invariants and fixed mechanisms. In this form, it ultimately led to a pragmatically useful learning mechanism for AI systems (maxim 10).

The first step on this journey (Newell & Rosenbloom, 1981) turned out to be rather impactful, with over 3,300 citations (Google Scholar, 2025), even though it was published as an invited book chapter—as were many of my early papers—rather than as a rigorously reviewed conference paper or journal article. Invited articles were a particularly amenable venue for exploratory work, although this paper did include many statistical analyses that examined the comparative accuracy of different functional forms—such as power laws, exponentials, and hyperbolics—as models for practice curves. It might have made a good journal article had it not been written in response to a specific invitation to Allen. However, journal articles were not so much the focus in AI in those days. Publication in books and conferences was much more the thing. There was always so much new to explore that taking the time to carefully produce and go through the comment-and-revision cycles typically required for journal articles was the exception rather than the rule. Journal articles were more the standard in cognitive psychology. But, without a focus on new experiments, this work might not have been a good candidate there, even with the impact it has since had in psychology and cognitive science.

As is clear from the number of citations, there was much follow up on this work. Much of it simply used the article to support the power law nature of practice curves, but others challenged whether practice curves were truly power law or whether our explanation for why they were power law was the correct one. Still, the citation numbers keep growing. I was invited to write a paper (Rosenbloom, 2006) for a special issue of *Tutorials in Quantitative Methods for Psychology*[12] in celebration of the twenty-fifth anniversary of this paper's "seminal" publication (Lacroix & Cousineau, 2006).[13]

I must admit, though, that just about all of the intellectual horsepower reflected in the original paper was Allen's, with me very much in a junior support role. Most of my effort involved developing and applying a statistical analysis package—which didn't exist in any standard form at the time—that would search the parameter spaces of power law and other curves to find the best fits to the data from the various experiments we were modeling.

The explanation for the power law of practice that we included in this paper was based on the existing notion of *chunking* in psychology, of combining multiple distinct memory elements into a new compound one (Miller, 1956). We adapted this to combine sequences of rule firings into new rules that in one step accomplished what originally required multiple steps (aside 15.2). This was ultimately developed into a domain-independent architectural mechanism in Xaps3, an architecture without activation despite still having "a" in its name.[14] But the process of getting there was one of the most challenging and frustrating periods in my research career. It seemed like there was at least a full year during which I made essentially no progress, spending much time pacing my office thinking, or walking around neighboring Schenley Park doing the same.

Aside 15.2: Exponentials versus power laws and implications for the intelligence singularity.

One general lesson from work on the power law of practice is that data that might appear exponential may be more accurately modeled as a general power law (maxim 30). An exponential requires a constant rate of improvement at each step, as with compound interest. With a power law, the rate of improvement decreases over time. With the power law of practice, the assumption was that it became necessary to learn larger and larger chunks over time, with these larger chunks getting rarer and rarer. Over a small enough time span, over which the decrease in the rate of improvement is small enough, such a curve may appear to be exponential.

Might a general power law actually turn out to be a better model for improvements in AI agents than the exponential hypothesized to underlie the intelligence singularity (section 4.3.1)? The two models would agree on there being portions of the curve that appear exponential, but the former would assume a slower rate of growth overall plus an asymptote. The skepticism expressed about the intelligence singularity in section 4.3.1 concerned whether there is sufficient headroom above human-level intelligence for continued growth at an exponential rate. Here, this can take the more explicit form of hypothesizing an asymptote; that is, that there is a limit on how intelligent it is possible for any agent to be that may not be that much above what we find in humans.

(continued)

Aside 15.2 (continued)

There may also be a slowdown in the rate of improvement, converting an exponential to a power law. In large language models, for example, this may be due to a slowdown over time in the rate of architectural improvements, or in the rate of increased availability of computational cycles, or in the (useful) data needed for training. Recent articles summarize concerns that this last possibility may be reality (Hu & Tong, 2024; Robison, 2024; Seetharaman, 2024), and there is increasing evidence that a power law may be at least part of the story (Hestness et al., 2017; Rosenfeld et al., 2019; Kaplan et al., 2020).

Not coincidentally, this was when computer Othello (aside 15.3) went from being a hobby to something I poured a lot of passion and energy into, becoming what I would later refer to as "what I did while unable to make progress on my thesis." A normal person would have abandoned the chunking problem and moved on to greener pastures long before, and a normal thesis advisor would have lost patience with a student who could not make progress on their topic. However, I am stubborn—or inflexible or obsessed or blind—enough that the thought of abandoning it frankly did not cross my mind; and Allen was understanding enough to tolerate this. This is a trait that has served me well at a variety of times, but not so well at others.

Aside 15.3: Computer Othello.

During my time at UCSD, in addition to taking courses, working on XAPS, and dabbling in psychology experiments, I began developing *Iago*, a program for the game of Othello—or, generically, "Reversi" (2024)—a simplified variant of Go. This was one of the few times in my career I was inspired by an application rather than an architecture. I was frustrated with the terribly simplistic program on the PDP-11 minicomputer in the LNR (Lindsay, Norman, and Rumelhart) lab, even though I was a complete novice with respect to the game at the time.

This effort started with the simple notion that, given what I had already learned about AI game playing during my initial years at CMU, it couldn't be hard to do much better than that program. However, it ultimately resulted after I returned to CMU in a world-championship-level program[15] written in the SAIL programming language, and my first journal article (Rosenbloom, 1982).[16]

I remember Allen Newell shaking my hand after this article was accepted without modification,[17] a form of acceptance I never afterward achieved. The CS faculty surprised me by passing me on what was called the "area qual" based on this effort. Prior to this, I was supposed to write an extended review article, a useful but not terribly exciting task, for the area qual.

It was also during this effort that I took some of the worst advice I ever received as a student—this advice was not from Allen—and listed two fellow students who worked with me on *Iago* in the acknowledgments rather than as coauthors. The resulting sense of guilt led to maxim 24, of being as generous as can legitimately be justified when assigning coauthorship. To follow up, I would like to apologize finally and formally to my fellow students at the time, Bruce Ladendorf and Charles Leiserson.

The easy part of this problem was figuring out how to build a chunking process that would compile experience over time into new rules that could replicate within a single step the consequences of that experience for similar later situations,[18] and thus reduce the time required in these later situations. The hard part was figuring out how to integrate this in with the rest of the architecture, in a domain-independent manner, so that new rules of the right sort would be learned automatically at the right times and be reused automatically, as appropriate, in later situations in lieu of the original processing. Making chunking impasse-driven—that is, when gaps became apparent in the knowledge needed for decision-making, chunks would be created based on the problem-solving that filled the gaps—ultimately yielded a simple, elegant solution to this problem (Rosenbloom & Newell, 1986). It also yielded the first major architectural *aha moment* of my career. This was a process that Allen guided but whose result was truly torn from my own soul. It also became central to my PhD thesis (Rosenbloom, 1983).[19]

The less traumatic portion of my thesis concerned a psychological phenomenon known as *stimulus-response (S-R) compatibility* (Fitts & Seeger, 1953). Simplistically, think of an elevator that can go up and down, and two buttons that can be used to call it to take you up or down. A compatible arrangement is one where the upper button calls the elevator to go up, and the lower button calls it to go down. An incompatible arrangement would reverse this ordering. Compatible arrangements enable faster performance and fewer errors. Although the phenomena around S-R compatibility were well known for over two decades, there was still no domain-independent theory of it at this point.

Noticing that across many experiments on this topic the differential time to perform the tasks could be predicted by the number of steps taken by simple programs hypothesized to underlie human performance on them,[20] I ended up developing an algorithmic theory of S-R compatibility that in a goal-oriented reformulation could be combined with chunking in Xaps3 to yield a cognitive architecture capable of modeling both practice and S-R compatibility (Rosenbloom & Newell, 1988). Without additional work it could also then model the interactions between these two phenomena, when compatibility conditions were repeated over time.

These results on S-R compatibility influenced work in human–computer interaction by Bonnie John, one of Allen's later graduate students (John et al., 1985; John & Newell, 1990) but did not have a long-term impact on my own path.

15.2 Soar

Transitioning out of the XAPS architectural lineage began with merging a reworking of the domain-independent version of chunking from Xaps3 into an evolution of the Soar[21] problem-solving architecture[22] that was separately being developed by John and Allen (figure 15.1[23]). They had developed a follow-on to Allen's earlier work with Herb Simon and Cliff Shaw on the General Problem Solver (GPS) (Newell et al., 1959), and the IPS project, which could use knowledge in the form of rules to define and control search in problem spaces, developing what they called a *Universal Weak Method* (Laird & Newell, 1983) that could exhibit many different search algorithms depending on the knowledge fragments, encoded as rules, that were available. They then enhanced this with a form of impasse-driven reflection, called *Universal Subgoaling* (Laird, 1986).

Key to implementing chunking in Soar was that both Xaps3 and Soar were built around rules and impasses, although the form of the latter in Soar was much more general. Where the two architectures principally differed was that Xaps3 leveraged these capabilities to support learning by chunking, whereas Soar leveraged them to provide general problem-solving. The combination of the two made a big splash in both AI and cognitive science,

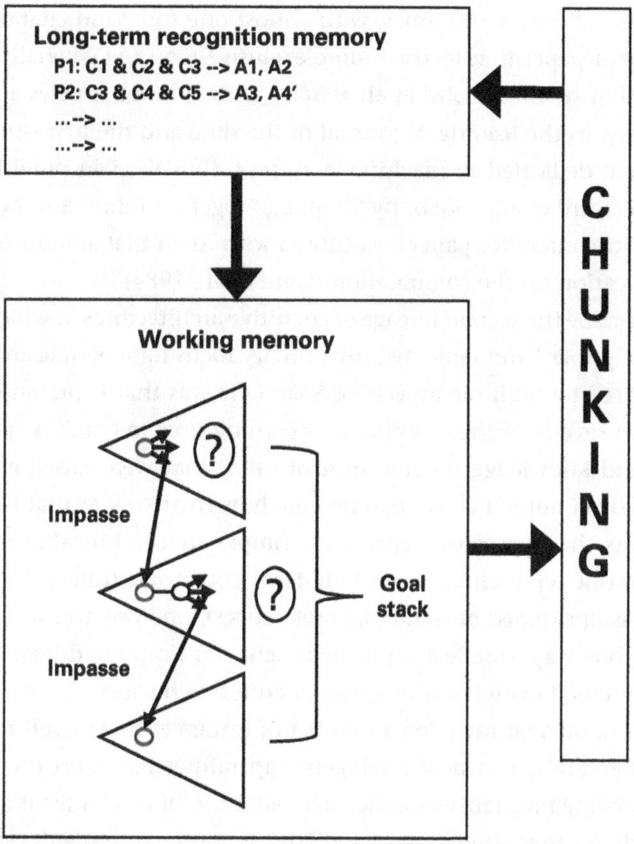

Figure 15.1
Early Soar, as a combination of a long-term recognition memory (rules); a short-term working memory in which a hierarchy of goals and problem spaces results from impasses; and a chunking mechanism that creates new rules based on traces of the rules that fire during impasses.

demonstrating a wide range of learning behaviors—not just simple practice learning—from the architecture plus two kinds of knowledge: domain-general knowledge about problem-solving, and domain-specific knowledge about tasks to be performed.

John took the lead in implementing this combination, as the implementation of Soar was really his baby, but we both contributed conceptually to the result. According to Google Scholar (2025), the first journal article on Soar as a whole that included this learning mechanism (Laird et al., 1987) has been cited over 4,100 times, with almost one thousand citations for a concomitant paper that focused more exclusively on the generality of the learning that resulted (Laird et al. 1986a). These journal articles appeared, respectively, in the leading AI journal of the time and the first issue of the first journal dedicated to machine learning (ML). We even published our first book (Laird et al., 1986b) by "stapling" together John's and my theses plus a short conference paper coauthored with Allen that amounted to the first publication on the combination (Laird et al., 1984).[24]

Soar became the second lineage of cognitive architectures in which I took a leadership role[25] and remained my primary focus for a decade and a half. I was excited by multiple aspects of Soar. One was that it provided direct answers to several of the questions concerning how to combine problem-solving and knowledge (in the form of rules) that were raised earlier by the IPS project but which we had no clue how to answer at that time. The second was that it was conceptually a simple architecture that elegantly combined one representation for long-term knowledge (rules), one way to make decisions (based on symbolic preferences), one way to learn (chunking), and one way to reflect on its own behavior (impasse-driven subgoaling), yet it could exhibit a wide range of critical behaviors.

The work on Soar included a number of instances in which listening to the architecture led to new intelligent capabilities based on interactions among existing mechanisms rather than addition of mechanisms intended specifically for them (maxims 16 and 20–22). John's and Allen's early work showed how many different problem-solving methods and search algorithms could arise from interactions among problem spaces, impasses, and knowledge. Adding chunking yielded many forms of learning (Steier et al., 1987) that might otherwise have required their own learning mechanisms, with data chunking (Rosenbloom et al., 1987)—as mentioned in chapter 14—perhaps the most unexpected such form. The same combination also yielded varieties of planning without explicit planning modules, with chunking storing plans piecemeal into rule memory (Laird & Rosenbloom, 1990; Rosenbloom et al., 1990; Rosenbloom et al., 1993).

Overall, from this conjunction plus appropriate knowledge, we were able to demonstrate a wide range of humanlike intelligent behavior in Soar, including problem-solving and planning, learning, knowledge-intensive behavior, natural language processing, and control of robots. In 1987, Allen took advantage of an invitation to give the "William James Lectures" (2024)

Figure 15.2
John Laird, me, and Allen Newell (left to right) at the Sixth Soar Workshop (1989) at the University of Michigan. In 2024, John hosted the forty-fourth such workshop. In case you are puzzled by the numbering, workshops were held at a very aggressive pace during the early years.

at Harvard to make a major statement about using cognitive architectures such as Soar as the basis for *unified theories of cognition* (Newell, 1990).[26] By the early 1990s, under the continuing leadership of John, Allen and me (figure 15.2), Soar formed the basis for two large books of collected papers, written by a community of researchers that grew to over one hundred individuals reaching out across disciplines, universities, and continents (Rosenbloom et al., 1993).[27]

During this period, based on personal considerations that included a desire to be closer to family and friends and to be somewhere there was more sun, I moved from CMU to Stanford, where I was an assistant professor of computer science and psychology for three years. This had seemed like an ideal position, back in California and at Stanford where I had thoroughly enjoyed my undergraduate years, and with a joint appointment in two top departments that I assumed would recognize and hopefully facilitate my desire to work on Soar across these two fields. Yet, despite the fact that I enjoyed being at Stanford and interacting with some very smart students while there, including three who completed a PhD with me—Andy Golding (CS), Dirk Ruiz (psychology), and Amy Unruh (CS)[28]—it proved to not be the best of fits. By my third-year renewal, it was clear that psychology was not interested in my continuing in that department, perhaps because my work did not seem closely enough tied to human cognition, with a complete lack of experiments on humans and only a rather abstract relationship to the modeling of human cognition. My fit within CS was somewhat better but still somewhat out of paradigm for the AI group there.

My position in CS at Stanford would have continued at least a few more years and I may have been able to create a longer-term niche for myself if I hadn't, for personal reasons, left in 1987 for the University of Southern California (USC) in Los Angeles.[29] At USC, I started as a project leader at its Information Sciences Institute (ISI) (aside 15.4), but then within a few years added the academic titles of research assistant professor and then (tenured) associate professor in the Computer Science Department.[30] I was later promoted to full professor. During my years at USC/ISI, I graduated four PhD students working either on Soar or on a topic inspired by it—Bonghan Cho, Jihie Kim, Soowon Lee, and Benjamin Smith.

Aside 15.4: LA and ISI.

The move to ISI was quick and easy, other than arriving in Los Angeles just in time for *Shake and Bake*, an earthquake on top of a heat wave, with our moving van trapped at the epicenter of the earthquake for a few days. Due to the housing frenzy at that time, our realtor talked us into putting a full-price offer on a house before it came on the market and without even having a chance to see inside, although the offer was conditional on seeing and approving the inside. We are still happily living in that same house over thirty-five years later.

Due to this shift being last minute, it was important that there be no delays due to academic appointment considerations; and I already knew, or knew of, a number of people at ISI. As one of the earliest Internet—or ARPANET, as it was then—sites, ISI was very much within the research community with which I was familiar (it was in fact one of the places at which I interviewed after completing my PhD in 1983). I originally planned to be at ISI for only a year or two, but it proved to be a wonderful place to do research, with great and friendly people, a supportive environment, and lovely views of Marina del Rey and the mountains around Los Angeles (figure 15.3[31]). So, I stayed and simply added on the academic appointment.

Figure 15.3
ISI takes up much of the tall building in this photo from 1982.

The Intelligent Systems Division (ISD) at ISI—of which I served as the deputy director for a decade—grew to be one of the largest AI groups in the world prior to the recent machine learning (ML) boom. The ISD name deliberately omitted "AI" because at the time of its founding AI was entering one of its down cycles—known as *AI winters*—where the field was out of fashion. This has, however, recently been rectified with a change in the name to the Artificial Intelligence Division (AID).

Soar evolved in various ways during this time (Laird & Rosenbloom, 1996) (aside 15.5), to a point where serious applications could be considered. When the opportunity—that is, the funding—arose to engage in a large-scale effort focused on building humanlike automated forces for military simulation and training, we threw all of our energy into it. This started with modeling pilots of fixed-wing aircraft (Tambe et al., 1995a). John Laird's group at the University of Michigan continued to explore this topic (Jones et al., 1999), while my group at ISI switched to rotary-wing aircraft (Tambe et al., 1995b)—that is, helicopters—and scaled up to groups of helicopters and their accompanying commanders (Hill et al., 1997) (aside 15.6).

Aside 15.5: More on Allen Newell and the Soar project.

Allen Newell died during this period, in 1992—a major loss for us all, and for the AI and cognitive science communities more broadly.[32] He was a major figure in founding both fields, as well as a winner of the Turing Award, a National Medal of Science, and much more.[33] Allen had a huge influence on my own development as a scientist, while serving as a model of what a scientist should be that I was never quite able to live up to. He was genial and generous, deep and creative, passionate about his work, and worked incredibly hard, good at both talking and listening, and cared about both people and organizations.

Within the Soar project, his leadership of the CMU component was replaced for a few years by Jill Fain Lehman, who had previously been leading the work on natural language processing in Soar (Lehman et al., 1991), before that site wound down. John Laird had himself moved on from CMU to Xerox PARC and then to a faculty position at the University of Michigan, where he remained working on Soar until his retirement in 2022. He has now spun up the Center for Integrated Cognition as a venue for continuing work on Soar and the common model of cognition.

Many graduate students were also trained within the Soar project, both during this time and later, moving on to successful academic or industrial careers, but most not directly continuing to work with Soar.

Aside 15.6: Visiting military bases.

This modeling effort necessitated visits to a range of military bases, such as Fort Knox, Kentucky (tank simulators); Fort Rucker, Alabama (helicopter simulators);

Fort Leavenworth, Kansas (US Army National Simulation Center); and Naval Air Station (NAS) Oceana, Virginia (WISSARD [What If Simulation System for Advanced Research and Development]—figure 15.4). We spent weeks at a time at some of these bases, as well as at the Applied Research Laboratory (ARL) at the University of Texas, Austin, including periods during which we needed to babysit our simulations full time as they ran around the clock as part of larger exercises.

Figure 15.4
Commander Dennis McBride (USN/DARPA) with me and John Laird, at the WISSARD facility at Oceana NAS (late 1990s).

Some of the lasting memories from these visits include how Fort Rucker sits in a region referred to as UCLA, although here it stands for the Upper Corner of Lower Alabama; how there was an Italian restaurant in the city of Leavenworth that served wonderful Kansas City Strip Steaks at ridiculously reasonable prices; and how F-14 Tomcats took off in a dramatic, and very noisy, manner at NAS Oceana.

This application also led to extending Soar in such directions as real-time behavior and multiagent systems. By using an instance of Soar to separately model each pilot of each aircraft, as well as commanders, and modeling abstractly typical communication among them, we were able to go beyond the previous state of the art of regimented Semi-Automated Forces (SAFs) to Intelligent Automated Forces (IFORs) and Command Forces (CFORs), with both individual and group behavior that was more autonomous and more humanlike. Watching these vehicles operate in this fashion did give the impression to key observers that they were controlled individually by humans rather than by AI systems. Ultimate success was measured by participation in several large Defense Advanced Research Projects Agency (DARPA) demonstrations in the late 1990s—Synthetic Theater of War (STOW) '97 and '98—and by John Laird spinning off SoarTech to continue with such efforts, along with other applications of Soar and related technologies.

A huge part of the success of the ISI component of this effort was the three researchers I was fortunate enough to hire to work with me: Milind Tambe went on to become a leader in multiagent systems and a chaired professor at Harvard University, Randy Hill went on to become the executive director of USC's Institute for Creative Technologies (ICT), and Jon Gratch went on to become a world leader at ICT in virtual humans and affective (that is, emotional) computing. However, by the end of the effort, I personally was rather a basket case, with ambivalence about the application and a moribund basic research program. I found myself dulling my mind by inappropriately playing solitaire on my computer for hours on end and engaging in other forms of escapism.

What followed was a sabbatical during which I attempted to model emotions in Soar, inspired by my own emotional issues, plus a sense that there was something important to be done with cognitive architectures in this area. However, except fittingly for a bit of a model of frustration, I failed to make any substantive progress on this topic. By this point I was having trouble in general finding any direction in which I could make further progress on Soar while retaining the constraint that it remain simple and elegant, even if this constraint was implicit at the time, and it felt like I was at a dead end without the internal resources to understand how to get out of it and to start something new (aside 15.7). This was the burnout that has already been mentioned several times.

Aside 15.7: More on John Laird and Soar.

John escaped what I felt to be a dead end by dropping the constraint that there be at most one module for each major capability, freeing himself to develop the late version mentioned in section 1.2.1 (Laird, 2012; Laird, 2022), with new memories, learning mechanisms, and other modules such as for mental imagery (figure 1.8).[34]

Soar has by now been under development for over four decades, making it one of the longest lasting, most thoroughly investigated, and well-known cognitive architectures.[35] The other architecture in this class is ACT-R (Anderson et al., 2004). Both are computational and theoretical architectures that play in both AI and cognitive science, although Soar is slanted more toward the computational/AI side and ACT-R more toward the theoretical/cognitive-science side.

John has also played a leading role in developing the area of AI and (video) games and later did the same for interactive task learning (ITL), a return to the original motivation for IPS but now with technology and ideas that could make a difference. He also found time to explore the use of Soar on a variety of robots.

We have remained good friends over the years—we started our research journeys together as new graduate students at CMU in 1976 and spoke remotely at each other's retirement celebrations on the same day in 2022. We grew somewhat apart during my interregnum, but he helped me transition back to work on cognitive architectures when I started work on Sigma, and we have been working closely together again on the common model of cognition.

15.3 An Interregnum

Fortunately, at this point Herb Schorr, ISI's executive director at the time, suggested I take a break from direct research to work on, and eventually lead, ISI's new directions efforts. This was less business development than trying to understand where the field was going and to help ISI move in such directions. Fortunately, it was one of the best possible cures for my burnout and the associated sense of being at a dead end from which there was no exit, as it involved an explicit mandate to think broadly in terms of both topics and collaborators.

In Allen Newell's terms, this turned into a decade-long diversion from the core desire of understanding and developing humanlike cognitive architectures; or, in my terms, a period when my research life was not ruled by architectural exploration. So, leaving the ISI Soar group in Randy Hill's and Jon Gratch's very capable hands,[36] I set off on a decade-long exploration of the future of computing and its related disciplines.

My most significant accomplishment during this period was to take the lead in writing a complex multischool proposal to the US Army that established USC's Institute for Creative Technologies (ICT), with an initial vision of applying academic computer science and entertainment expertise to enhancing military simulation and training (Rosenbloom, 2024). As Dick Lindheim—a Paramount VP and later ICT's first executive director—would say, it was about bringing story and character to military simulation to achieve *verisimilitude*—the appearance of being real—or, in other words, trying to create the technology for a real-life *Star Trek* Holodeck. My position in this effort was based on the earlier work on IFORs, plus time I spent afterward as a consultant to Paramount, working with Dick Lindheim and others on an ICT precursor called the *StoryDrive Engine* (aside 15.8).

Aside 15.8: Brushes with Hollywood.

Although a major part of ICT's mission involves leveraging entertainment expertise, my own brushes with Hollywood all predate this. Just for fun, I'll review several here. The first occurred in the runup to the work on the StoryDrive Engine, when Alex Singer, a director on *Star Trek: Voyager*, took three of us from ISI on a behind-the-scenes tour of the set, including a chance to mix with the cast on the bridge and to "hide" in the transporter (or was it an elevator?) while filming occurred just outside, on the bridge. We tried to avoid peeking out in a way that would interfere with shooting, but we still did apparently distract the cast.

Another day a bit later also remains vividly in my mind. It was sunny and warm, and I participated in a lunch hosted at the Getty Center, in one of their white marble buildings with a view out over the city (e.g., Allaire, 2023), to discuss the possibility of a collaboration between the Getty Information Institute (ere it was shut down) and ISI, before driving over to Paramount for a meeting on the StoryDrive Engine. Driving up to the iconic Paramount gates (Google Maps, 2023) on Melrose Avenue with the windows and sunroof open, it felt like a quintessential LA day. I even had an office on the Paramount lot for a time and learned some valuable lessons about the differences between the world of academia and that of entertainment, such as their respective foci on reality versus believability and the differences in how they approach negotiations.

A totally unrelated experience involved two Sunday dinners—the first one for Mother's Day—at a cozy Italian restaurant in Santa Monica, which to our surprise appeared to be where many stars took their families when their cooks had days off. Across these two nights, we saw Tom Cruise and Nicole Kidman, Tom Hanks and Rita Wilson, Dustin Hoffman, Steven Spielberg and Kate Capshaw, and Robert Wagner and Jill St. John. This last pair showed up near the end of the first evening, when one of us jokingly wondered out loud whether the new couple to appear at the door were also movie stars.

The final experience was trying out for *Jeopardy* at Sony Pictures in Culver City. I had enjoyed the show since childhood and always been generally good at it (and *Trivial Pursuit*) at home. So, one day I suppressed my concerns about being in front of an audience, left ISI early, and tried out for it. I was one of the few to pass the written test but bombed the mock game they held for us afterward. I was basically too nervous to quickly push the button at the right time or to smoothly move on to the next clue after responding to the previous one. Needless to say, I was not called back to be on the show.

The plans for ICT were much more ambitious than our earlier work on military simulation in terms of how humanlike the simulated entities were to be—not just behaving appropriately in battlefield scenarios in which the AIs were entirely embedded in vehicles but serving as *virtual humans* that could engage in face-to-face interactions with each other and with real people. USC offered me the opportunity to take a leadership role at ICT, but I declined partly because I wasn't interested in leaving ISI at that point—I had just begun work on new directions—partly due to not yet feeling ready to take on such a role (aside 15.9), and partly because my earlier work on military simulation and training was one of the factors in the burnout that I was just working my way out of. However, even without me, the ex-ISI researchers who moved to ICT—including Jon Gratch, Randy Hill, Stacy Marsella, Jeff Rickel, and Bill Swartout—still chose Soar as the basis for building their initial virtual humans (Swartout et al., 2006).

Aside 15.9: Maturation (maxim 26).

One of the things I've learned about myself over the years is that various aspects have matured at different rates. It wasn't until my fifties and sixties that I felt

(continued)

Aside 15.9 (continued)

sufficiently confident in my ability to lead anything beyond my own research group. Before then I was hobbled by my perceived shortfalls with respect to leaders I admired, and by my awkwardness in dealing with people.[37] As with the ICT experience, and my earlier time at ISI, I thus typically shunned opportunities to lead, more commonly and quite willingly instead taking on a background role as a deputy of some sort. As discussed further in chapter 19, it seems that it also wasn't until these later years that I had achieved the level of intellectual maturity and self-confidence needed to set my own large-scale research problems.

At the other extreme, I matured relatively early in reading and STEM (Science, Technology, Engineering, and Math) thinking. Other areas took an intermediate amount of time. It wasn't until my twenties and thirties that I learned to appreciate the outdoors in any real sense. As a child I would typically much prefer to sit inside and read than go outside and play. But I did eventually acquire a love of the natural world and have greatly enjoyed walking in it over the years, most particularly if there is low humidity and either a view or a grove of redwoods.

It wasn't until my forties and fifties that I truly began to appreciate art and literature, even with always having been a voracious reader. I don't think I was aesthetically, emotionally,[38] or culturally mature enough before then to get much out of either. I first got hooked by Jane Austen and Vincent Van Gogh, but my interests then expanded across eras and genres, although still with relative biases toward eighteenth and nineteenth century English literature and impressionist and postimpressionist art.[39,40] In addition to visiting art galleries wherever I traveled and reading much more broadly than I had previously, in somewhat manic phases I ended up collecting both old books and online art images, while educating myself further about each in the process. I'm afraid I'm still not mature enough though to get much out of most poetry or opera.

Other topics I explored during this period included automated building construction (aside 15.10); responding to the unexpected, in response to 9/11 (Arens & Rosenbloom, 2002); virtual teams composed of robots, agents, and people—aka RAP teams (Scerri et al., 2003); technology and the arts (including helping to recruit to ISI a group from the Getty Information Institute when it was shut down); translational medicine (helping with the informatics side in the early days of the proposal that led to the Southern California Clinical and Translational Science Institute); and high-performance computing (where I played a role in setting up USC's High Performance Computing Center). A range of collaborations were also explored with other engineering disciplines (e.g., biomedical, electrical, civil, astronautical, and industrial and systems), the sciences (e.g., geosciences and marine biology), medicine (e.g., biomedical imaging), and the arts.

Aside 15.10: Contour Crafting.

Automated construction—scaling up rapid prototyping to work for buildings, both terrestrial and extraterrestrial—in the form of contour crafting (figure 15.5) (Khoshnevis, 2012; "Contour Crafting," 2023), was one application that truly inspired me during this interregnum, so I was excited to collaborate with Behrokh Khoshnevis, a USC engineering professor who had invented the concept. However, I never had the insight to make substantive technical contributions on this topic. The challenge was clearly tied to (building) architecture, but there was no architecture of the sort to which my skills would apply. So, only my proposal-writing and teamwork skills ended up being leveraged. Fortunately, this effort continues to show great promise, with a spin-off—"Contour Crafting Corporation" (2024)—following through on it, and other groups imitating it around the world.

Figure 15.5
Contour crafting gantry robot with concrete nozzle in lab at ISI next to wall segment built using it.

Not surprisingly, some of the efforts I attempted to initiate during this decade never bore fruit, at least in the form intended. The most memorable was a proposal to the Department of Homeland Security for a center following up on the groundwork that Yigal Arens[41] and I had earlier laid on responding to the unexpected. The core idea here was to build integrated models of cities, including their infrastructure, buildings, and populations, for use in understanding threats and planning to avert or respond to them—an approach now known generally as *digital twins*. The idea aligned closely with the whole approach of architectural exploration, as well as with the integrated simulation approach that underlay both my 1990s work on military simulation and ICT's initial vision. However, at that time USC had no set procedure for handling funding opportunities with institutional limitations, and in this case only one proposal was allowed per institution. So, when a pair of more senior researchers convinced the leadership that their more eclectic approach was what was needed, our substantial existing work was set aside, and they took over the opportunity. They did offer us a limited role in their proposal, but at the time this whole thing felt like the kind of betrayal mentioned in aside 13.2, and I responded in a manner that was not totally professional. Yet, in retrospect, perhaps USC was not wrong, as their proposal was funded, initiating the USC Center for Risk and Economic Analysis of Terrorism Events (CREATE).

I spent a number of years as an associate director at ISI while leading new directions activities and progressively getting more involved in the overall administration of ISI. When Ron Ohlander retired as deputy director, I set my sights on replacing him, with the thought of eventually succeeding Herb Schorr as executive director, should he at some point retire. Considering trying for the position of executive director of ISI was a big step in maturing to the point where I thought that even with my various weaknesses, whether relative or absolute, I could still bring enough to the table to consider contributing significantly via a leadership role. Although nothing came of this, having gone through the process helped me agree some years later to go up for election as president of USC's faculty (aside 15.11).

Aside 15.11: A USC odyssey.

My feelings about USC have long been complicated. I grew up in Los Angeles as a UCLA fan—both of my parents attended UCLA, as did I for a year in high school. As a family, we also had strong ties to Stanford, and some to Cal (that is, UC Berkeley)—all of which made USC the enemy (maxim 29) to which we also had somewhat of a condescending attitude, given that it was not the greatest of universities through much of this time. When I first got an office on the main campus, I cringed to see the Trojan logo on the telephone in my office.

USC also later treated Sharon terribly (she worked at Children's Hospital Los Angeles as USC faculty). At a critical point, I got all of the way to the dean of engineering with a resignation letter in hand before he convinced me I could continue working at ISI—which I still very much enjoyed—and just ignore the rest of the university. I have wondered to this day, though, whether I should have stuck to my guns at that point.

Still, after working at USC long enough, I eventually bonded, albeit still somewhat ambivalently. The fact that it was becoming a very good university helped, as did the fact that the leadership who were directly involved in Sharon's situation left.[42] Leading new directions activities at ISI also had me interacting in interesting ways with various parts of the USC community. I even became, god forbid, a rabid USC football fan during the Pete Carroll era, to the disappointment of my wife and son, who remained diehard Stanford fans.

I dug myself in much further when, after two years as chair of the Engineering Faculty Council and finding that I was sufficiently comfortable and competent at it, I was asked to run for the university-wide position of president of the faculty (aka, president of the Academic Senate). To say yes, I needed to overcome continued ambivalence about USC and insecurities about being a leader, but it did turn out to be one of the best decisions of my life. It is also a good example of maxim 6, on the importance of service.

My year as president of the faculty (2017–2018) was spectacularly challenging, starting with the first of two scandals involving medical school deans; continuing with multiple further scandals, including a horrific one involving decades-long abuse of female students by a university gynecologist; and ending with a town hall that was so hostile—and, to my amazement, orchestrated to be so by a contingent of our fellow faculty—that I was in shock before moving on to lead an emergency meeting of the Academic Senate, during which we voted to call for USC's president to resign. Making it through this year required a close team comprised of the immediate past president—and my mentor in this—Paula Cannon (Keck School of Medicine), the academic vice president Yaniv Bar-Cohen (Children's Hospital), and me, backed up by the Senate's Executive Board.

With support from Paula and Yaniv, it was my job after the senate meeting to call the president with the news, and then to follow up as necessary with

(continued)

Aside 15.11 (continued)

Rick Caruso, the new chair of the USC Board of Trustees. Unlike my crisis responses in prior years, I have had no second thoughts about this one—what we did was necessary for USC to be able to move forward. Perhaps I could have dealt better with the town hall though, as I simply ended up passively listening, thinking that at least that would enable our colleagues to be heard and not having a better idea how else to respond, which might charitably be ascribed to a lack of experience at this sort of thing.

Figure 15.6

Leading the podium recession at USC's 2018 graduation ceremony. I'm smiling, even if it doesn't look much like it.

Despite all of this, for me personally, the year as a whole turned out to be one of the most fulfilling, and even joyful, experiences of my life.[43] Part of this was the wonderful people I worked with from across the university and how, in order to work effectively with them, I needed to tap into personal and interpersonal aspects of myself that I have traditionally suppressed in my professional life (see aside 15.12). Part of it was the surprising pleasure of participating centrally in various university rituals, such as carrying the university mace while leading the faculty past thousands of students and their families during graduation (figure 15.6). For someone who has always been bored by ritual—where nothing new happens, almost by definition—participating in this way was shockingly satisfying, and I found myself with a smile on my face the whole time.

One of the largest parts, however, was the chance to nudge the university in important directions, with the most critical being the chance to co-lead—with Jeff de Caen, president of the Staff Assembly—a task force that recommended how to revamp USC's culture and processes so as to fix not only the immediate issues that led to that year's scandals but also many that had festered for decades.

I did become deputy director of ISI, but even with the many years I had worked with/for Herb on new directions and as an associate director, we quickly learned that he and I had wildly divergent views at this point about both the nature of the position and the future of the institute, so I ended up leaving ISI rather suddenly, before completing a full year in this position. In essence, I gave him an ultimatum that he accepted as a letter of resignation. I do believe I was right to take a stand at this point, yet I sometimes wonder whether there was any other approach I might have taken that would have been more effectual for both me and ISI.

It was a difficult time overall at ISI, with morale that continued to worsen until Herb was ultimately replaced as executive director. When a search was then held for a new executive director, it considered only external candidates—something else I wasn't happy about at the time but, in retrospect, was likely the right approach given the overall situation there. For me personally, what followed once I left ISI was a year full time in the CS Department, wallowing too much in feelings of uselessness and a sense of hopelessness for the future while focusing largely on teaching introductory AI courses. I was feeling rather beaten down in general and was no longer sure what I might contribute. With respect to research, it had been a decade since I had led a project of my own—I was explicitly expected to avoid leading any projects of my own during this period—and during this recovery year away from ISI, the most I could do was just start to think about a few new directions I might personally consider pursuing.

One of these new directions was to follow up on some sketchy, back-of-the-envelope-style explorations I pursued while still at ISI, trying to understand whether there was any fixed coherence—that is, an architecture of a sort—behind the wide range of largely interdisciplinary efforts I had been involved with while leading new directions activities. As part of this follow-up, I taught a small graduate seminar on the topic during my year of full-time teaching. The other new direction was based on an idea that also came up during the interregnum, of merging the insights from symbolic cognitive architectures like Soar with what had been learned over the years about probabilistic reasoning, in the form of probabilistic graphical models (Koller & Friedman, 2009) and statistical relational learning (Getoor & Taskar, 2007), a hot topic at the time.

With these two largely undeveloped ideas in hand, but no research funding, it proved ideal to spend the following year on sabbatical. After some informal discussions with Randy Hill, who was by now ICT's executive director, and Bill Swartout, who was the director of ISI's Intelligent Systems Division—and one of the people who had originally attracted me to ISI—before becoming ICT's director of technology, about my plans, they offered to host my sabbatical at ICT and to supplement those funds with some unrestricted research funding. This proved to be by far the best sabbatical of my career, enabling me to initiate two new large-scale efforts.

The first idea developed into the relational model of computing, and the notion that computing was a great scientific domain (chapter 18). The second idea turned into the Sigma cognitive architecture (sections 1.2.2 and 15.4). This second idea also led to a long-term affiliation with ICT and one of the most productive decades of my career, supported largely but not solely by the all-too-rare boon of relatively stable basic research funding—from the US Army through ICT's core contract—that enabled exploring essentially whatever directions that I felt Sigma needed to go. So, although I did not move immediately to ICT when it was spun up, I did ultimately reap some of the benefits of my earlier work on the proposal when I moved there a decade later.

15.4 Sigma

For a number of years, Sigma went unnamed, as it wasn't clear whether the ideas being explored would yield a broad toolkit for building probabilistic cognitive architectures or an actual probabilistic cognitive architecture. It eventually was named[44] when it became sufficiently clear that the constraints implied by combining what was learned from Soarlike architectures and probabilistic reasoning led to a single lineage of architectures rather than a broad space of possibilities.[45]

Over the years, a set of four desiderata crystalized around this lineage (sections 1.2.2 and 5.4.3): (1) *grand unification*, which goes beyond the focus on high-level, central, symbolic thought processes typical of traditional cognitive architectures such as Soar to include low-level, peripheral, numeric, and perceptual-motor processes; (2) *generic cognition*, which spans both modeling natural human thought processes and creating artificial humanlike thought processes; (3) *functional elegance*, which looks to generate a broad range of capabilities and behaviors from the interactions among a small set of general mechanisms; and (4) *sufficient efficiency*, which constrains systems to run fast enough for relevant modeling and applications purposes. In retrospect, generic cognition might be reconsidered as humanlike cognition (chapter 1) and functional elegance as exuberance (section 5.4.2.1).

Some of these desiderata were inherited from Soar, whether they were explicitly stated there or merely implicit. Generic cognition and sufficient efficiency are both of this ilk. As previously mentioned, functional elegance is my personal take on Allen Newell's earlier maxim of listening to the architecture. Grand unification represents the largest shift, going beyond Soar's goal of unifying high-level cognitive functionality to include the low-level phenomena that it couldn't handle without separate modules but that seemed within reach in a functionally elegant manner via Sigma.

The early days of exploring Sigma felt much like the early days of exploring Soar, in that so many problems that previously appeared to have no appropriate resolutions suddenly became clear, with simple and elegant solutions. This is not to underplay the amount of conceptual and implementational work that had to be done over a number of years to bring this about, including coming up to speed on probabilistic graphical models[46]—something I continued with over the years but never completely mastered—and getting back into programming after a decade away from it.[47] But it was a very exciting time in which many of the previous barriers to progress—or at least the ones in my mind—came tumbling down and breathtaking vistas opened up.

In a significant sense, Sigma grew as a variant of early Soar, with only a single long-term memory, albeit with Soar's rule-based memory replaced with an extension of statistical-relational factor graphs (Rosenbloom et al., 2016b); only a single learning mechanism, albeit with Soar's chunking mechanism replaced by gradient descent learning (Rosenbloom et al., 2013); and decision and reflection capabilities much like Soar's, albeit with numeric preferences rather than the symbolic ones with which Soar had begun. Sigma's single long-term memory turns out to be general enough that it can still support the kind of rule-based procedural memory that Soar started with, but it can in parallel also support the two kinds of long-term declarative memories later added to Soar—for semantic and episodic knowledge—plus probabilistic reasoning, neural networks, and forms of perception and motor control. As mentioned in section 1.2.2, this all occurs over two architectural layers (figure 1.9)—one graphical and one cognitive—plus the use of skills and knowledge (figure 15.7[48]).

Sigma's single learning mechanism can support various forms of symbolic, probabilistic, and neural learning (Rosenbloom et al., 2013; Rosenbloom, 2012b; Rosenbloom, 2014; Rosenbloom et al., 2017). What it does not yet support, however, is the acquisition of new rules from experience,

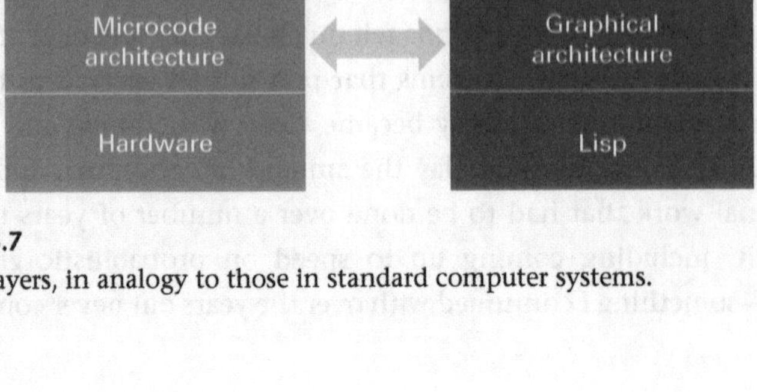

Figure 15.7
Sigma's layers, in analogy to those in standard computer systems.

such as is provided by Soar's chunking mechanism. As mentioned earlier, numerous attempts over the years to create a chunking mechanism in Sigma, whether a narrow version that only generates rules or a more general version that can produce all of the kinds of knowledge structures that Sigma encodes, have failed to pan out. This is another case of my extended stubbornness, but not one that has yet proven fruitful.

The development of Sigma was pursued by a much smaller group of researchers, featuring in addition to me, my deputy and colleague Volkan Ustun, several PhD students (aside 15.12) and programmers who made significant contributions—particularly Abram Demski, Himanshu Joshi, Sarah Kenny, and Dinesh Rajpurohit—a number of helpful collaborators, and summer interns from a variety of universities and differing educational levels. But together, leveraging combinations of Sigma's existing mechanisms plus limited extensions to the architecture and appropriate variable content—in the form of knowledge and skills—that are enabled by the architecture, the traditional symbolic capabilities of the early Soar versions were extended without additional modules to include such capabilities as mental imagery (Rosenbloom, 2011a; Rosenbloom, 2012c), simultaneous localization and mapping (SLAM) (Chen et al., 2011), object and speech recognition (Joshi et al., 2014; Joshi et al., 2016), distributed vector and holographic representations (Ustun et al., 2014), and neural networks (Rosenbloom et al., 2016b) and deep learning (Rosenbloom et al., 2017).

Aside 15.12: Working with students.

In my ten-plus years of developing Sigma, Himanshu Joshi is the only PhD student who did a dissertation directly on it, although Abram Demski recently completed a dissertation inspired by Sigma (see chapter 19). I also had one other PhD student—Lixing Liu—who was working more directly with Volkan Ustun (see section 15.4) on a topic that harks back to what we did in the 1990s with military simulation, but here the idea is to learn the behaviors based on graph neural networks and multiagent reinforcement learning (Liu et al., 2021). With the eight earlier dissertations that were either on Soar or inspired by it, this reveals a distinct paucity of PhD advising on my part over an academic career of nearly four decades—most particularly in the latter years—even if you deduct the decade of administrative work at ISI and the years in faculty leadership.

(continued)

Aside 15.12 (continued)

Part of this was using research funds that might have supported more students to "buy out" of some of my teaching each year.[49] Part of this in the latter years is likely due to a sense that I was nearing retirement and so should think twice about taking on new students, along with concern as to whether there were really jobs out there in cognitive architecture. Part of it is also that I was finding it increasingly difficult to find topics that were easy to split off from such a unified architecture for students to pick up quickly and make progress on. Part was likely simply a disinclination at this point. Whatever the reasons, I clearly did not shine in this area, in contrast to John Laird who has graduated many PhD students over the years on Soar.

John was also able to create a lively social environment for the Soar group at Michigan, whereas I always strove to keep a "professional distance" from my students, and even typically from my colleagues. I thought I had good reasons for this at the time, but on reflection, this seems to have been a serious mistake—I missed out on some great opportunities for friendship, or at least to know better those around me and to build a more welcoming environment for all of us.

Another way I fell short in working with my PhD students was in not understanding how my responsibility to them shouldn't have ended when they finished their theses. Throughout most of my career it didn't cross my mind that I should have been at their graduations to "hood" them, nor did I think about mentoring them in their post-PhD job searches. Whether this was lack of training in the faculty role, simple cluelessness, or indicative of any other failings on my part, it is something for which in retrospect I want to apologize to them.

I have also worked with some undergraduate and MS students, via internships, directed research, and more informal means, but even though I generally found it rewarding, I did not engage in a great deal of it.

Sigma has also been extended in recent years to incorporate aspects of theory of mind (Pynadath et al., 2013; Pynadath et al., 2014), attention and emotion (Rosenbloom et al., 2015)—returning to where I last failed with Soar—and motivation (Rosenbloom & Ustun, 2019). There was also to be an exploration of personality—on the "real" five-factor (or OCEAN) model that includes dimensions for openness, conscientiousness, extraversion, agreeableness, and neuroticism (McCrae & John, 1992), rather than Myers–Briggs—but it was put aside when I became ill.

By this time, we were striving for *humanlike autonomous social cognitive (HASC) systems* (Ustun et al., 2018) able to:

- **Perceive:** Observe the environment and others in it
- **Reason:** Draw conclusions from perception and knowledge
- **Act:** Choose and execute appropriate actions
- **Interact:** Communicate naturally (verbally and nonverbally) with others
- **Use Theory of Mind:** Model one's own mind and that of others
- **Be Affective:** Understand and exhibit emotional behavior
- **Be Motivated:** Have desires and the ability to choose among them
- **Exhibit Personality:** Maintain a distinct yet consistent character
- **Learn:** Adapt one's own behavior through experience

HASC systems were intended to support applications requiring virtual humans and other humanlike synthetic characters (Ustun et al., 2015; Ustun & Rosenbloom, 2015; Ustun et al., 2018)—Sigma's most promising area for application—but this work later evolved under Volkan's leadership to use other approaches for learning basic character and team behaviors, such as multiagent reinforcement learning, with the hope that Sigma's richness would still prove useful later (Ustun et al., 2021).

Whereas the implementation of Soar was very much John's baby in the early days, Sigma was mostly mine. That did change somewhat, however. Beyond limited early contributions by others in the group, there was more recent work to extend Sigma in languages beyond LISP and to take advantage of parallelism and special-purpose chips. Sarah Kenny, for example, explored leveraging parallelism in C for Sigma's inner loop. Himanshu Joshi has been exploring a fast reimplementation of a stripped-down variant of Sigma in C++. Jincheng Zhou explored a new formulation of Sigma in Python that was to better support continuous data while also ultimately supporting a version of chunking (Zhou & Ustun, 2021).

Looking back, I started exploring cognitive architectural lineages based on symbolic rules and neural-like activation in the late 1970s and ended up with an architecture that still features such a combination. The intervening years involved architectures exhibiting increases in the generality and scope of the mechanisms embodied, the varieties of skills and knowledge that could be expressed and processed, and the kinds of applications that could be pursued. From this personal perspective on the evolution of cognitive architectures, a form of history of AI can be found in aside 15.13[50].

Aside 15.13: A personal cognitive architectural perspective on the history of AI.

When I started as a graduate student at CMU in the mid-1970s, AI was still deeply in its foundational symbolic era, centered on such topics as AI languages, search, and knowledge representation and reasoning (KR&R). Cognitive architectures got their start in this era, as AI languages that not only integrated together capabilities such as search and KR&R but also added a theoretical aspect concerning how they combine to yield minds. The topic of learning was just beginning to have an impact late in this era, including in cognitive architectures.

The knowledge-intensive era, spearheaded by Ed Feigenbaum[51] and others, built on the symbolic era but centered on the knowledge required for effective task performance. The slogan was *knowledge is power*, which was backed up by the first successful wave of AI applications during the 1980s expert-systems boom. During this era, work on cognitive architectures was primarily considered interesting to the extent it did useful things with knowledge (such as in Rosenbloom et al., 1985).[52] This era ebbed as the limits of hand crafted, rule-based knowledge became apparent, leading to an *AI winter* during which interest in, and funding for, AI declined.

The next big era centered on probabilistic reasoning, which enabled shades of knowledge and the explicit representation of uncertainty in more principled ways. It also enabled learning of statistical relationships in ways that automated some of the manual knowledge acquisition of the previous era. Judea Pearl was the pioneer of this approach (Pearl, 1988), but when I arrived at Stanford, several of the next generation of leaders, such as David Heckerman, Eric Horvitz, and Stuart Russell, were senior graduate students there.

Cognitive architectures were largely considered irrelevant during this era, as all that was assumed to be necessary were the isolated learning and inference algorithms needed to convert probabilistic information into applications. As with the knowledge-intensive era, there were a number of successful applications. Although there was work on statistical learning in architectures (e.g., Langley et al., 1989), architectural researchers were slow to pick up on the deeper insights from this era. It was not until they were merged with the earlier ones from the symbolic era, in what was initially known as *statistical relational learning* (Getoor & Taskar, 2007), could attempts such as Sigma appear in which cognitive architectures were grounded in these ideas.

The current era centers on artificial neural networks (ANNs) and machine learning (ML). Early work (McCulloch & Pitts, 1943) predates AI's symbolic era, which is dated from the Dartmouth Conference in 1956. However, overinterpretations of a result (Minsky & Papert, 1969) that proved limitations on what could be done with the one-layer ANNs that could be learned at that point (Rosenblatt, 1958) effectively suppressed the approach for a decade. The modern ANN explosion was enabled by the backpropagation algorithm for learning multilayer ANNs (Rumelhart et al., 1986) but didn't truly take off until massive datasets, along with the massive computing necessary to train large ANNs on such data, became available.

For much of the ANN community, not only have architectures been irrelevant but so also has been symbol processing, except as emergent from ANNs. In the other direction, there have been a number of attempts to incorporate aspects of ANNs into cognitive architectures, from my own primitive ones with XAPS and Neuro-Soar (Cho et al., 1991) to the pervasive use of activation in architectures like ACT-R (Anderson et al., 2004) and hybrid architectures such as Clarion (Sun, 2016) and SAL (Jilk et al., 2008). Sigma has provided a new approach based on a tighter coupling of these capabilities. It is, however, possible to view recent work on ANNs as evidence that an increasing subcommunity is seeing the ultimate limitations of "pure" ANNs (Hu & Tong, 2024) and thus the need to combine these ideas with more structured (Zhou et al., 2019), symbolic (Silver et al., 2018), and architectural (Eliasmith, 2013; LeCun, 2022; Park et al., 2023; Sumers et al., 2023) aspects.

The recent ascendance of *generative AI* (e.g., Brown et al., 2020), with its unique combination of awe-inspiring generality—yielding an era that combines *knowledge is power* with what can be termed *learning is generality*—and dangerous hallucinations, looks likely to move us into a new era of interest in cognitive architectures. There is already much discussion as to whether such systems yield AGI, of how they might fall short, and of how they might be merged with cognitive architectures. Over the history of AI, shifts in focus from *narrow* intelligence that concerns individual capabilities to more *general* intelligence typically occurs when people get sufficiently frustrated with the limitations of the former to consider broader integrations.

My history with Sigma has likely ended, although not with a dead end as it did for me with Soar, as I still see significant untapped potential within its functionally elegant framework, including the possibility of serious applications. Although any such effort will need to be in the hands of others, my history with cognitive architecture research will hopefully continue on, but most likely in the more abstract forms found, for example, in the common model of cognition (section 1.2.3 and chapter 16), relevant bits of essential analysis (part I), and dichotomic maps (part II and chapter 17).

16 Common Model of Cognition: Searching for Consensus [Tier 2]

The previous chapter focused on three related lineages of cognitive architectures. Soar is one of the two most iconic architectural lineages overall—the other being ACT-R (Anderson et al., 2004)—having been under development for decades and having a wide range of applications. XAPS was more of a flash in the pan, although embodying an early attempt at combining symbolic rules with numeric activations and other aspects such as chunking that had a more lasting impact. Sigma has been an attempt to retain what still seems right to me about Soar, and to a lesser extent ACT-R, but to replace its core memory and reasoning components with a more modern and general approach based on an extended form of graphical models.

The common model of cognition, as introduced in section 1.2.3, is an attempt to abstract away from individual humanlike cognitive architectures to find a community consensus that may be incomplete but that captures what is agreed upon in a common language. Work on the common model has involved exploration of both scientific and social architectures. On the scientific side, which was the focus of section 1.2.3 and thus will not be of further concern here, the common model is itself a theoretical architecture, concerning the nature of humanlike minds. It can also be considered a meta-architecture that is about the space of humanlike architectures. In being unimplemented, it is not directly computational, but it falls within the class characterized in figure 2.2a, as a theoretical architecture that is not far from being computational.

On the social side, the overall effort began with a final presentation that Christian Lebiere gave to sum up a AAAI Fall Symposium on *Integrated Cognition* that he and I coorganized in 2013.[1] In this presentation, he introduced the notion of a *standard model of the mind* in analogy to the standard model of (particle) physics and mentioned a few aspects he believed from what he had seen at the symposium, and likely from his general background as well, should be in it. It frankly shocked us—in an extremely positive manner—that the participants in that symposium, who came from AI, AGI, cognitive science, robotics, and neural perspectives—and who worked on a wide range of architectures themselves—agreed with what he presented. This talk and the resulting reaction jointly inspired the broader effort that has turned into the common model of cognition.

The first step was agreeing to an interuniversity collaboration among Christian (CMU), John Laird (University of Michigan), and me (USC), and thus implicitly among the three architectures we represented at the time—ACT-R, Soar, and Sigma, respectively. While John was not a coorganizer of the 2013 symposium, he had been a participant and was soon included as a full partner in the overall enterprise. The social architecture here could remain fairly simple, but perhaps for that reason proved to be quite productive, resulting in a joint paper on the standard model (Laird et al., 2017) that again surprised us in terms of how much could be agreed upon, at least among the three of us and our architectures, after a significant amount of challenging interaction concerning what differences were real versus merely surface distinctions that masked deeper underlying similarities.

In one sense it should not be that surprising that some kind of consensus could be achieved over these researchers and architectures. We all received our doctorates in computer science at CMU, two (John and I) advised in parallel by Allen Newell, and the third (Christian) advised a decade and a half later by John Anderson. The three cognitive architectures are thus all part of a broad CMU architectural tradition. Soar, of course, owes a direct debt to Allen Newell's earlier thinking on cognitive architectures and their precursors (Newell & Simon, 1956; Newell et al., 1959; Newell & Simon, 1972; Newell, 1973a) and to his personal involvement with its early development. Sigma is in many ways a descendant of Soar, although it does bring in a radically different technology at its core. ACT-R is the most recent member of a lineage of architectures that earlier included HAM (Anderson & Bower, 1973),[2] ACT (Anderson, 1976), and ACT* (Anderson, 1983),[3] and which John Anderson began working on prior to joining CMU but continued at CMU since he arrived there in 1978.

Yet, it is notable how ACT-R and Soar—which started from rather different bases—turn out to be more alike now in how they map to the common model than either is to Sigma. The broader body of work considered in developing the consensus should further have helped prevent it from becoming too parochial. Still, much more was clearly required to reach a consensus across the broader range of relevant researchers and architectures in the field(s).

The next step on this path was holding a sequence of two additional AAAI Fall Symposia (Flenner et al., 2018; Adler et al., 2019) and the creation of a community mailing list that last I checked had 270 members. At the first of these two symposia we were excited to find near unanimity that what we had proposed in the paper was an appropriate starting point for developing a standard model of the mind. The biggest disagreement concerned its name[4]—there was a significant sense that it was too aggressive, with worries about the possibility of the kind of top-down pressure that was seen with the *standard theory* in linguistics (Chomsky, 1965). So, we conducted a multistage community-wide poll that resulted in changing the name to the *common model of cognition*.

We also started virtual working groups on eight topics for which we thought progress could be made with respect to the common model: declarative memory, emotion/mood/affect/motivation, higher-level knowledge, language processing, metacognition, neural/physiological underpinnings, perceptual/motor processing, and procedural memory. In an attempt to encourage a more distributed leadership to emerge, the three of us opted not to lead any of these working groups. Instead, we each served as resources in multiple of them.

Some of these groups got as far as producing reports on their topics for presentation at the second symposium. These were subsequently published in *Procedia Computer Science*. More recently, an online journal—the *Common Model of Cognition Bulletin*—has been created with a direct focus on papers related to the common model. By now, however, all of the working groups have gone quiescent, likely due to the common model being only a side interest at best for most researchers. So, Christian, John, and I began to rethink what to do next.

We started by recruiting Andrea Stocco (University of Washington) as a fourth member of our interuniversity collaboration. His background has ties to ACT-R, having done a postdoc with John Anderson at CMU, plus he brings in a cognitive neuroscience perspective that was largely lacking (although Christian does have some relevant background). The four of us now hold weekly Zoom meetings to discuss how best to move forward in a manner that will maintain the community-consensus aspect while leveraging our joint commitment to the effort.

From these conversations we have so far been able to organize the first of what will hopefully be a continuing sequence of specialized workshops on the common model; this one in June of 2022 on emotion and the common model,[5] which led to the beginnings of a potential consensus on this topic that we attempted to extend and refine before taking it back for further discussion with the community. Although this extension did lead to a proposal for a consensus it has not yet actually led to a consensus (Rosenbloom et al., 2024).

We also were able to publish a short, general-interest piece on the common model and AGI (Rosenbloom et al., 2022) and coorganized—along with David Reitter (Google DeepMind)—a 2023 AAAI Fall Symposium on Integration of Cognitive Architectures and Generative Models. Although not directly about the common model, it did play a substantive role in several of the presentations.

Still, as the old saying goes, "the proof of the pudding is in the eating." There is much more to be done in developing and using the common model, and in achieving consensus across broader communities, including even those not directly related to cognitive architectures. The ultimate success and worth of the common model will be determined by how all of this plays out and how useful it ends up being. However, I think there is a good chance that this effort will have a larger and more enduring impact than any of our individual cognitive architectures.

17 Dichotomic Maps: Scope, Similarities, Differences, and Combinations [Tier 3]

As was seen in part II, a dichotomic map is a theoretical architecture that leverages the cross product of n dichotomies—that is, this-or-that distinctions—to structure and help to understand an n-dimensional space of possibilities. Whereas the common model is driven to identify the essential *similarities* among humanlike *cognitive architectures,* dichotomic maps are instead driven to identify the key *differences* among *technologies* that provide core memory/knowledge, reasoning and learning capabilities more broadly across AI and related disciplines. Still, the full exploration of dichotomies yields insight into both similarities and differences, as a function of which poles are shared and thus of the L1 distances among the technologies, but at a smaller granularity than is provided by the common model.

This exploration into dichotomic maps grew out of a sequence of attempts to understand and explain how Sigma spanned such a broad range of capabilities from a small set of general mechanisms. This started more than a decade ago with a paper on how to build architectures for virtual humans that were *broad spectrum* (making creation of simple virtual humans simple while enabling extension to arbitrary sophistication as needed), *tightly integrated* (yielding much more than the sum of its parts), and *functionally elegant* (as already discussed for Sigma; (Rosenbloom, 2011b). This work considered how a grab bag of dichotomies relevant to such architectures might be bridged, and how Sigma supported such bridging.

The broad-spectrum aspect of this earlier work morphed later into a rejected conference paper that proposed Sigma's approach of extended factor graphs as *a/the language for AI* that could yield broad functionality from a simple and theoretically elegant base (Rosenbloom et al., 2019b). This idea was inspired by combining the notion of a *cognitive hourglass* (Rosenbloom, 2009)—a way of thinking about Sigma that arose in analogy

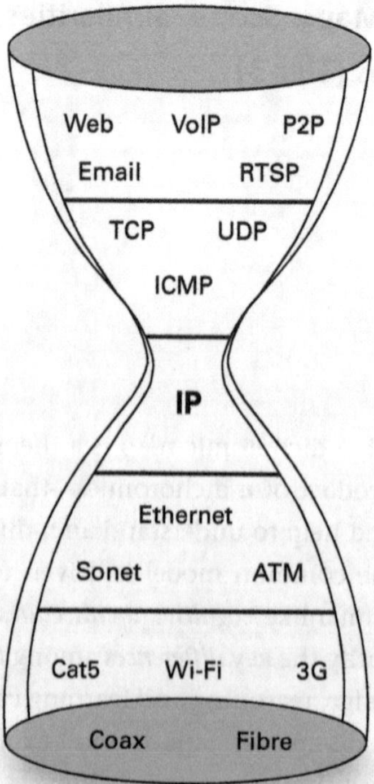

Figure 17.1
Internet hourglass.

to how the Internet Protocol (IP) provides a narrow waist of the *Internet hourglass* (figure 17.1[1])—and the long history of work in general on languages for AI.[2]

The failure to get traction publication-wise with this approach led to backing off from it to focus more on specific dichotomies and their interactions. This was done in two pieces. The first considered how the structure of many cognitive architectures could be understood in terms of *(a)symmetry × (non)monotonicity*, an earlier take on (a)symmetric × (un)controlled. The second considered how the traditional but coarse-grained dichotomy of symbolic versus neural could be understood in terms of *(sub)symbolic × (a)symmetric × (non)combinatory*, where the first dichotomy was an earlier take on (meta)data and the last was an earlier take on (un)controlled. The notion of opposite locations in a map is grounded generically in this work.

The initial attempts to publish these explorations at general cognitive science and AI conferences also failed (Rosenbloom, 2019b; Rosenbloom et al., 2019c), but repurposing them for specialized conferences more directly related to cognitive architectures ultimately succeeded (Rosenbloom, 2019a; Rosenbloom et al., 2019a). I've laid out this particular publication saga in some detail because of how well it illustrates the methodological and publication issues raised early in this memoir.

As ultimately published, these two explorations were similar in nature and shared the (a)symmetric dichotomy, but it wasn't until later that I was able to figure out how to combine them into a single theoretical architecture. The biggest conceptual stumbling block was that the former focused on the structure of individual systems and the latter on the structure of whole disciplines. The solution in part II manages to distill the original four dichotomies down to three while clarifying the difference between architectural and disciplinary maps, with the former serving as theoretical meta-architectures for integrated systems and the latter as a component of metascience (section 1.1) that can be used to model entire disciplines.

The work described in part II greatly expands on these earlier articles. This has involved going further into what other dichotomies might be considered as variants of the three core ones, and how the resulting clusters should be labeled, ultimately settling on (a)symmetric, (un)controlled, and (meta)data; how dichotomies might be spanned to yield the best of both poles; and an exploration of criteria for evaluating technologies and integrated systems (chapter 7) and how they could be leveraged in understanding the trade-offs involved in both individual dichotomies and cross products of them. This all helps in understanding and comparing both individual technologies and integrated systems across AI and related disciplines. It may also ultimately provide one path toward considering all possible forms of intelligence. However, as articulated in chapter 12, much remains to be done.

This expansion originally led to an eighty-page draft that I was thinking might end up as a journal article, or possibly a monograph or even a full book; that is, I was following maxim 13 again here, of writing what I wanted to say and only then worrying about how it might be published. Once the reflections that now form part I began to accumulate, my thoughts gradually turned toward publishing this material as one part of a larger book, which is ultimately what has happened.

18 Relational Model: Computing as a Great Scientific Domain [Tier 4]

As mentioned in section 15.3, the relational model grew out of attempting to make sense of the diversity of topics, all at least loosely affiliated under a computing banner, explored during my interregnum. On first blush, there was no coherence across them other than that they all involved computing in some manner. However, as I explored different possibilities, a deeper congruity started to emerge in which the various topics exhibited characteristic relationships among particular scientific domains. This insight—it was an extended *aha* moment—first led to the development of the *relational model*, which provides a theoretical architecture delimiting the kinds of relationships that can appear between domains, and then the notion of *great scientific domains* as the primary slot fillers (that is, the variable content) in these relations, accompanied by the identification of computing as one of these domains (Rosenbloom, 2012a).

Like dichotomic maps, the relational model has a dyadic aspect and can be used in mapmaking. This yields a high-level commonality between the two architectures, but the details differ significantly. First, the dyads in the former are built from pairs of technological traits, whereas in the latter they are built from pairs of scientific domains or disciplines, typically what came to be called great scientific domains—the physical, life, social, and computing sciences. Second, the relationship underlying the dyads in the former is effectively *opposite*, whereas in the latter it is either *implementation* or *interaction*. Third, rather than maps based on cross products of dyads, they are built here by crossing relationships with domains.

Figure 18.1[1] shows a simple way to view the relational model over the three traditional great scientific domains. In it, the domains are abbreviated as their initials—P, L, and S—with the generic notion of a relationship

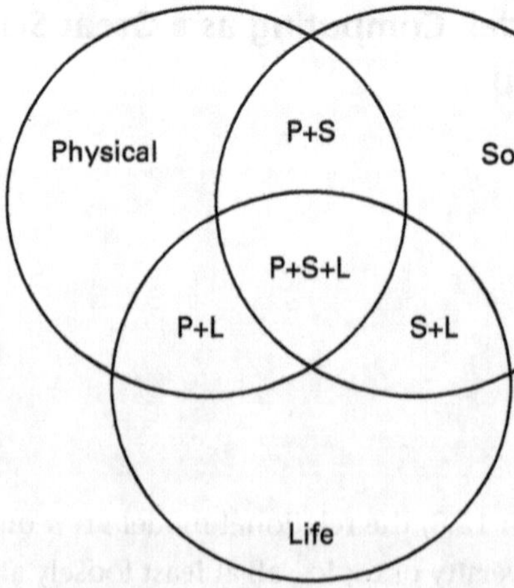

Figure 18.1
Overlaps among the traditional great scientific domains, yielding monadic, dyadic, and triadic regions.

denoted by +. The figure includes monadic regions in which each domain is considered in isolation (and its name is spelled out in full), dyadic regions that combine two domains, and one triadic region that combines all three. For example, physics, biology, and psychology fit respectively within the monadic regions of P, L, and S; climate change, cognitive neuroscience, and molecular biology fit respectively within the dyadic regions of P+S, S+L, and P+L; and ecology fits within the triadic region of P+S+L.

If the generic relationship denoted by + is refined into implementation (denoted by /) versus interaction (denoted by ↔, →, or ←, depending on the direction(s) of influence involved in the interaction), a richer and more informative representation can result for the dyadic and triadic disciplines: climate change becomes P↔S, to represent both the human impact on climate change (P←S) and the impact of climate change back on humanity (P→S); cognitive neuroscience becomes S/L, to represent how the mind is implemented by the brain; molecular biology becomes L/P, to represent how life is based on molecules; and ecology becomes (P↔L)↔S, to represent the interaction among all three domains.

When computing is added to the picture, figure 18.1 swells into figure 18.2[2], embodying a range of possible overlaps, and thus relationships, with the other domains. Theory and algorithms are classical examples of

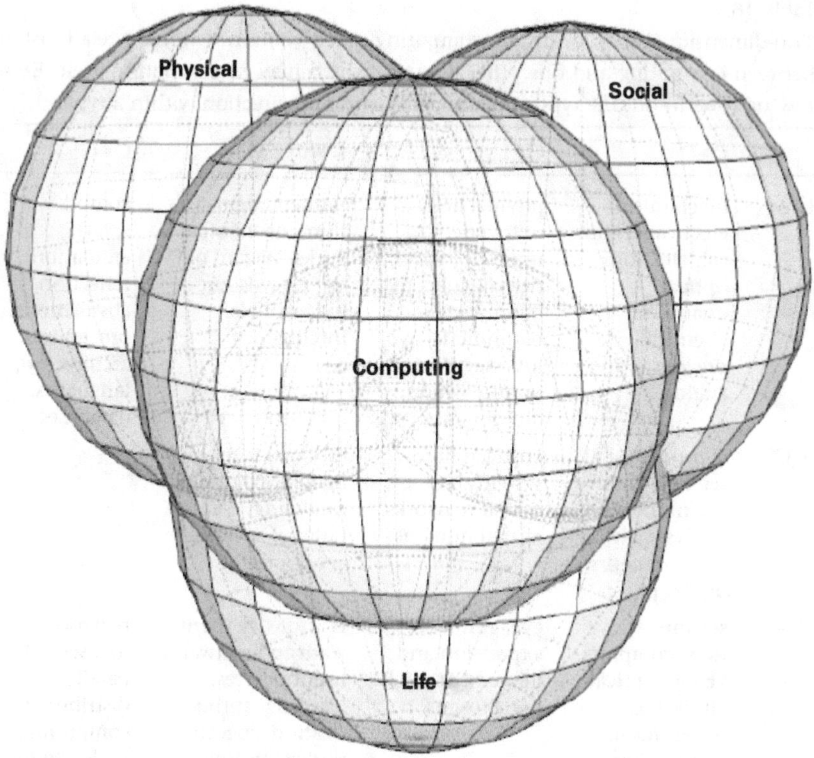

Figure 18.2
Three-dimensional representation of the four great scientific domains and their overlaps.

monadic computing. The abundance of dyadic possibilities—involving computing plus one other domain, which may include itself—can be seen in table 18.1[3,4] The one additional bit of representation here is Δ, which refers to any or all of the domains (P, L, S, and/or C).

It was the development of a very early version of this map, and how it was able to structure and relate so much of what was going on, or had gone on, in computing broadly, that got me excited about the potential of the relational approach to help understand computing as a discipline. To highlight just two specific examples that are both based on relationships between the social and computing domains (S+C), but in very different ways, artificial intelligence implements the social domain by computers (S/C), whereas human–computer interaction involves a bidirectional

Table 18.1

Two-dimensional map of dyadic computing. Each column denotes a relationship between computing and one other domain, which may be computing itself. Each row narrows in on one type of relationship and one direction within a type.

	C+P	C+L	C+S	C+C
C/Δ	mechanical, electromechanical, electronic, optical, quantum, chemical, precomputing, analog	phylogenetic, ontogenetic, immunological, neural, collective, biomolecular, biomimetic, evolutionary, autonomic, swarm, biomimetics	human cognition, mechanical Turks, Wizard of Oz, crowdsourcing, artificial intelligence	compilers, OS, emulation, reflection, abstractions, procedures, architectures, languages, databases
Δ/C	computational science, graphics, numerical analysis, virtual environments, digital physics	artificial life, systems biology, neural networks, virtual humans	artificial intelligence, cognitive modeling, computational sociology, econometrics	
$C \leftarrow \Delta$	sensors, scanners, computer vision, optical character recognition, localization	eye, gesture, expression and movement tracking, biosensors	IT industry (hardware and software), input devices, learning, authorization, speech understanding	networking, security, parallel and distributed computing, grids, clouds,
$C \leftrightarrow \Delta$	robots, informatics	brain–computer interfaces, assistive robotics, biomedical informatics	computer and software engineering, human–computer interaction, full immersion, games, natural language, social robotics	multirobot and multiagent systems, automated programming and debugging, self-monitoring
$C \rightarrow \Delta$	locomotion, fabrication, manipulation	bioeffectors, haptics, sensory immersion	presentation, screens, printers, graphics, speech generation, cognitive augmentation, computational thinking	

interaction between the social and computing domains (S↔C). Both can thus be found in the same column but are distinguished by their rows.

Beyond dyadic relations, mapmaking gets more difficult, but the *metascience expression (ME)* language, which has at its core the notations that have already been introduced, enables specifying complex representations among multiple domains, as well as the possibility of individual domains playing multiple roles within a single expression. For example, one way to represent a robot is as l/(P↔C), where the lowercase "l" reflects a simulation of life rather than actual life. Building on this expression, a human enhanced with a prosthesis that contains an embedded computer can be denoted as L/(L↔l/(P↔C)), where the prosthetic—that is, a particular form of robot—interacts with the living body to implement a new form of life that is the combination of the two. We can also think of this new form of life as a *cyborg*, although if the mind were also to be augmented as well as the physical body, more terms would be needed that include S and at least one more C.

Up until this point, this discussion of the relational model has assumed that there are four great scientific domains, the three traditional ones—the physical, life, and social sciences—plus the computing sciences. Most people likely agree with the first three without really needing to think about it, unless perhaps they are of the contrary sort that want to deny that the social sciences do science.[5] One of the things that surprised me though when I began this work was that, even with the long history of dividing up the sciences in this manner, there was no standard definition that explains why these three fit into this class while no other areas do.

Using the relational model as a means of understanding the field of computing implicitly assumes that computing is of a kind with these other three, but many question whether computing is even a scientific discipline—as opposed to "merely" a form of engineering—not to mention a great scientific domain on a par with the other three. Making this claim required a further excursion into philosophy of science—or metascience, as discussed in chapter 17 and section 1.1, and as implied by the name of the ME language—to hypothesize the existence and nature of great scientific domains, and to understand whether the resulting definition applies to computing.

The definition I ultimately settled on for a great scientific domain was the *understanding and shaping of the interactions among a coherent, distinctive, and extensive body of structures and processes*. This definition amounts to a post hoc rationalization, as the nature of the domain of computing was considered along with the three traditional domains, so an extra degree of skepticism is also in order here (see section 5.3.1.1 on the paradox of predictivism). But the broader hope is that it still provides a useful new perspective. Although this definition was derived years before essential analysis was formulated as a definitional method, the process of defining great scientific domains roughly fits that model, with the four domains serving as the exemplars from which the essence was extracted.

The combination of understanding and shaping implies that each domain—not just P and C but also L and S—intertwines forms of science and engineering, cutting the overall intellectual pie at the top level by domain rather than by science versus engineering (maxim 9). The domains are then distinguished by the characteristic structures and processes—or, as defined in section 4.1.1.1, the mechanisms—that they study and manipulate, with computing in particular concerning *informational structures* and *processes that transform this information* (Rosenbloom, 2010b). This yields a broad definition of computing as a whole—again via a form of essential analysis or the reflective method—that includes not only monadic C topics, but all topics in which C participates, whether science or engineering, academic or industrial, research versus applications, hardware or software, or based on static information or dynamic processing (maxim 8). Both this paragraph and the previous one provide examples of maxim 7, on breaking down walls rather than building them.

This way of cutting the pie provides, independent of the relational model, a novel theoretical architecture for science and engineering, grounded in revised definitions of them—yielded by additional applications of essential analysis—that extends to include a variety of disciplines not conventionally considered part of either, such as mathematics (part of the understanding side of C); the humanities, law, business, and education (all part of S); and the practice of medicine (part of the shaping side of L).

When I gave a talk some years ago on the relational model at CMU, one of the strongest objections was to the idea of mathematics being part of computing. There are a range of very serious reasons for such an objection, but it is a direct consequence of the model, based on how the domain of computing is defined broadly in terms of a focus on informational structures. As mentioned in section 5.2.1.6, alternative names could be considered for the domain, such as information sciences or mathematical sciences—which was the name of my undergraduate major—but I chose computing sciences because it emphasizes not just the structures, but also the transformational processes that are applied to these structures. Mathematics essentially becomes a static, formal component of the broad domain of computing.

This does not, however, necessarily imply that mathematics should structurally be part of either a department or school of computing. As discussed in section 1.1, stronger methods tend to reduce the acceptability of weaker methods, even in areas where the stronger methods don't apply. Although there does exist theoretical computer science, which does thrive on proofs, much of computing is experimental, a strong method in its own right but weaker than proofs. So, incorporating mathematics into a computing unit would likely be problematic. In fact, some computer science departments started as parts of mathematics departments but split off early on for such a reason. Including mathematics within computing may do no favors to mathematics either, given how it might reduce its autonomy and prominence, as well as highlighting its more limited access to external research funds. Still, despite these pragmatic reasons for keeping the two separate, the conceptual relationship makes sense.

At first, I considered the notion that computing was a great scientific domain as simply a *lemma*, or subsidiary point, that was necessary for consistency among the slot fillers in the relational model. It took Peter Denning, one of our leading thinkers about computing as a scientific discipline (Denning, 2005; Denning, 2013), to point out to me that this idea was important in its own right, and possibly even more important than the relational model itself. Since the four domains are in effect variable content within the architecture defined by the relational model and the ME language, perhaps it is not surprising that with my architectural focus I missed this until it was pointed out to me. Once it was pointed out, however, it resulted in an inversion of how the book was written that summarized this exploration (aside 18.1).[6]

Aside 18.1: *On Computing.*

Writing *On Computing: The Fourth Great Scientific Domain* (Rosenbloom, 2012a) required significant work spanning two sabbaticals, although with an unusually small-time gap between them, plus much of the time in between. I started with some initial ideas on how to proceed from notes I had jotted down while at ISI, plus an earlier article on this topic (Rosenbloom, 2004)[7] and the graduate seminar I taught in 2007, but it still ended up as one of those efforts in which much of the requisite research occurred during the writing.

I originally proposed to MIT Press, who published the book, the image in figure 18.3 for its cover, based on overlapping subimages that iconify the four great scientific domains. Computing is represented here as a Turing machine, with the physical domain an image of two black holes interacting, the life domain an image of DNA replication, and the social domain an image of an artist looking in a mirror to paint a self-portrait. MIT Press countered with the image in figure 18.4[8], wanting something simpler that would "pop" on a shelf. It is essentially an abstraction of the subimage I had proposed for the physical domain. Although my proposal was richer and more accurate, it definitely did not pop as did theirs (particularly in color).

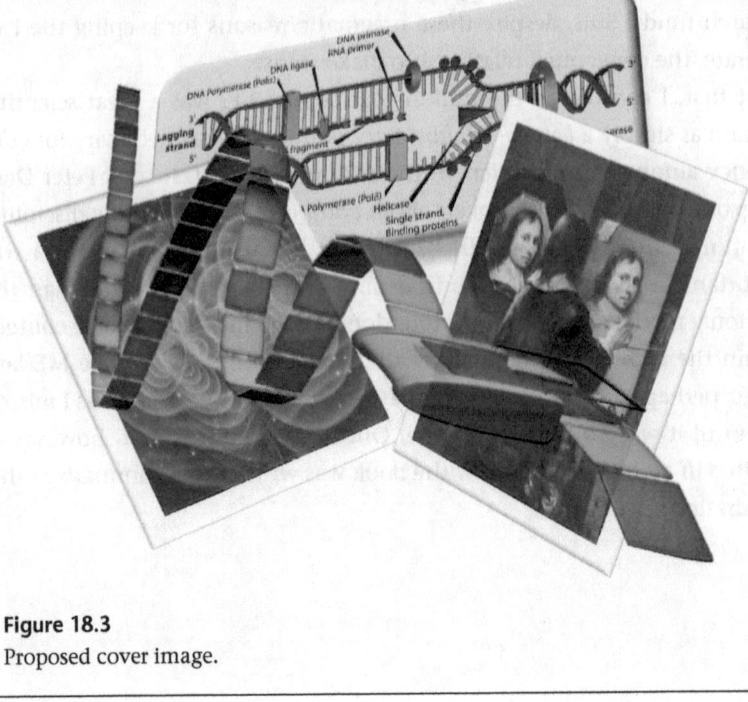

Figure 18.3
Proposed cover image.

Figure 18.4
Final cover image.

I felt at the time, and still feel, that the book captured a bunch of novel insights that led me to reorient my thinking in multiple ways as I explored them. In brief summary, such insights included: the nature of great scientific domains, including their essential intertwining of understanding (i.e., science) and shaping (i.e., engineering); how science/understanding and engineering/shaping should be defined and thus also what each should comprise; computing sciences as the fourth great scientific domain; how the relational model could provide an interdisciplinary understanding of the breadth and structure of the computing sciences; and the potential of the relational model to extend beyond this to understanding a wider range of human intellectual endeavors. The relational model itself is the architecture that captures these insights.

Yet unfortunately but perhaps not surprisingly, *On Computing* did not have the impact I had hoped on the broad computing community. It did get a very nice review in *Nature* (Gilbey, 2012), one of the top general science journals/magazines, and has been cited eighty-four times (Google Scholar, 2025) on top of the citations garnered by various short articles on this general topic, but I can't say that the field as a whole has paid much attention. Perhaps, when I am long gone, the field will find some of these ideas useful as it reflects more on itself. At the moment, much of the attention is limited to members of the small subcommunity focused on understanding the nature of computing as a discipline, and a science, such as Peter Denning—who got interested in the work early on, encouraged and mentored me on it, and coauthored an article with me on it (Denning & Rosenbloom, 2009)—Matti Tedre (2015), and Subrata Dasgupta (2016). This level of impact still matters, but my original ambitions were much greater.

In contrast, the book somewhat surprisingly did have an impact on how we introduced computing to undergraduate students at USC for several years (aside 18.2). Very early ideas that led to the book also inspired a restructuring of how we talked about the different research areas within the CS Department at USC—providing a novel, coherent way of characterizing it—but this too was abandoned after a few years because it was too eccentric to make sense to most people without a lot of associated explanation.

Aside 18.2: Teaching the relational model of computing.

I was surprised, or perhaps shocked is a better term, when in 2013 the chair of CS—Gaurav Sukhatme—suggested basing a new required freshman course on *On Computing* and offered to coteach it with me. I was of course gratified that he recognized something useful in these ideas for CS education and, furthermore, that he found them critical enough to work with me on exposing all of our undergraduates to them in this manner. But, at the time, I had been thinking more modestly in terms of an upper-division elective. My one previous experience teaching this material was the graduate seminar I had taught years before.

We did create and coteach this course, pairing *On Computing* with *Computing for Ordinary Mortals* (Amant, 2013), with the goal of providing our freshmen a broad framework for understanding computing—as much more than just programming—that could carry them through the remainder of their education and beyond.[9] The student evaluations after the first semester showed that there were indeed some students who got out of it what we hoped they would, but there were also a surprising number who absolutely hated it.[10] We continued to refine the course over several years, but it never made it to where enough of the students saw its value. I comforted myself with the hope that looking back later even the skeptical students might understand its utility to them, but ultimately the department revised and downsized the course to a more conventional focus on algorithmic/computational thinking.

In a bit of a contrast, the notion of computing sciences from this work seems to have had an impact more broadly on thinking about informatics education in Europe, with Denning and Rosenbloom (2009) cited as one of three complementary characterizations of *informatics*—the European term for a broader field than what is considered computer science in the United States—by The Committee on European Computing Education (2017).

One other interesting impact of the relational model has been in the area of *digital humanities*, which concerns the interactions between computing and the humanities (Nyhan et al., 2013). I was introduced to this topic when a USC student, Max Feinstein, came to talk to me about applying the methodology of *critical code studies* (Marino, 2020), a part of digital humanities that treats computer programs like literature to be analyzed for cultural context and significance, to the code that implements Soar. I must admit to having been quite puzzled initially by the idea, not being familiar with how scholars in the humanities approach their subject matter. However, after attending a workshop at USC on digital humanities, I decided it would be fun to see what would happen if I applied the relational model to such a new and relatively unformed area. So, I set out to explore in some detail the dyadic relationships possible between computing and the humanities, with a focus on understanding and structuring the existing topics within the field and on determining if there were any notable gaps in current work that might be worth proposing for future investigation.

It was indeed satisfying to discover that the relational model could bring some useful structure to the field, resulting in a journal article on the topic (Rosenbloom, 2012e).[11] This work attracted sufficient interest within the digital humanities community to be republished in the first reader aimed at helping to define the field (Nyhan et al., 2013). Multiple translations of this reader now exist, including into Chinese and Russian.

There is certainly much that can still be done with the relational model. Exploring its implications more broadly and deeply across computing comprises one aspect of this. But so too does exploring its aptness and utility across the rest of the intellectual world that does not include computing as a component, although I do not have the expertise myself in these other domains to take the lead in such an effort.

Several people have asked over the years as to whether the relational model can be used prescriptively—for example, to identify previously unconsidered areas of computing worth exploring—rather than just retrospectively as a way of understanding what already exists. Although the potential to do this was part of the original motivation for the model, and I have taken a few stabs at attempting to locate such examples—such as by systematically generating, and endeavoring to interpret, all relatively small expressions that include C, as well as by exploring what might be missing from the conventional scope of digital humanities—this mostly remains an intriguing area for future consideration.

19 Conclusion to Part III

Looking back on my career from today's perspective yielded the bipartite theme found in chapter 13 and which is repeated here:

1. *Substantive*: Mind and its encompassing disciplines.
2. *Methodological*: Seeking insight via architectural exploration.

Together this pretty much sums it up in as simple a manner as possible, although based on figure 1.5 it is also possible to refine this a bit further, to where the architectural explorations are partitioned into those that are scientific and those that are metascientific (or reflective). For a more personal summary, see aside 19.1.

Aside 19.1: Personal summary.

I haven't been able to devise a comparably concise summary of my personal life, yet there are some clear pieces. The most important have been the strength and stability that my family has provided throughout my life (aside 13.1) and the impact my nature (aside 13.2) has had on essentially everything. A second is that if you start with the notion of multiple forms of intelligence—as discussed in section 4.1.4.3—in my life the differing rates at which these have matured has been striking (aside 15.9). Perhaps related to this has been an improved ability over the decades to deal with major crisis points, or at least to be comfortable with how I've dealt with them (aside 15.11).

My career was not particularly a planned one (aside 19.2), except perhaps for an early long-term commitment to my grand challenge (maxims 14–15), which in thematic terms becomes *seeking insight into mind*—in particular, complete humanlike minds. The remainder of the substantive subtheme, rather than deriving from a grand challenge, evolved late in my career more from frustration about the current state of things in areas in which I was working and a sense that deeper principles and forms of order would be discernible if I were to let myself thoroughly explore the metascientific possibilities. This in turn led to essential analysis, dichotomoic analysis, and relational analysis; and ultimately in section 1.1 to their synthesis into the reflective method.

Aside 19.2: Long-term planning and unexpected illness.

My one concrete longer-term plan outside of a proposal or commitments to a degree program was created recently and was only peripherally related to research. When I was elected academic vice president of USC's Academic Senate in 2016, it was actually a three-year commitment, with automatic progression to president of the faculty and then immediate past president of the Academic Senate. My plan started with those three years, followed by a year of sabbatical and a year back as normal faculty (a requirement of the sabbatical). After these five years, I intended to decide whether to carry on—if I still had sufficient ideas and funding—or retire.

This plan played out as expected for 3.5 years until I noticed in October 2019, shortly after Elaine and I had returned from visiting Michael in the Washington, DC area, where we had tramped on foot for miles around the district and northern Virginia, that I was having trouble simply walking across USC's flat and relatively compact campus without frequently stopping to catch my breath. At that point I made the mistake of going directly to a specialist—a pulmonologist since I was having a breathing problem—who suggested I see a cardiologist. After several months and too many heart-related tests, I was finally diagnosed with *acute myeloid leukemia (AML)* by my internist—who I should have seen first—based on a simple blood test.

The "acute" label was well earned here, in terms of the rapidity of the disease's onset. I was in terrible shape by the time I saw an oncologist and was quickly admitted to Saint John's hospital in Santa Monica, where I had been born sixty-five years earlier.[1] The last one-and-a-half years of my planned five years therefore ended up being dedicated instead to three months in the hospital—two at Saint John's and one at the City of Hope in Duarte—for multiple treatments, including chemotherapy and a stem cell transplant, followed by an extended recovery period. Treating AML is supposed to be a marathon rather than a sprint, but to me it felt more like a triathlon.

Happily, I have been in remission since after the first month of therapy in early 2020. I largely avoided delving into the AML survival statistics during the treatment period, as I knew they were rather grim for someone in my situation; but looking back at them now, I can feel extremely fortunate to have made it as far as I have, with a good chance of continuing to enjoy what I can only view as bonus years.

Elaine and Michael were both indispensable throughout this entire process.[2,3] But, of course, the pandemic hit while I was in the middle of the three months of treatment. For us, it meant among other things that I could have no visitors for many weeks in the hospital. This was a particular challenge given the condition I was in through parts of the treatment, but at least it also meant that driving out to City of Hope for follow-up visits after I was discharged could occur without having to fight normal LA traffic. I can't recommend having cancer in the time of Covid, but it was better than having severe Covid itself. Still, as I was wheeled out through the lobby on my way home after recovering "sufficiently" from the transplant, a standing ovation from the staff—as I have only seen on TV for Covid survivors—brought me to the edge of tears. It enabled me to fully grasp how I had survived a high-risk trip through hell.[4]

The initial combination of frustration and a sense that I could do better was also what earlier led to the work on Othello; yet, whereas it was clear throughout that effort that Othello would remain a diversion, the later work grew from what initially appeared to be another diversion—as it was in fact described in *On Computing*—into a far-reaching extension of the substantive subtheme. Overall, the choice of substantive topics was thus determined by a combination of curiousity—whether of the abstract intellectual sort or the sort induced by more concrete frustrations—plus local judgments concerning whether the problems were sufficiently important and amenable to my making significant contributions.

The choice of methodology was also largely not planned, instead typically emerging from a combination of the demands of the substantive subtheme with the biases induced by my personality. The specifics of the substantive subtheme were themselves biased by my personality, as well as by various environmental contingencies. This combination of influences has yielded an eccentric research path that only occasionally intersected with the mainstream of either AI or cognitive science. I would often try to understand the latest topic that was animating the mainstream, and to see what if any implications it might have for my own work—as with the incorporation of probabilistic graphical models at the core of Sigma based on a sense that they captured a much more general approach to many of the problems I had been dealing with, or failed to deal with, in Soar, and the subsequent extension of these graphical models to deep neural networks—but even then I would adapt the work for uses that were typically uninteresting to the mainstream.[5] My research community, whether concerning mind or disciplines, was most often therefore rather small, when in fact such a community existed at all.

As was shown earlier in figure 13.6, and as revisited in a slightly modified form in figure 19.1[6], my research path as described in this memoir ultimately yielded a career that can be modeled via a four-tiered architecture. At the core (tier 1), as at the core of my career, was work on a sequence of three specific cognitive architectural lineages—XAPS → Soar → Sigma—that combined both computational and theoretical aspirations. The common model of cognition (tier 2) expands outward to explore a consensus theoretical architecture for modeling humanlike minds. Dichotomic maps (tier 3) expand further outwards to explore two kinds of theoretical architectures (as can also be found in figures 1.3 and 1.5): architectural maps covering the structure of individual cognitive/A(G)I architectures (and the common model); and disciplinary maps covering the wider range of

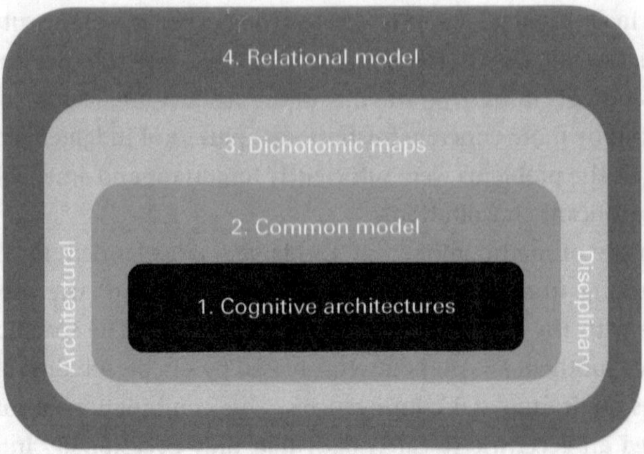

Figure 19.1
Reprise of substantive subtheme from figure 13.6, with tier 3 split into architectural and disciplinary maps.

technologies found in AI (and cognitive science). The relational model (tier 4) expands all of the way out to not only include a theoretical architecture for the great scientific domain of computing, but also potentially for all intellectual endeavors.

At a coarse grain, tiers 1 and 2 concern mind, tier 4 concerns the discipline of computing (and beyond), and tier 3 in some sense bridges these two rather distinct perspectives in providing both architectural maps of mind and disciplinary maps of AI and cognitive science. Digging even deeper, we can begin to explore what relationship, if any, exists between each adjoining pair of tiers. Starting with the innermost pair, there is a simple relationship given that the common model (tier 2) is an abstract consensus over the humanlike architectures found in tier 1. The relationship between tiers 2 and 3 depends on using architectural maps for understanding the common model, and thus humanlike architectures more broadly. This has the potential to supplement the common model itself by explaining more fundamentally, in terms of dichotomies and criterial trade-offs, why humanlike architectures combine the technologies they do, as well as to extend the consideration beyond humanlike minds.

Considering tiers 1 and 2 plus the architectural maps from tier 3, the bulk of my "mind" career can be seen as comprising two large synthetic explorations of computational architectures—Soar and Sigma (with XAPS more minor)—and two large analytic explorations of theoretical architectures—the common model and architectural maps. Each of the latter two can be seen as reflective or meta explorations, although the common model also maintains aspects of normal science. Both Soar and Sigma clearly have theoretical aspects with respect to AI and cognitive science, but the computational aspect has been dominant for both. Figure 19.2 was developed with John Laird for our joint presentation in accepting the *Herbert A. Simon Prize for Advances in Cognitive Systems* in 2018 and shows the overall trajectory of my tiers 1 and 2 in context with his work on the same two tiers and Allen Newell's (and Herb Simon's) work on tier 1.

Tiers 3 and 4, with the focus now narrowed to disciplinary maps in tier 3, can be brought together via an abstract expression from the ME language—Δ + Δ—representing two domains of any sort (Δ) connected by any relation (+), assuming that the domains can, as necessary, be arbitrarily narrow. Applying the ME language to narrower subdomains was already anticipated in the work on digital humanities, where H was introduced

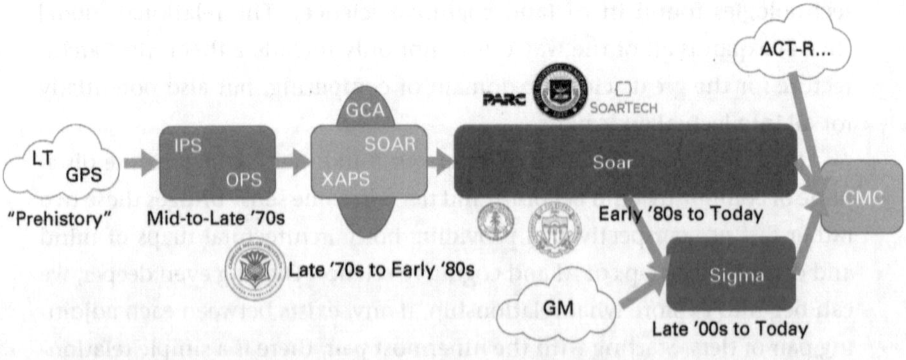

Figure 19.2
Research trajectory on cognitive architectures, developed jointly with John Laird. The sequence of organizational affiliations below the Soar box are mine, whereas those above it are John's. LT (the Logic Theorist) (Newell & Simon, 1956) and GPS (the General Problem Solver) (Newell et al., 1959) were critical to where we got our start but are prehistory from the perspective of our own trajectories.

as the humanities portion of S. It may then make sense to add an *opposite* relation—perhaps denoted as ⊕, based on the similarity between it and *exclusive or*—to the relational model to represent dichotomies.

As mentioned in section 6.2, although I didn't notice it at the time, it also turns out that the two relations at the heart of the relational model—implementation and interaction—map rather well onto the two key approaches to spanning dichotomies that were earlier referred to as *reduction* and *addition* (Rosenbloom, 2011b). Reduction concerns how elements at one level can be reduced to combinations of elements at the level below, while implementation takes the dual perspective of how a combination of elements at one level can yield elements at the level above. Addition adds more mechanisms that interact with existing ones to enable a wider range of capabilities overall. Given this mapping, rather than adding *opposite* as a relation type, given how different it is in nature from the other two, these two tiers might instead be brought together by applying the relational model to explain how dichotomies can be spanned.

Extending beyond the cognitive-architecture research trajectory from figure 19.2 to view my career as a whole on a chronological basis yields a triphasic structure spanning three phases—*early*, *middle*, and *late*. The early phase, roughly covering the first two decades, largely aligned with Allen

Newell's notion of desires and diversions. The substantive subtheme—that is, the desires—was laser focused on mind, with work in tier 1 on XAPS and Soar via a methodological focus on cognitive architecture. During this phase, which was clearly the most impactful objectively with respect to the broader research community, I could succinctly be considered a cognitive architecture researcher. This phase did include one major diversion—on Othello—plus the large-scale application of Soar to military simulation that in a sense blended desires and diversions. This early phase also coincided with the peak in my leadership with respect to the AI community.[7]

The middle phase was my interregnum—or *midlife (research) crisis*—which lasted roughly a decade. This phase was athematic with respect to my research career as a whole, although more narrowly it did have a focus on initiating interdisciplinary new directions via a rather eclectic methodology.

The late phase, lasting over a decade and a half so far, has been a time of rejuvenation, continued growth, and surprising efflorescence. Research on mind continued in tier 1, with Sigma, while also extending into tiers 2 and 3. In addition, the substantive theme grew beyond a focus on mind itself to include mind-encompassing disciplines, with work in tiers 3 and 4. As well, during this phase the methodological subtheme fully ripened, whether thought of in terms of architectural exploration or the reflective method, and leadership with respect to the USC community peaked. Although the early phase was the most impactful one in a traditional academic sense, this late phase has in its own way been highly innovative—at least in my own judgment—and perhaps the most productive. It serves as a prime example of late maturation (aside 15.9) and of how not all of one's best ideas are limited to one's youth (maxim 27). A recent article on late bloomers captures this notion well (Brooks, 2024).

I have also realized, and accepted, quite late in writing this memoir that it was not until this last phase that I was truly mature enough intellectually, and perhaps embued with sufficient intellectual self-confidence, to set new, large-scale, research problems for myself (maxim 26). The big early problems, such as chunking and Soar were set by Allen Newell, and for the latter John Laird as well, although I did contribute significantly to the pursuit of both while also devising many relatively smaller problems to which I could contribute. The work on Othello did originate with me, as did the extension of the work on practice to include stimulus-response compatibility and the interaction between these two, but these were more small to medium-sized efforts that did not start with any grand ambitions. And thus my career continued for a number of years while working on Soar and a range of small to medium ideas.

Yet, the more recent work on Sigma, dichotomies, and computing—plus more recent reflections that started with chapter 14 in this memoir and continued through part I of this book—are examples of problems I would not have been able to set for myself earlier, or for the most recent reflections even have had sufficient self-awareness to pull off. The earlier inability to set new large research problems for myself may in fact have been a large, previously unacknowledged—even to myself—contribution to the sense of burnout discussed earlier in this memoir.

This caveat notwithstanding, my research over the years reached a level of influence—with the work on Soar clearly being the most impactful—that I have been elected a fellow of three professional societies—the American Association for the Advancement of Science (AAAS),[8] the Association for the Advancement of Artificial Intelligence (AAAI), and the Cognitive Science Society. I have also received a few additional awards for, among other things, work on Soar (the *Herbert A. Simon Prize for Advances in Cognitive Systems* in 2018, jointly with John Laird), Sigma (the *Kurzweil Award for Best AGI Idea* in 2011 and the *Kurzweil Award for Best AGI Paper* in 2012), the relational model (*USC Phi Kappa Phi Faculty Recognition Award* in 2012–2013), and the rethinking of the physical symbol systems hypothesis (the *Springer Prize for Best Paper at AGI-23* in 2023). My research has not, however, (yet?) reached the level of influence that would make me an obvious candidate for the top honors in any of the broader fields.

Moving forward, I don't currently anticipate doing more serious programming, and so am unlikely to continue with Sigma, unless the temptation to give it a personality becomes too great. Fortunately, two students have investigated ideas for what might come after Sigma—Abram Demski[9] pursued an approach in his dissertation that involves an even more general layer below Sigma's extended factor graphs, based on weighted logic programming (Eisner & Filardo, 2011), that can implement them and much more (Demski, 2022); and Jincheng Zhou, a USC student at the time who started working with us as a sophomore, investigated a generalization of Sigma that deals better with continuous data and holds out the possibility for a new approach to chunking (Zhou & Ustun, 2021). These are both computational architectures, with implementations in Julia and Python, respectively.

What I do anticipate continuing to work on, via the new life balance described in aside 19.3, are the noncomputational theoretical architectures found in tiers 2 and 3, along with the additional kinds of reflections found in part I. Whatever the directions might end up being, the point will always be searching for insights with potential impact; and to do so via architectural explorations of topics I care about, and that leverage my strengths while working within or around my weaknesses and limitations.

Aside 19.3: A new life balance.

As I began my recovery from AML, it felt in many ways like I was starting over from scratch, with very limited physical, mental, and social abilities. (Following up on the photos in figures 13.2–13.5, figure 19.3 shows my last two professional headshots, as they appeared on my web page, prior to falling ill.) As I gradually came back, I was able to start thinking and writing in the mornings—I am a morning rather than an evening person—and have been exercising twice daily to try to overcome the physical damage from the treatments and concomitant immunosuppressants.

Figure 19.3
Professional headshots from 2011 and 2016.

(continued)

Aside 19.3 (continued)

For a number of years, I have felt that my ideal day would include working in the morning, with this stretching out longer only when necessary or desirable, and the remainder of the day available for family interaction, reading, walking/exercising, and satisfying other needs and desires. In effect, I gradually built up to this during my recovery period, structuring my days around my personal life rather than my work life, with work restricted to a specific time frame. In these terms, retirement needn't mean an end to work, particularly work that I enjoy, but an inversion of its role in my life that yields a new life balance moving forward.

One huge side benefit of having had to depend on assistance from Elaine and Michael throughout this recent period has been how it has strengthened and enriched our mutual bonds (no, not the financial sort). I understand that many families went through something similar during the pandemic, but for us this effect has been multiplied, further enriching this new life balance.

20 Final Thoughts

Reflecting may be endemic to later periods in life, or induced by traumatic events, or simply be a natural approach to dealing with the world. In my case, it has been some combination of the three that has led to this book, an attempt to capture and explain much of what I think I have learned over my career of architectural exploration that goes beyond the technical details involved in building cognitive architectures. This includes reflections on cognitive architectures and key concepts relevant to them. It also includes reflections on the range of fields within which cognitive architectures are studied. Going even further, it includes reflections on aspects of my life that have both preceded this career and proceeded alongside it, many of which have had a strong influence on my career's trajectory.

Part I reflected on four key concepts from the study of mind, attempting to clarify each by stripping it down to an essence plus a space of variations. *Architectures* are fixed mechanisms that when combined with variable content yield systems. Two key variations here concern whether the architectures are computational or theoretical. Cognitive architectures are the relevant specialization of architectures when considering minds and are typically both computational and theoretical. When combined with knowledge and skills, cognitive architectures yield intelligent systems. *Symbols* are distinct atoms, or placeholders. Whether symbols designate, an aspect that has traditionally been definitional for them, becomes here a key variation rather than part of their essence. A second key variation here is whether quantitative metadata is associated with the symbols in support of probabilistic, neural, and hybrid approaches to intelligence.

Intelligence, as conventionally considered, becomes a nested trilevel hierarchy of alternative concepts that can be characterized from the inside out as immediate, cumulative, and full-spectrum. Variations here then correspond to a space of approximations to these three levels. Alternatively, a simpler essence for intelligence may be conceived of around the notion of adaptivity. *Theories* are where the notion of designation that was stripped out from the essence of symbols returns to center stage, in the form of aboutness (and metasystems). Key variations here stem from many of the aspects traditionally assumed to be necessary for something to be a theory—such as correctness, generality, and principledness—but which are now considered optional.

Each of these four concepts has been explored in a broad sense, but always with an eye for understanding the mind, and even more particularly for understanding cognitive architectures. The implications of these analyses have varied widely, from the nature of deeply hybrid systems and evaluation criteria for theories to AI ethics and free will. The hope with each of these has been to help provide insight, clarity, and coherence, even in those cases when the implications tend to be more on the speculative side.

Part II shifted the focus from concepts to technologies, and to the integrated systems that are to either model or be minds. Three essences—cast as the core dichotomies of *(a)symmetric*, *(un)controlled*, and *(meta)data*—came into focus here, with explorations of each individually and of how they interact to structure both disciplines and architectures/systems. As category labels, each core dichotomy organized a larger set of dichotomic variations found across AI and related disciplines. As dimensions of the space of technologies found across these disciplines, the core dichotomies yielded maps within which technologies could be located, compared, and contrasted. Useful notions, such as whether technologies were identical or opposite with respect to a map, along with a distance metric that such maps induce, also came to the fore. Complementing these disciplinary maps, architectural maps enabled understanding the structure and completeness of architectures and integrated systems of interest with respect to the overall space delimited by the dichotomic dimensions, while enabling comparisons among them in terms of what they both include and omit.

The question of how to evaluate cognitive architectures has been a long-term concern in the field, with little substantive progress having been made over the years. It may, or may not, have been a surprise then that evaluation criteria ended up playing a significant role in several ways in parts I and II. Extensive lists of evaluation criteria showed up in part I for theories (chapter 5) and in part II for technologies and integrated systems (chapter 7). Abstractly, this split corresponds to the distinction between theoretical and computational architectures mentioned above, but with an overlap between the two that occasionally yielded conflicts in terminology usage. The criteria in part II took on a particular prominence in helping to understand trade-offs between poles of the core dichotomies and, beyond this, in helping to compare and contrast both pairs of locations within disciplinary maps and pairs of architectures/systems across architectural maps.

Part I also listed aspects of full-spectrum intelligence that were then used in evaluating both LLMs and Sigma (chapter 4). This can thus also be seen as providing a third list of evaluation criteria, but here concerning aspects/capabilities necessary for achieving this form of intelligence. As such this list applies to architectures and integrated systems intended to provide humanlike intelligence, whether theoretical or computational. Complementing this was a fourth rather small list of evaluation criteria introduced early on to help explain why the evaluation of cognitive architectures is not just about controlled experiments (chapter 2).

Part III explored my career, based on a substantive theme of mind and its encompassing disciplines, and a methodological theme of searching for insight via architectural exploration. This part also examined my life more broadly, with much of this geared toward understanding how key aspects of it shaped my career, but with other aspects included to further flesh out the overall picture. Part III has provided the third version of a memoir I began while recovering from a serious illness. Some of the background material that was developed for the earlier versions of this memoir proved also to be useful as background for parts I and II, leading to promotion of this material from part III to the overall introduction in chapter 1.

Both chapter 1 and the more focused introduction to part III provided ways of viewing my career as a whole in terms of a particular combination of research topics and methodologies—reflected in figures 1.5 and 19.1, respectively. Due to when it was conceived, the version in figure 19.1 omitted consideration of the material in part I of this book, but it did include part II plus my decades of work on the grand challenge of developing cognitive architectures that aid in understanding the human mind and building artificial minds. It also included work on the common model of cognition; an interregnum that involved exploring the future of computing and its interactions with other domains of intellectual endeavor; and two large-scale efforts that came out of this interregnum: the development of the Sigma cognitive architecture, and an exploration of computing as a great scientific domain plus the use of relational analysis in understanding it.

As partially apparent in figure 19.1, and more fully developed in figure 1.5, three distinct forms of analysis—*essential*, *dichotomic*, and *relational*—formed the foundation for the reflective phase of my career that is the focus of this book. The first two forms of analysis were central to parts I and II, respectively, whereas the third was central to chapter 18 of part III. All three provide novel tools for understanding disciplines and systems, with the latter two also inducing semi-formal languages—the *criterial advantage (CA)* and *metascience expression (ME)* languages—that yield succinct means of summarizing and comparing topics within their scope. As part of bringing these three forms of analysis together in this book, and reflecting on them jointly in the process, a more general *reflective method* has been identified that can be seen to subsume these more specialized methods.

From a personal perspective, the exploration and summarization of the reflections in this book proved to be a rather challenging but correspondingly rewarding effort. For you the reader, the hope is that attempting to assimilate these reflections will have proven useful, or at least thought provoking, on multiple levels.

Appendix A: Technologies (Part II)

Many technologies relevant to core processing across AI and related disciplines were mentioned in part II but mostly not explained. The purpose of this appendix is to provide brief, informal explanations of them. This process begins in table A.1 with a succinct summary of this range of technologies. Although the table is densely packed, it has no pretensions to being exhaustive. Some existing technologies may effectively be combinations of those listed while others may simply be missing. What this table is instead to provide is a simple but useful structuring over the technologies that are then explained in what follows.

Each row in this table starts with a category of technologies. These categories are not necessarily intended to be orthogonal, and some may also seem a bit odd, but all together they provide some useful structuring. Following the category labels are one or more listings, separated by semicolons, of modifiers of these categories. Each comma-separated listing of modifiers specifies alternative, but not necessarily orthogonal or exhaustive, variants of its category. If there is a natural ordering to such a listing, it is used; otherwise, it is simply presented in alphabetical order. Further decompositions of alternatives are provided in parentheses. Many modifiers within a row can be combined freely—such as symbolic constraint propagation and numeric constraint satisfaction—but this need not always be the case.

In the following, the technologies in the table are briefly explained in an informal manner intended to convey the gist of the ideas rather than formal definitions. Much of this should be easily recognizable to anyone broadly familiar with the disciplines. However, there are aspects that are idiosyncratic to how the terms are used in part II.

Table A.1
Summary of the range of technologies appearing in part II, characterized as modifiers or variants of relevant category labels

Category	Modifiers/Variants
Band	Biological, Cognitive
Constraint	Programming, Satisfaction (Propagation, Conditioning); Numeric (Soft), Symbolic (Hard)
Control	Reaction (Policies), Deliberation (Action Selection and Execution, Cognitive Cycle), Reflection (Problem Space Search, Planning); Numeric, Symbolic, Hybrid (Probabilistic); Motor, Skill; Influence Diagrams
Emotions and Affect	Proto-Affect (Reactive), Primitive (Routine), Cognitively Elaborated (Reflective)
Functions	Numeric, Symbolic, Hybrid; Monotonic, Nonmonotonic
Knowledge	Facts, Frames, Instances, Cases, Analogies; Ontologies, Knowledge Graphs; Logical Sentences, Constraints
Learning	Bayesian (Solomonoff), Neural Network (Backpropagation, Hebbian), Structure (Composition (Chunking (Data), Distillation, Macro), Proceduralization); Control (Reinforcement (Deep; Model-Based)), Model; Knowledge Acquisition (Declarative), Skill Acquisition (Procedural); Fast, Slow; Numeric, Symbolic; Knowledge-Level, Symbol-Level; Induction (Decision Tree, Genetic Algorithms/ Programming, Inductive Logic Programming, Neural Network, Solomonoff); Discriminative, Generative; Supervised, Unsupervised
Logic	Propositional, First-Order, Probabilistic; Monotonic, Nonmonotonic; Horn Clauses, Situation Calculus
Markov	Decision Processes, Hidden Models, Networks
Memory	Declarative/Knowledge (Semantic, Episodic), Procedural/Skill (Rules), Working (Blackboard, Global Workspace); Explicit, Implicit; Numeric, Symbolic, Hybrid
Neural Networks	Autoassociative (Asymmetric (Autoencoder), Symmetric (Hopfield, Boltzmann, Self-Attention)), Heteroassociative (Feedforward, Recurrent (Gated, Long Short-Term), Attention); Embedding; Structured/ Relational (Convolutional, Encoder-Decoder (Autoencoder, Transformer), Generative Adversarial, Graph, Local Connectionist, Recursive); Hyperdimensional Computing/Vector Symbolic Architectures, Tensor Product Representations
Physics	Quantitative, Qualitative

Category	Modifiers/Variants
Probability and Statistics	Function-Based (Arithmetic Circuits, Sum-Product Networks), Model-Based (Probabilistic Graphical Models (Bayesian Networks, Factor Graphs, Markov Networks)); Propositional, First Order (Statistical Relational AI)
Processing	Reaction, Deliberation, Reflection; Sub/Nonsymbolic (Numeric (Dynamical Systems)), Symbolic, Hybrid (Probabilistic); Automatic, Controlled; Thinking (Fast, Slow); Parallel, Serial; Tensor, Graph; Modular, Non-modular; Cognitive, Noncognitive; Conscious/Explicit, Unconscious/Implicit; Central (Knowledge Representation and Reasoning), Peripheral (Sensory-Motor Behavior); Action-Centered, Non-Action-Centered; Iterative, Anytime (Speed-Accuracy Trade-offs)
Programs	Declarative (Constraint), Procedural; Numeric, Symbolic, Hybrid (Probabilistic); Code (Executable, Source)
Reasoning	Causal (Association, Intervention, Counterfactual); Shallow, Deep; Rule-Based, First-Principles; Function-Based, Model-Based; Conjunctive, Disjunctive; Numeric, Symbolic, Hybrid (Probabilistic); Truth Maintenance (Assumption-Based, Justification-Based); Analogical, Case-Based
Search and Planning	Knowledge, Problem Space (Problem-Solving); Local Optimum (Greedy, Gradient Ascent/Descent, Hill Climbing), Global Optimum; Exhaustive, Heuristic (Probabilistic (Monte Carlo Tree)); Goals, Utilities; Goal-Oriented, Decision-Theoretic

- A *band* is a sequence of contiguous timescales for human thought that covers three orders of magnitude.
  - The *biological* band covers the organelle, neuron, and neural circuit timescales, at approximately 100 μs to 10 ms.
  - The *cognitive* band covers the deliberate act, operations, and unit task timescales, at approximately 100 ms to 10 sec.
- A *constraint* limits the values that can be jointly assigned to a set of variables.
  - Constraint *satisfaction* is a form of declarative *programming* (see later) that seeks compatible assignments to a set of variables given a program specified in terms of a set of constraints.
    - Constraint *propagation* directly reduces sets of possible variable values based on possible values of other variables and the constraints among the variables.

        - Constraint *conditioning* induces a search over possible choices of values for the variables.
    - *Numeric (soft)* constraints can be approximately satisfied.
    - *Symbolic (hard)* constraints are all-or-none with respect to satisfaction.
- *Control* determines how processing proceeds.
    - *Reaction* amounts to an immediate response to changing circumstances, as for example determined by a *policy* that directly specifies actions that are preferable in particular situations.
    - *Deliberation* amounts to a sequence of one or more decisions involving *selecting and executing actions*, whether internal or external, without the need for reflection. Each decision involves a single *cognitive cycle* of memory access followed by a choice.
    - *Reflection* amounts to stepping back from the immediate decision-making situation to consider more broadly what it is best to do.
        - *Problem space search* and *planning* accomplish this by exploring the impact of potential future sequences of actions on goal achievement, with planning also concerned with yielding the sequence of steps—that is, the *plan*—that yields goal achievement.
    - *Numeric, symbolic,* or *hybrid* control considers whether there are quantitative, qualitative, or both aspects in deciding how to proceed.
        - *Probabilistic* control is a special case of hybrid control where the quantities are probabilities.
    - *Motor* control concerns how external actions are directed.
    - *Skill* concerns an extended pattern of controlled behavior.
    - *Influence diagrams* add decision (and value/utility) nodes to probabilistic graphical models to enable choices to be made.
- *Emotions* (or *affect*) concerns feelings.
    - *Proto-affect* comprises the *reactive* component of feelings.
    - *Primitive* emotions comprises the *routine* (or *deliberative*) component of feelings.
    - *Cognitively elaborated* emotions comprise the *reflective* component of feelings.
- *Functions* compute output values from inputs.
    - The outputs may be *numeric, symbolic,* or *hybrid* in analogy to the comparable forms of control.

- A *monotonic* function either increases or decreases in value across its domain but not both. It may also remain constant through regions of the domain.
- A *nonmonotonic* function can increase and decrease in value across different regions of its domain, as well as remain constant over other regions.

- *Knowledge*
  - A *fact* is an individual relational statement that may be true or false.
  - A *frame* is a set of facts about a single entity.
  - An *instance* represents a specific thing.
  - A *case/analogy* is a preexisting instance that can be adapted to current problems that share essential commonalities.
  - An *ontology* is a set of terms, definitions, and relations that characterize a topic area.
  - A *knowledge graph* is a network of facts about sets of entities, possibly accompanied by ontologies over them.
  - A *logical sentence* is a statement in a logic that expresses facts.
  - A *constraint* (see earlier).
- *Learning* modifies a system based on experience to (hopefully) improve its performance in future situations.
  - *Bayesian* learning leverages Bayes' theorem to determine the most likely hypothesis to explain a set of observed examples.
    - *Solomonoff* induction is a complete yet incomputable approach to calculating the posterior probability of any explanation given a set of observations.
  - *Neural-network* learning acquires values for weights in a neural network (see later) based on observations.
    - *Backpropagation* modifies weights in a heteroassociative neural network by propagating errors backwards through the network.
    - *Hebbian* learning modifies weights locally at synapses in an autoassociative neural network based on the activations arriving in forward and backward directions.
  - *Structure* learning creates new structures from existing components.
    - *Composition* combines a sequence of procedural structures into a single larger procedural structure.

- *Chunking* is a form of composition applied to sets of rules that execute during resolution of an impasse.
  - *Data chunking* leverages chunking to simulate the acquisition of declarative knowledge within procedural memory.
- *Distillation* is a form of composition applied to sets of neural networks.
- *Macro* learning combines sequences of operators into new operators.

- *Proceduralization* incorporates elements of declarative memory into elements of procedural memory.

- *Control* learning adapts to make better choices.
  - *Reinforcement* learning acquires policies by propagating rewards and evaluations backward over sequences of states and action executions.
    - *Deep reinforcement* learning expresses its policies as heteroassociative neural networks.
    - *Model-based reinforcement* learning includes models of the domain as input to the learning process.
- *Model* learning creates representations of situations and actions.
- *Knowledge acquisition* learns knowledge (in *declarative* memory).
- *Skill acquisition* learns skills to be performed (in *procedural* memory).
- *Fast* learning involves innate, automatic learning processes.
- *Slow* learning involves deliberate or reflective learning strategies.
- *Numeric* learning updates metadata.
- *Symbolic* learning updates data and its associated structures.
- *Knowledge-level* learning increases the amount of knowledge ascribable to an agent as reflected by its ability to use the knowledge in rational decision-making toward goal achievement.
- *Symbol-level* learning may improve the performance of an agent—for example, by speeding up its ability to return answers—but does not increase the knowledge ascribable to it.
- *Induction* enables making generalized predictions about unseen items based on items that have been seen.
  - *Decision tree* induction learns hierarchical decision-making models.
  - *Genetic algorithms/programming* learn solutions/programs by searching over populations of possibilities for ones that perform well on

problems of interest, while using operations such as crossover and mutation on existing populations to generate new populations.
        - *Inductive logic programming* learns logical programs from examples.
        - *Neural networks* learn network weights from examples.
        - *Solomonoff* induction (see earlier).
    - *Discriminative* learning enables making choices among alternatives.
    - *Generative* learning acquires distributions over alternatives.
    - *Supervised* learning supplies both inputs and outputs for learning.
    - *Unsupervised* learning either does not provide outputs for learning or fails to distinguish them from inputs.
- *Logic* provides a formal means of representing and reasoning about knowledge in terms of a language, its syntax and semantics, and its rules of inference.
    - *Propositional* logic concerns the truth values of grounded facts and logical relations among them.
    - *First-order* logic adds notions such as variables and quantifiers to propositional logic in service of expressing more general facts.
    - *Probabilistic* logic extends normal (symbolic) logic to include the probabilities of truth values.
    - *Monotonic* logic never reduces what can be inferred as more knowledge is added.
    - *Nonmonotonic* logic may lose some existing inferences as more knowledge is added.
    - *Horn clauses* are logical sentences that take the form of rules.
    - *Situation calculus* simulates the retraction of inferences—that is, a form of nonmonotonic logic—by introducing situation variables into monotonic logic.
- *Markov* denotes technologies in which predicting uncertain futures need only consult the present and not the past.
    - *Markov decision processes* model past-independent sequential decision-making under uncertainty.
    - *Hidden Markov models* model past-independent sequential non-decision-making processes under uncertainty.
    - *Markov networks* (see under probability and statistics)
- *Memory* holds information for later reuse.

- *Declarative* memory is a long-term memory for factual information (i.e., *knowledge*).
  - *Semantic* memory holds general information about the world.
  - *Episodic* memory holds information about sequences of situations that have been experienced.
- *Procedural* memory is a long-term memory for information about how to control behavior (i.e., *skills*).
  - *Rules* are condition-action structures that encode actions to be performed when conditions are met.
- *Working* memory holds, and shares across modules, declarative knowledge that is retrieved/generated by modules such as declarative memory, procedural memory, and perception. It is often transitory, but not always.
  - A *blackboard* is a variation of working memory found in systems constructed of sets of independently operating experts.
  - A *global workspace* is a variation of working memory focused on distributing information globally in the brain.
- *Explicit* memory is examinable by cognitive processing.
- *Implicit* memory plays a role in determining behavior but is not examinable by cognitive processing.
- *Numeric* memory has meaning primarily embodied in metadata.
- *Symbolic* memory has meaning primarily embodied in data.
- *Hybrid* memory has meaning embodied in a combination of metadata and data.

- *Neural networks* are abstract interconnected models of natural neurons that can learn to perform a wide range of tasks.
  - *Autoassociative* networks use the same variables for both input and output.
    - *Autoencoders* are *asymmetric* autoassociative networks for which the output has the same variables as the input.
    - *Symmetric* autoassociative networks consist of a single bidirectional network in which each variable can act as both an input and an output.
      - *Hopfield* networks and *Boltzmann* machines are two classical examples that it is not necessary to distinguish here.
      - *Self-attention* networks are like attention networks (see below) but relate elements within a single sequence rather than across two sequences.

- *Heteroassociative* networks use different variables for input and output.
  - *Feedforward* networks pass information in one direction with no cycles.
  - *Recurrent* networks simulate time sequences by nominally connecting outputs back to their inputs, although this may be done without loops by including multiple stages, where outputs at one stage are connected to inputs at the next stage.
    - *Gated recurrent units* (GRUs) and *long short-term* memories (LSTMs) are both forms of recurrent networks that add additional units and connections to provide access to more of the past.
  - *Attention* networks fully remove the Markov assumption found across the layers of feedforward networks and the stages of recurrent networks in order to learn explicitly what prior aspects are relevant.
- *Embedding* enables symbols and relations among them (aka structures of them) to be represented as patterns of numbers in vectors.
- *Structured/relational* networks reflect the structure of (at least some of) the domain in the structure of the network.
  - *Convolutional* networks embody a 2D structure to support visual processing.
  - *Encoder-decoder* networks combine a sequence of two feedforward networks with a bottleneck between the two to yield encoding in the first network and decoding in the second.
    - *Autoencoders* (see earlier).
    - *Transformers* combine self-attention networks, encoders, and decoders.
  - *Generative adversarial* networks combine networks that generate candidates with ones that evaluate them.
  - *Graph* and *local connectionist* networks use links between elements to model relations in the domain.
  - *Recursive* networks leverage tree structured graphs in which the weights are shared across levels.
- *Hyperdimensional computing,* or equivalently *vector symbolic architectures,* use high-dimensional vectors and algebraic operations to represent structural information.
- *Tensor product representations* leverage tensors to represent relational structures.

- *Physics* concerns the understanding of how the physical world works, in terms of fundamental concepts such as matter, energy, and force.
    - *Quantitative* physics is the standard approach to physics based on equations grounded in continuous mathematics.
    - *Qualitative* physics abstracts quantitative physics by quantizing what would otherwise be continuous values.
- *Probability and statistics* concern the randomness and likelihood of events.
    - *Function-based* approaches can predict the probability of one output variable from the evidence provided for input variables. *Arithmetic circuits* and *sum-product networks* are two closely related variations on this.
    - *Model-based* approaches yield distributions over the probabilities of all variables from the evidence provided for any subset of the variables.
        - *Probabilistic graphical models* are model-based approaches based on graphs of nodes and links. *Bayesian networks, factor graphs, and Markov networks* are three variations that differ in terms of the types of nodes they use and whether the links are directed or undirected.
    - *Propositional* approaches are grounded, with each structure representing only a single instance.
    - *First-order* approaches are lifted, with each structure representing a set of instances. *Statistical relational AI* is a model-based first-order approach to statistical reasoning over symbols.
- *Processing* in this context concerns *reaction, deliberation,* and *reflection* plus a wide range of additional variations on these.
    - *Sub/nonsymbolic (numeric), symbolic,* and *hybrid (probabilistic)* processing here are analogous to the comparable control types.
        - *Dynamical systems* are numeric approaches based on systems of differential equations.
    - *Automatic* processing proceeds without explicit reasoning about what to do next.
    - *Controlled* processing involves explicit reasoning about what to do next.
    - *Thinking* involves the processing of knowledge in a very general sense.
        - *Fast* (*System 1*) thinking is considered to be fast, automatic, parallel, unconscious, emotional, and instinctive.

- *Slow* (*System 2*) thinking is considered to be slow, controlled, serial, conscious, rational, and learned.
- *Parallel* processing can perform multiple activities concurrently.
- *Serial* processing can only perform one activity at a time.
- *Tensor* processing organizes computation around the regular structure of a tensor, typically yielding computations that may be easy to parallelize.
- *Graph* processing organizes computation around the irregular structure of a graph, typically making it hard to parallelize.
- *Modular* processing leverages distinct bodies of specialized material for each type of activity.
- *Nonmodular* processing leverages a single general body of material across all, or at least many, types of activities.
- *Cognitive* processing involves symbolic reasoning.
- *Noncognitive* processing involves sub/nonsymbolic calculation.
- *Conscious/explicit* processing is accompanied by awareness.
- *Unconscious/implicit* processing proceeds without awareness.
- *Central* processing concerns *knowledge representation and reasoning* and is typically considered to be explicit, cognitive, and modality independent.
- *Peripheral* processing concerns *sensory-motor behavior* and is typically considered to be implicit, noncognitive and modality specific.
- *Action-centered* processing (or subsystem) controls internal and external actions.
- *Non-action-centered* processing (or subsystem) maintains general knowledge.
- *Iterative* processing repeats over and over again.
- *Anytime* algorithms/processing can be stopped at any point and produce an answer that typically improves as more time is allowed, yielding *speed-accuracy trade-offs*.

- *Programs* are specifications of computational behavior.
  - *Declarative* programs specify behavior indirectly in terms of what is to be computed.
    - *Constraint* programs are structured as sets of constraints that should be satisfied (see earlier).
  - *Procedural* programs specify behavior directly in terms of how computation is to proceed.

- *Numeric, symbolic,* and *hybrid (probabilistic)* programs are defined in analogy to the similarly denoted control types.
- *Code* amounts to programs for the control of computers.
  - *Executable* code can be directly run by the computer hardware.
  - *Source* code is what is written by programmers and compiled into executable code.

- *Reasoning* comprises inferring conclusions from premises.
  - *Causal* reasoning concerns relationships between causes and effects.
    - *Association* focuses on probabilistic causal reasoning.
    - *Intervention* considers the causal impact of action execution.
    - *Counterfactuals* consider hypothetical causal chains.
  - *Shallow* reasoning is based on simple associations between premises and conclusions.
  - *Deep* reasoning leverages a deeper understanding that underlies the relationship between premises and conclusions.
  - *Rule-based* reasoning uses condition-action structures for reasoning.
  - *First-principles* reasoning leverages theories (aka models) about domains of interest.
  - *Function-based* reasoning computes a single conclusion unidirectionally from the premises.
  - *Model-based* reasoning can yield conclusions in any direction over an explanatory model.
  - *Conjunctive* (*and*) reasoning considers situations where multiple things must be true at once.
  - *Disjunctive* (*or*) reasoning considers situations where only one out of several possibilities must be true.
  - *Numeric, symbolic,* and *hybrid (probabilistic)* reasoning are defined in analogy to the similarly denoted control types.
  - *Truth maintenance* tracks dependencies during reasoning.
    - *Assumption-based* truth maintenance tracks dependencies of conclusions on assumptions.
    - *Justification-based* truth maintenance tracks dependencies among conclusions.
  - *Analogical* and *case-based* reasoning transfer conclusions from similar past situations.

- *Search and planning* explore spaces of possibilities to find elements, or sequences of elements, that match or maximize specified criteria.
    - *Knowledge* search finds relevant elements in long-term memory based on the contents of working memory.
    - *Problem space* search considers sequences of actions and states that connect initial states to ones that match goals. *Problem-solving* can be viewed as a synonym for problem space search.
    - *Local-optimum* search finds the best element according to some criterion that is reachable by moves among adjacent elements that never decrease the criterion. *Greedy* search, *gradient ascent/descent* search, and *hill climbing* are all synonyms or variants.
    - *Global-optimum* search finds the best element according to some criterion in the full space of possibilities.
    - *Exhaustive* search examines the entire space of possibilities.
    - *Heuristic* search uses knowledge about the problem domain to order or prune the choices made during the search. In this usage, no distinction is made based on whether or not the knowledge is guaranteed to be correct.
        - *Probabilistic* search uses a probability distribution to stochastically guide the choices made during the search.
            - *Monte Carlo tree* search stochastically examines a set of paths from the current state to the end of a two-person game.
    - *Goals* are symbolic criteria to be achieved.
    - *Utilities* are quantities to be maximized.
    - *Goal-oriented* search or planning seeks situations that achieve goals.
    - *Decision-theoretic* search or planning seeks situations that maximize utilities.

Appendix B: Maxims (Part III)

As discussed in chapter 13, this appendix distills the wisdom—structured as maxims plus minimal elaborations on them—I would try to impart in the context of the memoir laid out in part III. Some of these maxims were explicit early on while others remained implicit for much of my career until becoming clarified by later reflection. Not all of them will make sense for everyone, nor do they all work for every career style (as Allen Newell also qualified the maxims in *Desires and Diversions*). Some may also be trite, but they have all mattered to me and my career. They are enumerated here, loosely grouped by topic, and with a final grouping that revisits some of Allen's maxims in terms of their relationship to my career.

Insight and Impact

1. *It is the insight that matters.*[1,2] It is a necessary truth that before an insight is applied, it must be validated—whether theoretically, experimentally, or computationally—to an extent appropriate to the application, but it is paramount not to lose the centrality of the insight itself in the weeds of how it is validated.
2. *Struggling in the weeds can be essential for insight.* Whether the weeds are in early exploration or later validation, remaining above—or out of—them can mean losing a major source of insight. This may appear to contradict the previous maxim but is instead intended to complement it.
3. *Implementation is critical for understanding computational architectures.* This can be viewed as a special case of the previous maxim.
4. *Making an impact requires combining compelling visions with attention to detail.* This is likely trite, but essential nonetheless, and it applies in all phases of research.

Problems, Disciplines, and Venues

5. *Pledge your primary allegiance to problems rather than to disciplines.* Work to the extent you can in whatever disciplines, departments, institutes, or organizations facilitate your progress on your problems of choice. Although this risks making you neither fish nor fowl, and thus complicating your career path, it has its own rewards, and if you succeed, everything will turn out well.
6. *Service to your disciplines/organizations is an essential complement to problem-focused research.* This can be a win for both you and your disciplines/organizations and may help you toward being both fish and fowl.
7. *Break down walls rather than build them.* This is my own variation on the now common phrase "build bridges rather than walls." It is particularly helpful when trying to survive and thrive while exploring interdisciplinary problems.
8. *Define disciplines broadly rather than narrowly.* This is related to the previous maxim but suggests constructing any walls that must be built as far as can be workable from a discipline's nominal center. In terms of the relational model, don't just include the discipline's monadic region within the walls, but also dyadic and polyadic regions. Accept that the result will be overlaps among disciplines, and that the walls themselves will therefore be rather weak, as they should in fact be.
9. *Great scientific domains include both science (understanding) and engineering (shaping).* An effective intertwining of both should become more and more important as they continue to become more informative and essential for each other across all four domains.
10. *There is power in blending AI and cognitive science in exploring minds.* Although related to all three of the previous maxims, this one is much more specific, emphasizing the utility of judiciously combining the functional and experimental evidence yielded by these two fields, as well as both science and engineering, in exploring the space of human-like cognitive architectures that are relevant to both fields. The use of *judicious* here implies not being a slave to the details of either field but leveraging what is most critical from each.
11. *Stronger research methods within a discipline push out weaker ones.* Strength here is the level of veracity that a method can guarantee. This maxim is true even in those areas of a discipline in which the stronger methods

within that discipline don't (yet?) apply. Erecting new disciplinary boundaries that wall off the latter areas can sometimes help keep them active, but the walled off areas then risk becoming insular and irrelevant.

12. *Distinguish between absolute and relative method strength.* Use the strongest methods available for the problem at hand (i.e., attend to relative strength) but do not shy away from important problems just because there are not strong methods available, at least at present, for them (i.e., disregard absolute strength). Absolute strength is critical when assessing the veracity, and considering the application, of any results, and may even rightly play a role in determining funding levels; however, it should not be a bar—or even lead to derogation of topics or researchers—where it is not attainable.
13. *Be idea driven in writing papers rather than venue driven.* In other words, figure out what you want to say and use that to drive where you try to publish the result rather than the other way around.[3]

Problems, Challenges, and Hypotheses

14. *Commit to solving a problem that is big enough to make a difference.* But be sure it is a problem you deeply care about, have the skills to contribute to, and enjoy the day-to-day aspects of working on.
15. *Dedication to a grand challenge can provide focus and long-term joy.* The times at which major advancements—and associated *aha* moments—become apparent can yield particularly heady experiences. However, they can be few and far between. So, if you are to pursue such an approach, it is best to be okay with delayed gratification (while, as in the previous maxim, finding sufficient short-term reward in the day-to-day aspects).
16. *Milk an ambitious hypothesis for all it is worth.* The more ambitious the hypothesis the more committed you need to be to it. It can take years, or even decades or more, to understand and derive implications from, and to validate or invalidate truly ambitious hypotheses. There is a deepening of understanding that occurs in the process, however, that is invaluable.
17. *Be willing to make radical changes when there are compelling reasons to do so.* This maxim complements the previous one. You have to be willing, and to give yourself permission, to abandon ambitious hypotheses when it is clear they no longer have legs or when the case for an alternative hypothesis becomes overwhelming.

18. *Focusing on a long-term grand challenge helps avoid being sucked into research fads*. But, as a result, your work may at best go through cycles of being in and out of fashion, depending on how well it matches the current zeitgeist. When it is unfashionable, resources to pursue the work can be difficult to obtain and publication can also be difficult.
19. *Create your own niche.*[4] Given your skills, experience, expertise, history, knowledge, personality, and perspective (see maxim 25) find—or, more realistically, create—and pursue problems and/or approaches that no one else can, or even would consider, focusing on. This gives you a chance of making a unique contribution that only you could have made rather than constantly looking over your shoulder in fear of the competition getting to the finish line first. Such problems and/or approaches are moreover likely ones that you would particularly enjoy pursuing.[5]

Functional Elegance

20. *Always keep your eye out for elegance and beauty, and for what simplifies, unifies, and explains*. I think of this as the physicist's mindset on science, but for me it applies across the board. I seem to be wired in such a manner that without such a focus, I feel like I'm stuck in a quagmire and get little satisfaction from what by other metrics might count as progress.
21. *Seek similarities before considering differences.* In other words, look for as much simplicity and elegance as possible before acknowledging that there must be messiness. There is also a relationship here to Occam's razor.
22. *Seek explanations before settling for descriptions*. Descriptions—even mathematically rigorous or statistically accurate ones—may be pragmatically useful but settling for them when deeper explanations may be within reach is a lost opportunity for true insight.
23. *Dig deeply into proposed -otomies*. Traditional dichotomies may conflate multiple distinct concepts that typically align but that can be better understood when teased apart. Higher-order -otomies, in their turn, may be better understood as cross products of simpler dichotomies.

More Personal

24. *Always be as generous as can legitimately be justified with coauthorship*. Don't let other considerations get in the way of this.
25. *Understand your own strengths and limitations*. Leverage the former as much as you can, even when they are side effects of a broader

"abnormal/negative condition." Although understanding and accepting your limitations can be quite trying, attempting to overcome or work around them can be extraordinarily valuable.

26. *Capabilities and interests mature at different rates both within and across individuals.* Some mature young, others take decades, and others may never mature. It isn't terribly productive to try to force them before their time.
27. *Not all of one's best ideas are limited to one's youth.* Continued maturation can lead to significant innovation even late in life. Some of what I consider my best ideas were developed late in life.
28. *Family is the most important consideration in the big decisions.* However, how to prioritize with respect to the smaller decisions is a more subtle problem, particularly if you want to get a lot done.
29. *Early prejudices should be left to childhood.* In addition to my early prejudice against USC, I also had ones against the military (growing up in the 1960s and having a lifelong aversion to violence, whether real or cinematic), German cars (a hangover from stories about WWII and the Holocaust), and living in Beverly Hills (reverse LA snobbery). My destiny, however, has included spending most of my career at USC, being funded largely by the Department of Defense (although experiencing more and more internal conflict the closer the work came to actual battlefield relevance), owning a Volkswagen at one time, and living near Beverly Hills for much of my life.

Miscellaneous

30. *Consider the possibility of a general power law whenever you think you see an exponential.* Even if things seem exponential locally, in the bigger picture there may be slowdowns in the rate of change over time as well as initial and final curve flattening.

Revisiting Some of Allen Newell's Maxims

Some of these are summarized from memory of what Allen Newell used to say and others are reproduced from Laird and Rosenbloom (1992). The point of having them here is to discuss how they played out in my own career.

- *There will be diversions; make them count.* This applies most directly to my interregnum, which was a decade-long diversion from work on cognitive architectures. I was able to make it count locally—that is, on its own—not only by what grew out of working on the various new directions but also by developing the relational model and the notion that computing was a great scientific domain. But I was also able to make it count in returning to work on cognitive architecture with a kernel of a new idea and the explicit understanding that the only way I felt I was making progress was when I was guided by functional elegance (maxims 20–23). This was the core of what I tried to do with Sigma, providing much of the oomph behind its first three desiderata. It was also implicit in why I felt boxed in earlier while working on Soar rather than feeling free to take the road John Laird later did of adding more memories and learning mechanisms to the architecture. It is furthermore what I have tried to do with the various noncomputational theoretical architectures.
- *To do interdisciplinary work you must become a professional in all of the relevant disciplines.* I did manage to uphold this maxim throughout much of my career, particularly with my work across AI and cognitive science. However, at various times, I did cut corners, including in working with Himanshu Joshi on speech understanding in Sigma, and in engaging in philosophy of science. I did of course read on both topics but did not find the time nor the interest to approach professional competence in either. This may very well have hurt both efforts as well as leading to a sense of failing to meet standards, whether my own and/or Allen's, but at this point I just need to accept this.
- *Listen to the architecture.* This was Allen's way of saying that you should treat the architecture as a source of knowledge, much like the phenomena of interest is already treated. To me, the part of this that has always mattered most is the exhortation to explore what the architecture implies about how to implement a new functionality before jumping in to add a new mechanism for it. As such, it ties in with maxims 20–22. More generally, my notion of functional elegance is a variant of this earlier maxim, although instead of being a heuristic for how to proceed it is a characterization of the goal you want to achieve. In this form it applies to a broader range of situations, including for example identifying a small set of dichotomies that structure AI and cognitive science.

- *Deep scientific ideas are exceedingly simple. Others usually see them as trivial.* This focus on simplicity can be seen across my career in both the desideratum of functional elegance that I have striven to satisfy in my work on cognitive architectures and in what I have tried to accomplish with the various analytical methods. The implied concern with triviality has also arisen a number of times, both in my own mind and in the minds of others.
- *Preserve the insight and deepen it. Deep ideas transform themselves beyond imagining.* This can be viewed as a forerunner to maxim 16. It is reflected in my work on cognitive architectures, particularly Soar and Sigma, on which I spent more than a decade each attempting to deepen my/our understanding. It can also be seen in the work on the relational model, which went from back-of-the-envelope calculations to short papers to a full book-length treatment that required multiple years and sabbaticals to complete.
- *New things get started by evolution or chance, not design.* In writing my memoir, part of the intent was to lay out the circumstances that led to the choices made over my career. Much of the driving force behind these choices can be seen as evolutionary or as a product of chance. Still, I would add two more factors when these choices concern starting new things—the ability to discern what is new and of potential import when it appears and the confidence in it and in oneself to pursue it.
- *We are all flawed—we all carry our burden of incapacities.* Another part of the intent behind writing my memoir was to lay out a whole person, with both strengths and flaws, and to understand how their combination led to the trajectory of my career. This maxim can also be seen as relating to maxims 25, 26, and 29.

Notes

Preface

1. Part III represents the third version of this story I have attempted to tell. The first version (Rosenbloom, 2023a) is the longest, including a variety of introductory material and personal details. The second version (Rosenbloom, 2025) is an edited-down professional version that eliminates much of the personal detail. The version here is edited down instead by moving selected material from the memoir itself to earlier in the book when it has turned out to provide useful background for discussions prior to part III.

2. *An Apologia.* I have occasionally been frustrated, particularly when I was a young researcher but even at times since then, when witnessing some senior researchers make claims about fields of which they know little or fail to ground their claims appropriately in the relevant literature. Such claims seemed both lazy and irresponsible, even when not apparently downright incorrect. However, as I've aged, and (hopefully) grown in perspective while also growing more limited in what I can accomplish, I have been increasingly tempted to follow such paths to "put out there" ideas that I believe may be of some use to others even when they fall short of normal standards of academic rigor and responsibility. I hope I am not blinding myself to the weaknesses of such an approach, but I do try to make the attempt in a spirit of humility, with the understanding that these ideas can be nothing more than thought-provoking speculations in their current form. Perhaps this can be considered the *privilege of the superannuated.* Much of this book may be considered to exist under this privilege.

Chapter 1

1. I have also been trained as a computer scientist and, to a lesser extent, as a cognitive psychologist. Part III highlights some of my other contributions to these disciplines, including my work on a broad understanding of computing as a science (Rosenbloom, 2012a) and early work on the power law of practice and

stimulus-response compatibility (Newell & Rosenbloom, 1981; Rosenbloom & Newell, 1988).

2. In this sense, figure 1.2 is actually a bit misleading in implying that this part of the book concerns only the three levels of subject matter included in it, rather than all five levels that will later be seen in figure 1.5. Although this limitation is largely true, the reflections on the nature of theories actually extend across all five of the levels shown in figure 1.5.

3. Essential analysis is also used in this work to define computing, science, engineering, and great scientific domains. More generically, it can also be seen to play a role in identifying the kinds of relationships that exist across such domains.

4. Although essential analysis does not induce, at least at present, a corresponding language, it is not a coincidence that both dichotomic analysis and relational analysis do. The close ties between analytical methods, architectures, and languages is a thread that will surface in various ways throughout this book.

5. Figure created by Efbrazil. Reused from https://commons.wikimedia.org/wiki/File:The_Scientific_Method.svg under the Creative Commons Attribution-Share Alike 4.0 International license (https://creativecommons.org/licenses/by-sa/4.0/deed.en#ref-appropriate-credit).

6. From *The Soar Cognitive Architecture* by John E. Laird, 2012, p, 17. Copyright 2012 by The MIT Press, 2012. Reprinted with permission.

7. Adapted from Rosenbloom et al. (2016a), p. 16, under Creative Commons Attribution-Noncommercial-Noderivatives 4.0 International license (https://creativecommons.org/licenses/by-nc-nd/4.0/deed.en). Converted from color to grayscale with a few more arrows added.

8. Adapted from Rosenbloom et al. (2016a), p. 14, under Creative Commons Attribution-Noncommercial-Noderivatives 4.0 International license (https://creativecommons.org/licenses/by-nc-nd/4.0/deed.en). Converted from color to grayscale.

9. This section is largely based on section 4 of (Rosenbloom, 2025). It is reused here under a Creative Commons Attribution-Noncommercial-Noderivatives 4.0 International license (https://creativecommons.org/licenses/by-nc-nd/4.0/deed.en).

10. Functionalities that are absent, or almost completely absent, in the current version of the common model include episodic memory (e.g., Kelley et al., 2018), emotion (e.g., Larue et al., 2018), metacognition (e.g., Kralik et al., 2018), natural language processing (e.g., Kelly & Reitter, 2018), perception, and action.

11. Drawn by Andrea Stocco. From Rosenbloom et al. (2022), under Creative Commons Attribution-Noderivatives 4.0 International license (https://creativecommons.org/licenses/by-nd/4.0/). Converted from color to grayscale.

Chapter 2

1. This chapter is an adapted/expanded version of an earlier conference paper (Rosenbloom, 2022a), which was itself an adaptation of an early version of chapter 14 of this book that later appeared in Rosenbloom (2023a; 2025). The most significant differences in content from the conference paper stem from the addition of the section on architectural constraint and the enhancement of the discussion of theoretical computational architectures.

2. In Rosenbloom (2022a), variable content was referred to as *variations*. However, in the larger context of this book, this latter term is best left for use in characterizing the content of a definitional space.

3. The same may also be true beyond computational architectures, to all transformational architectures, but the focus here is narrower.

4. From Rosenbloom (2025), figure 2, under Creative Commons Attribution-Noncommercial-Noderivatives 4.0 International license (https://creativecommons.org/licenses/by-nc-nd/4.0/deed.en). Converted from color to grayscale.

5. From Rosenbloom (2025), figure 3, under Creative Commons Attribution-Noncommercial-Noderivatives 4.0 International license (https://creativecommons.org/licenses/by-nc-nd/4.0/deed.en). Converted from color to grayscale.

6. Newell actually proposed a different perspective on this, that the entire implementation was the theory but that parts of it—most particularly those that would otherwise be considered implementation details—are just more approximate than others (Rosenbloom & Laird, 1993).

7. There can also be analog simulations that exploit a variety of physical phenomena to simulate other phenomena, but they are not of especial relevance here.

8. Hypotheses represented as complex mathematical models may also implicate such an understanding process, even when they are not active in the sense here of directly interacting with the phenomena. Instead, understanding these models, by deriving implications mathematically from them, can play an important role in the theoretical side of the scientific process that is similarly missing from figure 1.6.

9. One notable exception is the argument by Oberauer and Lewandowsky (2019) for distinguishing between discovery-oriented/exploratory research versus theory-testing/confirmatory research in psychology as part of the solution to what is known as the theory (or replication) crisis.

10. A much more thorough analysis of ways to evaluate theories in general, along with the implications of this for cognitive architectures, can be found in section 5.4. From a different perspective, chapter 7 explores evaluation criteria for AI technologies and integrated systems that are thus also relevant to cognitive architectures.

Chapter 3

1. This chapter is an adapted/expanded version of an earlier conference paper (Rosenbloom, 2023b) that won the Springer Prize for Best Paper at AGI-23. The most significant addition is section 3.5, concerning other variations on the physical symbol systems hypothesis. The analysis here has also been reformulated in terms of the reflective method.

2. A similar notion can also be found, for example, in Sun (2024).

3. Overviews of the use of LLMs in the construction of broader agents can be found in Wang et al. (2023) and Xi et al. (2023).

4. One of Allen Newell's maxims mentioned in appendix B was, "Deep scientific ideas are exceedingly simple. Others usually see them as trivial." I would expect he felt this was true of the PSSH.

5. Oddly, a similar analysis of Turing machines reveals that the locations on the tape are themselves symbols that declaratively designate the symbols written on the tape.

Chapter 4

1. This chapter is an adaptation of (Rosenbloom, 2023c).

2. A similarity in spirit can also be discerned with respect to a recent blog post by Yoshua Bengio (2023) that provides a sequence of "definitions, hypotheses, and resulting claims about AI systems."

3. There is also an interesting contrast with the approach taken by Rosen (1991), which mixes formal and informal aspects. While the latter includes a cascade of definitions, it is structured more as an extended discourse rather than a terser sequence. He also ultimately builds up to a definition of life, rather than of intelligence. However, in the process he also defines such terms as mechanisms and machines in a way that is intended to show that they are not fully up to exhibiting life, thus perhaps raising questions about the general assumption underlying this chapter, that we can commingle consideration of intelligence in both natural and artificial systems. Explicit refutations of Rosen's claim do exist, however—such as Chu and Ho (2006)—so this argument will be set aside here, as will also be the broader philosophical and scientific controversary over whether machines can actually think.

4. As in section 2.2, the original use of the term *variations* in Rosenbloom (2022a) has been replaced here with *variable content*. The former term continues to be used for the elements of definitional spaces.

5. I was pleasantly surprised when I later looked up the dictionary definition of "Understand" (2024) to find the very first entry to be nearly identical—"to perceive the meaning of." Even more elaborated definitions, such as "function in ways that

are causally explainable by appeal to its semantic properties" (Titus, 2024), seem to be built around this basic notion.

6. Please forgive the use of a teleological form here when considering evolution.

7. See, for example, Pezzulo et al. (2024) for a discussion of the implications of the lack of sensational grounding in LLMs. It is important to note, however, that this critique is not the same as the broader argument that no physical symbols systems can be grounded in this manner (section 3.2).

8. The two main threads in the literature on assigning rights, or moral status, are based either on properties, as here, or on relations to others, particularly to humans. This latter notion is not a focus here, although the idea of socialized agents (section 4.1.3.9) and its bearing on the determination of rights and responsibilities does bear some relationship to it. The notion of culturally significant objects (section 4.3.4.8) also bears some relationship to it.

9. In an early AI perspective on this, Arkin, Ulam, and Duncan (2009) propose a constraint-based agent as the governor for weaponized military robots.

10. Long et al. (2024) provides a thoughtful consideration of *AI welfare*, a topic that closely relates to the question of the rights of AI systems. There are interesting points of contact with the discussion here. In particular, they propose two specific aspects to consider, *consciousness* and *robust agency*, when considering AI welfare. The first is closest to what has been termed here self-aware affective sensates, whereas the second is closest to self-aware rational agents.

11. When I posted a version of this section on my LinkedIn page, Yoav Shoham pointed out that it echoes a pithily apt quote from the novelist Isaac Bashevis Singer: "We must believe in free will, we have no choice."

Chapter 5

1. A similar discussion could have also appeared in chapter 3, concerning whether the topic yielded by the essence of an *atom* should still be labeled as *symbol* rather than some other existing or new term. As here, it was felt most appropriate there to go with the existing topic label that engendered the analysis.

2. Please note that this usage of "ground" is quite distinct from the notion of grounding found in section 4.1.

3. Along these lines, according to Rosenbloom and Laird (1993), "Newell felt that a theory is adopted by an individual or community if it is useful to them. If it is not useful—because it is wrong in some way that matters to them, or because they do not trust or understand it, or because there is some other theory that is more attractive—then they will abandon it (or never even pick it up)."

4. There has been prior work identifying syntactic, semantic, and pragmatic views of theories (e.g., Savage, 1990; Mormann, 2007). According to this distinction, the focus on aboutness in this chapter would most aptly fit within the semantic perspective. However, here this triad instead maps onto three compound dimensions along which theories can vary, with section 5.4 in turn mapping them onto three general evaluation criteria for theories.

5. Although this notion is now challenged, albeit quite speculatively, by the notion that computation actually underlies physics, and thus everything else as well ("Digital physics," 2024).

6. Generative models are unique among human-made systems in their duality of extraordinary breadth, of both function and domain, and extreme challenges with respect to fidelity. The only other (nonhuman?) system that would seem to come close in breadth is the World Wide Web, but it is intrinsically less complete as a mechanism that can directly provide answers to questions; instead requiring humans to ferret out the desired answers from the potentially relevant pages that are returned.

7. For LLMs, this traditional notion of explanation needs to be distinguished from explanations the model itself may generate explicitly as text. In the latter case the explanation exists wholly within the explanandum (i.e., the output) of the model rather than providing a metatheoretic explanation of the explanandum in terms of the explanans (i.e., in terms of the generative model and the prior words input to it).

8. *Explanation-based generalization* (Mitchell et al., 1986), which later was reconceptualized as *explanation-based learning*, made this connection explicitly in the name of the method.

9. Keas (2018), for example, describes twelve *theoretical virtues*, most of which map at least roughly onto criteria included here, although several do raise the possibility of further extensions. Similarly, building on even earlier work, Douglas (2013) introduces four groups of *cognitive values*, while emphasizing justifications for their inclusion.

10. Lieto (2021) defines a *Minimal Cognitive Grid* for use in evaluating "cognitive artificial systems" that includes analogues of both accuracy (performance match) and power (generality) plus a third criterion—functional/structural ratio—that sounds much like exuberance but actually turns out to be something quite different.

11. I don't believe I have invented this notion but have not been able to track down a specific reference. The closest I have come is a notion in Newell (1990) that unified theories of cognition shift the discussion from discrimination among theories—essentially a focus on accuracy—to an approximative one that focuses more on cumulation.

12. This is actually a rather subtle question for procedural memory. The procedures themselves cannot be retrieved as propositions into working memory but, if they are expressed as rules, they may be seen as combinations of propositions to a human examining them.

13. Varma (2011) alternatively refers to exuberance in cognitive architectures as *parsimony*.

Chapter 6

1. Some of the material in this part is derived from section 5 of Rosenbloom (2025).

2. The phrase "dissecting AI" can be found in a recent news story (Science Daily, 2018), but the focus there was on deriving insights into the human brain by dissecting artificial neural networks rather than on understanding the field and its technologies more broadly.

3. Although Hopfield networks (Hopfield, 1982) and Boltzmann machines (Ackley et al., 1985) are truly symmetric, other forms of autoassociative neural networks—such as autoencoders (Kingma & Welling, 2019)—are structured asymmetrically, as back-to-back heteroassociative neural networks. As introduced in section 6.2 and discussed more thoroughly in section 8.4, this latter approach can be interpreted as simulating a symmetric form via an asymmetric one.

4. Addition and reduction turn out to align well with *interaction* and *implementation*, as they are used in the relational model (chapter 18) to understand the relationships among the great scientific domains that underlie topics in computing and beyond, suggesting a more fundamental role for a generic combination of horizontal (addition/interaction) and vertical (implementation/reduction) relationships in understanding a variety of topics.

Chapter 7

1. As mentioned in section 4.1.3, Wang (2019) referred to a definition of intelligence based on such functional capabilities as a *functional definition*, in contrast to a *principled definition* based on a single principle. Here we can see a contrast between grounding generality in such a functional definition versus problem generality, which he referred to as a *capability definition*. Kotseruba and Tsotsos (2020) also used a similar distinction in their overview of cognitive architectures.

Chapter 8

1. Technically, "IL" could be reduced to "I" here, as learnability is the only form of improvability considered here, but it will be kept as the former throughout simply to emphasize that this is learning.

Chapter 13

1. Large segments of this part are based on Rosenbloom (2025), with the main additions being the more personal aspects. They are reused here under a Creative Commons Attribution-Noncommercial-Noderivatives 4.0 International license (https://creativecommons.org/licenses/by-nc-nd/4.0/deed.en).

2. I never could get into neuroscience and the study of the brain, as opposed to cognitive science and the study of the mind. Early on, understanding the brain did not appear to yield much insight into the mind, even with the former obviously "implementing" the latter. As the decades passed, what was learned about the brain became increasingly relevant, but I was still not inclined to learn much about it, likely because it seemed irremediably messy (see aside 13.2). So, it plays a relatively minor role overall in this memoir.

3. Reprinted with permission. © The System Source Computer Museum (https://museum.syssrc.com/artifact/exhibits/220/).

4. Large segments of this part are based on Rosenbloom (2025), with the main additions being the more personal aspects. They are reused here under a Creative Commons Attribution-Noncommercial-Noderivatives 4.0 International license (https://creativecommons.org/licenses/by-nc-nd/4.0/deed.en).

5. A broadening of the standard notion of computer science to include all of how computing is understood and applied (Rosenbloom, 2012a). More on this can be found in chapter 18.

6. I couldn't resist a few ancient references in recognition of my interest in both reading and collecting old books.

7. I began including snippets of personal material in my book *On Computing: The Fourth Great Scientific Domain* (Rosenbloom, 2012a), which is the focus of chapter 18. This was an unusual step for what was largely a technical (and philosophical) book, but the intent was to provide a useful and engaging background for the more technical foreground. One reviewer commented "*On Computing* is an unusual, and welcome, mix of conventional academic text and personal odyssey." (Gilbey, 2012). Here, such material is deliberately more pervasive.

8. To reduce the temptation to either intrude on the privacy of others or to complain about perceived slights, I will largely eschew anecdotes and negative comments focused on others except when particularly germane.

9. This is a significant departure from my normal writing style, which minimizes footnotes and endnotes to reduce distraction.

10. Classically, this is a period between rulers, but here the term is used in the more general sense of a "pause or interruption in continuity" ("Interregnum," 2024), in particular, among cognitive architectures.

11. As hard as it is to imagine now, Stanford did not offer an undergraduate major in computer science at that time.

12. Although its name was hyphenated at the time, as Carnegie-Mellon University.

13. Fortunately, I was supported by a National Science Foundation (NSF) graduate fellowship at the time so there were no funding issues with switching universities and fields for a year.

14. This aside reflects my attempts to actualize maxim 25, of understanding one's own strengths and weaknesses.

15. While, in contrast to the five factor personality model mentioned in section 15.4, Myers–Briggs is not a scientifically validated personality model, it is included here because of how it does seem to succinctly capture my intellectual personality.

16. A recent guest essay in the *New York Times* seems to capture well much of this sense of myself (Szalavitz, 2023).

17. During my early years as a graduate student at Carnegie Mellon, I played darts frequently in the computer science lounge with my fellow students. Partly due to an intensional tremor that forced me to carefully time my throws to when they were aligned with the board, and which got worse under stress, and partly due to this desire to be less competitive, I modified the home-grown program used to track how well everyone did at darts so that it would ignore attempts to record my scores.

18. This last may be due to positive side effects of being bipolar and OCD, both of which I was diagnosed with earlier in life, although neither is terribly severe, and either the former has moderated, or I've learned to handle it better over the years.

19. There may have been some hero worship here as well, although I would have been unlikely to have admitted it at the time. I tended to be driven by a largely unconscious but occasionally explicit need to outdo her academically—part of the overall competitiveness already mentioned—through much of our early years.

20. From the Clear Creek High School *Den*.

21. From the University High School *Chieftain*.

22. Two painfully illustrative examples from my graduate student days were: my last-minute inability to do my duty as best man at my best friend's wedding in terms of leading the toast of the bridal couple; and my frantic/panicked call to Elaine back in California the night before I taught my first class as a graduate student at CMU.

23. It has been an interesting contrast to watch Michael grow up, with an ease and facility with writing, an apparent complete lack of self-consciousness in front of audiences—even seeking out and enjoying opportunities to act on stage—and an abiding enthusiasm for teaching. Whereas my father was the more natural mathematician, Michael is the more natural teacher. Perhaps, between the two genealogically, I ended up being the more natural researcher.

24. Cluelessly, I didn't understand until years later, from an introduction of me before a talk I was to give at UCLA, that I was considered one of that year's graduating "stars," possibly explaining why Allen suggested I consider faculty positions.

25. This was a junior position on the *research faculty*; a nontenured faculty track where financial support came from external research contracts and grants rather than from teaching.

26. Upon finishing my dissertation in 1983, I was just about to accept a tenure-track offer from the University of California, Berkeley (UCB) when Allen suggested I stay at CMU and team with him and John Laird on the Soar Project. After some hard thought, I agreed, ultimately setting the course for the rest of my career. When, within one more year, both Elaine—who had joined me in Pittsburgh to do her medical internship, residency, and fellowship—and I felt an overwhelming need to get back to California, UCB was no longer a possibility, so I picked the other main option I had perseverated over the previous year and joined Stanford's faculty. John left for Xerox PARC at the same time. He just reminded me at my retirement celebration that Allen was so high on the potential of Soar and so worried about the impact of our leaving on it that he told us it was in the national interest for us to stay in Pittsburgh.

27. In a manner, this memoir is structured analogously to his talk, but with the theme functioning structurally as "desires" and the rest functioning structurally as "diversions."

28. The proximate trigger, however, was my sister Kate's *Swan Song* (Pohl, 2021) as she retired from a career that included extensive work on the UCSC Genome Browser (2023). It got me thinking about trying something similar myself at this point.

29. As a memoir, this work is inherently self-centered, self-indulgent, and exhibitionistic, not traits greatly admired in academic writing. It is also presumptuous given that I have not reached the level of eminence that would lead a broad audience to want to understand my work and life in this form. But it felt like the right thing to do, I enjoyed doing it, it should be useful for me moving forward, and parts of it will hopefully be of use to others—all motivations that are compelling to me at this point in my life, and with the last one the primary rationale for including it in this book. After focusing on technical papers throughout my career, it has also been an opportunity to leave something more personal behind in written form.

30. From Rosenbloom (2025), figure 1, under Creative Commons Attribution-Noncommercial-Noderivatives 4.0 International license (https://creativecommons.org/licenses/by-nc-nd/4.0/deed.en). Converted from color to grayscale.

31. Essential analysis can also be seen as a variation on maxim 21, of seeking similarities before considering differences.

32. A lineage of architectures, as the phrase is used here, is a sequence of architectures that typically starts with one simple yet concrete version and evolves over time—normally via human programming rather than natural variation and selection—into successively more advanced versions. All of the architectures within such a lineage typically share a single name, just with added modifiers or increasing version numbers over time.

33. It is only recently, with early work on this memoir, that I have begun to think of the common model and the higher tiers as involving architectures, leading ultimately to the general thoughts on them that can be found in chapter 2.

Chapter 14

1. As mentioned in an endnote in chapter 13, my father was a more natural mathematician, with a PhD in electrical engineering. His father—Abraham Rosenbloom—was also an engineer, with a BS in mechanical engineering, making me a third-generation engineer.

2. This aversion to messiness could be due to my OCD nature, as mentioned in another endnote in chapter 13. This may also be a key factor in why I am driven to seek architectures that can capture any order lurking beneath the messiness found in complex phenomena.

3. As mentioned in an endnote in section 2.2, variable content was referred to as *variations* in an earlier version of this memoir.

4. For example, one of the biggest misses in work on Sigma, despite having considered doing it for some time, has been the lack of any attempt at bringing it together with a major "knowledge graph" (2024).

5. Chapter 5 explored beauty and simplicity in more detail from the perspective of theories. Chapter 7 also explored simplicity and elegance in the context of evaluating technologies and integrated systems.

6. The painter Helen Frankenthaler said something similar about art: "Let the picture lead you where it must go."

7. This, perhaps surprisingly, also dovetails with the reuse and adaptation of what already exists in the evolution of biological systems.

8. Maxims 16 and 20–22—milking the existing architecture/hypothesis for all it is worth while focusing on: simplicity in terms of the number of architectural mechanisms, similarity between new and old capabilities, and creating explanations in terms of (combinations of) existing mechanisms rather than immediately jumping to add new mechanisms that are disjoint from existing ones—provide further perspective on such decisions.

9. I am somewhat tone deaf in reality. Although I can hear notes, I can't distinguish keys or even major versus minor. Even worse, unless the words of a song are front and center, as in folk music—or I really concentrate—I mostly can't hear words in music. "Hamilton" (2024) to me is unfortunately just sound. This has led me to prefer genres like folk music and other forms of mid-twentieth-century music, where I can hear the words, and baroque music, where the sound patterns are interesting enough in their own right. Unfortunately, this same limitation also affects my ability to make out words when spoken with accents that my brain interprets as musical, thus making collaborating with some foreign students difficult.

10. This paper won the 2012 Kurzweil Award for Best AGI Paper, earning me $1,000 of Ray Kurzweil's money.

11. "I do not pretend to start with precise questions. I do not think you can start with anything precise. You have to achieve such precision as you can, as you go along." (Russell, 2009).

12. Herbert Simon was a Nobel laureate in economics, Turing Award winner, and Allen Newell's PhD advisor (and collaborator for many years), and thus my academic grandfather.

Chapter 15

1. The recent emergence of large language models (e.g., Brown et al., 2020) that are trained over vast bodies of text on many subjects has begun to blur the distinction given the capabilities they can exhibit from experience with such bodies of text (section 4.3.2). The work to build more complete architectures around them (e.g., LeCun, 2022; Park et al., 2023; Sumers et al., 2023) also provides further movement in this direction.

2. That first winter in Pittsburgh, of 1976–1977, was the coldest on record (Only in Your State, 2022) and the second was one of the snowiest. I made it through the first winter with a corduroy coat I had worn at Stanford. Fortunately, I lived on campus and so did not need to walk too far. By the second winter, I had wised up at least a little bit and bought a down jacket.

3. Lenat would later acknowledge that "For the first five years, nothing good came out of it" ("Eurisko," 2024).

4. I also recall us spending much of our meeting time discussing a name for the system that was to be developed. I just ran across a letter to my parents from fall 1976 in which I said "The biggest controversy so far is over the name of the program. So far things like DOVE, HUMOR, EUREKA, ISIS, DELPHI, and SYD have been proposed."

5. I would have started working with Allen immediately, but he was on sabbatical my first year at CMU.

6. One of the archaisms I have picked up from too extensive reading of old books, which I have "fun" with by contraposing to modern technical terms. I know it is a nerdy thing to do, but more literary nerdy than tech nerdy.

7. I did at least pass all of my PhD qualifying exams during this period, including the notorious sixteen-hour exam that was free to roam across all of the areas comprising computer science; as a result, I received my MS in CS. CMU CS at that time had no required courses. Instead, "all you needed to do" was to pass these challenging exams. You only participated in courses to the extent you thought they would help in passing the exams.

8. These attempts included taking three psychology courses—one in high school, one at UCLA, and one at Stanford—plus spending one summer doing independent reading on the topic, but none of them yielded what I sought. In retrospect, it is clear that this was because none touched on cognitive psychology, and at the time, I didn't even know that it existed or how to look for it.

9. Geoff was a visitor at UCSD that year like me, although he was post-PhD. My interactions with him there amounted to my first serious introduction to the world of neural networks, even if this was before the development of either backpropagation or Boltzmann machines. Geoff later served on my thesis committee at CMU, before going on to win a Turing Award for coleading the deep learning revolution, becoming known as a/the "Godfather of AI," and then winning a Nobel Prize in physics with John Hopfield. I may be unique in having both an original "father" of (symbolic) AI and a later "godfather" of (neural) AI on my committee, as well as having thesis committee connections to two "AI" Nobel laureates, Herb Simon (economics), my academic grandfather through Allen Newell, and Geoff Hinton (physics). A lot of reflected glory here.

10. Allen later mentioned that at the time he had thought it was likely I wouldn't return to CMU after being back home in California for a year. However, not returning was another one of those options I never really considered.

11. John and I went on to be close collaborators and best of friends over the years. More on John can be found later, in aside 15.7.

12. This journal has since been renamed *The Quantitative Methods in Psychology*.

13. In 2017, I was at a conference in Warwick, UK, that combined cognitive modeling with mathematical psychology and was surprised to learn over beers on an outdoor patio that attendees from the latter community, with which I was previously unfamiliar, knew my name well from this book chapter that was published thirty-six years earlier. They appeared amazed to finally meet the person behind the previously mysterious name they knew so well.

14. Activation was eliminated in moving to Xaps3 due to the experience with chunking in Xaps2 that led to formulating the *crypto-information constraint* that bans using information in architectural decision-making that is not available to the knowledge and skills on top of the architecture (Rosenbloom, 1986). Activation in Xaps2 was crypto information in this sense, although it is not so in either neural networks or probabilistic graphical models, and thus also not in Sigma.

15. Iago won the Santa Cruz Open Machine Othello Tournament (Frey, 1981). It ran at CMU on a "large" DECSYSTEM-2060 with 512K 36-bit words of primary memory running at <2 MIPS while being controlled remotely via a terminal in Santa Cruz by my sister Kate, who lived nearby.

16. This required creating a new Othello font for both a very early laser printer—the Xerox Dover—of which CMU had one of the few ever built, and a digital typesetter, to illustrate board positions and entire games.

17. An early forerunner of the "Paul Hollywood handshake" on the *Great British Baking Show*?

18. Although this did get considerably more complicated in Soar, inspiring many revisions over the years of how chunking worked there. Figuring out how to do this for chunking in Sigma proved even more difficult, ultimately becoming a major stumbling block.

19. The work toward my thesis was done on a DEC Vax-11/780—the later big brother of DEC's popular PDP-11—running Unix, plus a Xerox Alto, a pivotal precursor of the Apple Macintosh, that Xerox PARC had donated to CMU.

20. These programs were written in a semi-formal language developed for this purpose called Artless.

21. *Soar* was originally the acronym *SOAR*—State, Operator, And Result—but it later morphed into a proper noun.

22. Section 1.2.1 provides a more basic, technical introduction to Soar. Beyond this, a "gentle" introduction to Soar can be found in (Lehman et al., 1998). There is also an early introductory video (Lehman et al., 1994) with me as the rather wooden presenter, plus more detailed recorded tutorials due to John. More up-to-date references on Soar, from after I left the project, can be found in section 1.2.1 and aside 15.7.

23. Figure redrawn with permission from Young, Ritter, and Jones (1996).

24. This paper was nominated for the publisher's prize at the Fourth National Conference on Artificial Intelligence.

25. The very first version of Soar, from before I joined the project, was built around Xaps2, providing a second although more transient connection between the XAPS and Soar lineages, before Xaps2 was replaced by OPS5 as Soar's rule system.

26. Mitch Waldrop, a senior writer at *Science* Magazine, followed up on this with two articles that explained Soar and the idea of a unified theory of cognition to the broader scientific community (Waldrop 1988a,b). One unexpected result was a phone call I received one day from Francis Crick, codiscoverer of the structure of the DNA molecule, who had begun to think about consciousness and wanted to understand more about Soar.

27. Soar's impact, and the accompanying work we did on symbolic cognitive architectures more broadly, ultimately resulted in my last name—along with John's and Allen's—making an appearance in English literature. Mark Haddon, the author of *The Curious Incident of the Dog in the Night-Time* (Haddon, 2003), later wrote *A Spot of Bother* (Haddon, 2006), in which a pretentious character pretends to read a copy of the philosopher Daniel Dennett's *Consciousness Explained* (Dennett, 1991) at a café. Haddon quotes a snippet from Dennett's book that includes the citation "Newell, Rosenbloom, and Laird, 1989."

28. My time as research faculty at CMU was too short for the two PhD students whom I worked with there to continue with me as their advisor, but I did end up later coadvising two CMU CS PhD students—Milind Tambe and Bob Doorenbos—who had started working with Allen before he became ill. Milind's name will come up again very shortly.

29. The move to Los Angeles, where I grew up but had not lived for many years, was driven by Sharon's career. Having finished her medical training, there were no positions open in her specialty in the Bay Area. We had earlier survived a long-distance relationship for three years, but at this point I never considered staying in the Bay Area while she moved to LA (maxim 29).

30. Coincidentally, when I started in CS at USC, two of my fellow faculty—George Bekey and Ed Blum—were old-time colleagues of my father's, with the latter also being a long-term family friend. George and my father were PhD students together in electrical engineering at UCLA. Ed and my father were part of a close-knit group of (former) TRW employees whose families essentially grew up together and spent every Thanksgiving for many years in a group of nearby cabins in Yosemite National Park. My car's "vanity" license plate shows Yosemite Valley in the background with COG ARCH as the "number."

31. Reprinted with permission from William Swartout. Converted from color to grayscale.

32. I was in transit at Saint Louis's Lambert International Airport—traveling on TWA in those days—when I was paged and learned that Allen had died. I quickly rerouted to Pittsburgh to help deal with things there. I ended up taking on the job of calling many of his colleagues to let them know what had happened and stayed on to speak at his memorial service in the (now long gone) Skibo Auditorium at CMU.

33. He also played foundational roles in other areas of computer science, such as computer architecture (Bell & Newell, 1971) and human computer interaction (Card et al., 1983).

34. Included in this were new declarative learning and memory modules that essentially replaced data chunking as Soar's story. It turned out that although Soar could do declarative learning and memory without new modules, it required interrupting other activities to do so, rather than enabling them to proceed in parallel, as they needed to do.

35. Soar is mentioned in the introduction to the leading AI textbook (Russell & Norvig, 2010)—"The work of Allen Newell, John Laird, and Paul Rosenbloom on SOAR (Newell, 1990; Laird et al., 1987) is the best-known example of a complete agent architecture."

36. In my desperation to move on at this point, I probably left Randy and Jon without sufficient funding options, one of the reasons I assume they moved to ICT not long after. Whatever the cause, though, this shift did seem to work out well for both of them. By this time, Milind had accepted a tenured faculty position on campus.

37. There has also seemed to have been a gradual shift over the decades from an awkward self-consciousness to an empowered self-awareness that has enabled me to better cope with troublesome personal issues and thus to handle leadership positions more comfortably.

38. Oddly, I matured early in my ability to choose a wonderful significant other, starting to see Elaine when I was seventeen.

39. The appearance of my name in an English novel, as mentioned in an earlier endnote, thus provided a rather pleasing form of "coming full circle," from a fascination with reading such works that began relatively late in life to my name appearing in one.

40. When combined with my taste in music, I tend to think of myself as living in the present while working in the future and recreating in the past.

41. Yigal was the director of ISI's Intelligent Systems Division at the time while also leading a big push in *digital government*, concerned with leveraging digital technologies broadly across government agencies and activities.

42. Hopefully "Stockholm syndrome" (2024) was not a major factor.

43. I did also receive the *USC Distinguished Faculty Service Award* in 2019 for these efforts.

44. The name *Sigma (Σ)* was inspired by a self-referential acronym for something like "*S*igma is an *I*ntegrated (and/or *I*ntelligent) *G*raphical *M*odel *A*rchitecture," but based on the Soar experience, it was considered a proper noun rather than an acronym from the start. It was also intended to connote a *significant change* from what had come before plus Sigma's *integrative* aspect and the notion of *addition* that is central to the sum-product algorithm (Kschischang et al., 2001) underlying Sigma's graphical models. Kim LeMasters, ICT's creative director at the time, helped choose the name based on a set of raw possibilities I had generated for consideration.

45. I contacted the journal *Artificial Intelligence*, the leading AI journal and the venue for several of the foundational articles on Soar in the 1980s and 1990s, about the possibility of publishing the foundational paper on Sigma there but received the feedback that these types of papers were no longer of interest to them. In contrast, the journal *Artificial General Intelligence* was quite willing to consider it, so that is where it ended up (Rosenbloom et al., 2016a).

46. My biggest breakthrough in understanding how to apply such models to building a cognitive architecture came from reading a paper on factor graphs (Kschischang et al., 2001), a very general formalism whose breadth of applicability was made readily apparent in that paper. This was a true *aha* moment for me that was driven by the work of others—in this case from electrical engineering—rather from my own explorations.

47. As a short cut, and to retain some level of comfort while changing so many other things, I went back to the use of Lisp here rather than investigating any of the newer languages that might have been more appropriate by this time. I always enjoyed Lisp because it naturally captured symbolic processing and using it was incremental and interactive. I learned Lisp as an undergraduate, from a course taught by John McCarthy, one of the four "fathers" of (symbolic) AI—along with Allen Newell, Marvin Minsky, and Herb Simon—who invented the language and had a surprisingly formal way of teaching it as a programming language, perhaps due to him also being the one to introduce logic into AI for automated reasoning. John later became a colleague when I joined Stanford's faculty.

48. From Rosenbloom (2025), figure 8, under Creative Commons Attribution-Noncommercial-Noderivatives 4.0 International license (https://creativecommons.org/licenses/by-nc-nd/4.0/deed.en). Converted from color to grayscale.

49. Although I did learn to enjoy teaching and found satisfaction in doing it well, a full teaching load never seemed compatible with the kinds and amounts of research I wanted to do nor with the involvement I wanted to have with the various research

institutes to which I belonged. And, although I have the greatest respect for those who have teaching as a critical part of their mission, this simply wasn't me.

50. More thoughts on the past, present, and future of AI can be found in an earlier two-part interview (Rosenbloom, 2013).

51. Ed was the main CS/AI person who recruited me to Stanford as faculty after my PhD.

52. In this work, Ed appreciated how chunking in Soar could automatically compile a shallow rule-based system from experience in running a deeper model—also in Soar—of how task problems could be solved. In other words, it was a demonstration of learning across the (a)symmetric dichotomy (chapter 8).

Chapter 16

1. This symposium arose from an interaction between Christian and me at an earlier conference on computational cognitive modeling during which we lamented how, although there were many conferences that touched on aspects of cognitive architectures, there was not one that accepted work on all of the relevant threads and perspectives.

2. Gordon Bower was John Anderson's dissertation advisor at Stanford and John Anderson was the third member of my dissertation committee at CMU, after already counting Allen and Geoff Hinton (with Jaime Carbonell being the fourth). Gordon had earlier taught me mathematical psychology as an undergraduate and thereafter wrote me a recommendation letter for graduate school. My first attempt at a research project was with him, based on this class, but it unfortunately failed to yield any insight. Still, it seems likely that Gordon played a role in the psychology portion of my faculty appointment at Stanford.

3. I—and I think John Laird as well—participated in his seminar on ACT* shortly after this book came out.

4. Providing an odd harkening back to my earliest days with Eurisko.

5. An earlier working group report on how emotion might relate to the common model can be found in Larue et al. (2018).

Chapter 17

1. Figure created by Xander89 (Original), OlivierMehani (Vector). Reused from https://commons.wikimedia.org/wiki/File:Internet-hourglass.svg under the Creative Commons Attribution-Share Alike 3.0 Unported, 2.5 Generic, 2.0 Generic and 1.0 Generic license (https://creativecommons.org/licenses/by-sa/3.0/deed.en).

2. This idea was developed in parallel with the similar one, of an *interface layer* for AI based on Markov Logic (Domingos & Lowd, 2009).

Chapter 18

1. *On Computing: The Fourth Great Scientific Domain*, by Paul S. Rosenbloom, p. 28, ©2012 MIT, by permission of the MIT Press.

2. *On Computing: The Fourth Great Scientific Domain*, by Paul S. Rosenbloom, p. 31, ©2012 MIT, by permission of the MIT Press.

3. *On Computing: The Fourth Great Scientific Domain*, by Paul S. Rosenbloom, p. 177, ©2012 MIT, by permission of the MIT Press.

4. Although not metric in the same way as dichotomic maps, I have still found it useful to think of this as a map of the dyadic components of the great scientific domain of computing rather than simply as a table.

5. An example denying that psychology is a science can be found in the comments on our piece in *The Conversation* on the common model and AGI (Rosenbloom et al., 2022).

6. I originally intended to write a book for the general reader who might be interested in a broad perspective on computing, but after trying out a few chapters on noncomputing folks, it was clear I didn't know how to pull that off, so I backed off but still with the hope of at least being able to reach college students interested in computing.

7. In this earlier article, I proposed restructuring computer science and engineering "rationally" so that the former actually focused on science and the latter on engineering, an idea that got me into hot water with colleagues in computer engineering—a discipline focused on computer hardware, architecture, and the software closest to the hardware—who were concerned I was suggesting either that they did no science or should do no science. With the newer focus on great scientific domains, this all shifted strongly toward the idea that it is best to intertwine science and engineering rather than creating hard-and-fast disciplinary boundaries between them.

8. *On Computing: The Fourth Great Scientific Domain*, by Paul S. Rosenbloom, cover image, ©2012 MIT, by permission of the MIT Press.

9. I had earlier given a presentation on this potential reorientation of student thinking as part of a special session on *Re-centering Computer Science* at the 2006 ACM SIGCSE (Special Interest Group on Computer Science Education) conference, in an era when CS education was being rethought because of low enrolment following the dotcom bust. It was quite a different time from where we are now.

10. You can see one student's comment in an Amazon review of the book (Customer Review, 2015).

11. The discussion of research methodology in this article—reflecting some of the amateur philosophy of science I needed to engage in while writing *On Computing*—has, it turns out, found its way back into the philosophy literature (Tallant & Andow, 2020), perhaps entitling me now to claim a very small amount of professional status in that field.

Chapter 19

1. My oncologist at the time had arranged for me to be admitted directly to the cancer floor at Saint John's. However, by the time we reached the hospital, we learned of a Catch-22 imposed by the insurance company. For urgent cases they had an "expedited" approval process that would take "only" several days to approve such an admission; but I was too sick by this point to wait that long. For a "true" emergency, they insisted on a (re)diagnosis by ER doctors before admission; but by then my immune system had collapsed to the point where it would have been extraordinarily dangerous, even prepandemic as it was, to be exposed to whatever else might be present in the ER that day. So, I was advised to loiter outside the ER (and the hospital itself) while awaiting a last-minute insurance appeal. Fortunately, with encouragement from some key folks at USC, the appeal was granted in time for me to be admitted directly to the cancer floor that day.

2. Georgetown University, where Michael was working when I got ill, shut down in-person activities while he was visiting us for spring break of 2020, so he ended up simply staying with us while working remotely for more than a year.

3. A wide network of family, friends, and colleagues also supported me remotely throughout the process, as I sent out regular updates and they responded with both heartfelt messages and humorous bits that helped buoy my spirits.

4. A final tie to old books here with an allusion to Dante Alighieri's *Inferno* (1472).

5. The latest such example was the AAAI 2023 Fall Symposium mentioned in chapter 16 on *Integration of Cognitive Architectures and Generative Models*.

6. From Rosenbloom (2025), figure 18, under Creative Commons Attribution-Noncommercial-Noderivatives 4.0 International license (https://creativecommons.org/licenses/by-nc-nd/4.0/deed.en). Converted from color to grayscale.

7. In alignment with maxim 6, I, for example, served as councilor and conference chair for the American Association for Artificial Intelligence (AAAI)—which is now the Association for the Advancement of Artificial Intelligence (AAAI)—and as chair of the Association for Computing Machinery (ACM) Special Interest Group on Artificial Intelligence (which was SIGART then but is SIGAI now).

8. This one was especially satisfying from a personal "childhood" perspective, as I recall my father regularly reading *Science* magazine, the main publication of the AAAS.

9. My interactions with Abram on how to extend Sigma's graphical models to neural networks (Rosenbloom et al., 2016b) drove home how much more natural it is for me to think in terms of mechanisms in contrast to a/his more mathematical and semantical approach to the same problem(s).

Appendix B

1. In a cruder (Bill) Clintonesque form, this maxim might be phrased as *It's the insight, stupid.*

2. Paul Allen Newell, Allen Newell's son, upon reading this list of maxims, suggested that in any such list there should be one that takes priority over all of the others. In the process, he brought to mind how in *Desires and Diversions*, Allen had mentioned his father's, and thus Paul's grandfather's, favorite example—*Keep your eye on the main chance.* After a bit of thought, it became clear that if I were to pick only one from my list, it would be this one concerning insight.

3. Although, the differential formatting requirements across the range of potentially relevant conferences and journals can make post hoc reformatting once a venue has been selected a tedious occupation, and one that may feel like rather a waste of time.

4. This can be considered as part of the creed of the outsider, or even of the eccentric. It is a maxim I only identified quite late in the writing of this memoir, and which only truly took shape later in my life. It can be considered as my own personal twist on a Herb Simon maxim—of, as a scientist, finding your own "private" part of the universe to mine that no one else is looking at—that I heard of indirectly through Allen Newell. On the other hand, it does sound generically like the title of a self-help work.

5. Combining this maxim with others in this appendix should help ensure that what you do is important as well as uniquely personal. The risk, though, is that if your niche eventually proves to be unproductive, or even irrelevant, then you may not only work in obscurity during your lifetime, but your life's work may ultimately and correctly be consigned to oblivion. You may go down in history, if history remembers you at all, as a crackpot rather than as an eccentric or even as an incomparable.

References

Ackley, D. H., Hinton, G. E., & Sejnowski, T. J. (1985). A learning algorithm for Boltzmann machines. *Cognitive Science, 9*(1), 147–169.

Adler, A., Dasgupta, P., Depalma, N., Eslami, M., Freedman, R. G., Laird, J. E., Lebiere, C., Lohan, K., Mead, R., Roberts, M., Rosenbloom, P. S., Senft, E., Stein, F., Williams, T., Wray, K. H., Yaman, F., & Zilberstein, S. (2019). Reports of the 2018 AAAI Fall Symposium. *AI Magazine, 40*(2), 66–72.

Agüera y Arcas, B., & Norvig, P. (2023). Artificial general intelligence is already here. *Noēma*.

Aha, D. W., Kibler, D., & Albert, M. K. (1991). Instance-based learning algorithms. *Machine Learning, 6*(1), 37–66.

AHDEL. (2000). *The American heritage dictionary of the English language*. Houghton Mifflin.

Alighieri, D. (1472). *La Comedia di Dante Alleghieri*. Johann Neumeister & Evangelista Angelini da Trevi.

Allaire. (2023). Helicopter view of the Getty Center. *Ancient Glass Blog of The Allaire Collection*. https://ancientglass.wordpress.com/3b-museums-glass-collections-and-exhibitions/j-paul-getty-museums-glass-collection/#jp-carousel-11264

Alphonse, E. (2004). Macro-operators revisited in inductive logic programming. In R. Camacho, R. King, & A. Srinivasan (Eds.), *Inductive logic programming (ILP 2004)*, Lecture Notes in Computer Science, *vol 3194*. Springer.

Amant, R. (2013). *Computing for ordinary mortals*. Oxford University Press.

Anderson, J. R. (1976). *Language, memory and thought*. Lawrence Erlbaum Associates.

Anderson, J. R. (1982). Acquisition of cognitive skill. *Psychological Review, 89*(4), 369–406.

Anderson, J. R. (1983). *The architecture of cognition*. Harvard University Press.

Anderson, J. R. (1995). *Learning and memory: An integrated approach.* Wiley & Sons.

Anderson, J. R. (2007). *How can the human mind occur in the physical universe?* Oxford University Press.

Anderson, J. R., Bothell, D., Byrne, M. D., Douglass, S., Lebiere, C., & Qin, Y. (2004). An integrated theory of the mind. *Psychological Review, 111*(4), 1036–1060.

Anderson, J. R., & Bower, G. H. (1973). *Human associative memory.* Lawrence Erlbaum Associates.

Anderson, J. R., & Lebiere, C. (1998). *The atomic components of thought.* Lawrence Erlbaum Associates.

Anderson, J. R., & Lebiere, C. (2003). The Newell test for a theory of cognition. *Behavioral and Brain Sciences, 26*(5), 587–601.

Apollo 10½: A Space Age Childhood. (2025, January 10). In *Wikipedia.* https://en.wikipedia.org/wiki/Apollo_10_1%E2%81%842:_A_Space_Age_Childhood

Architect. (2025, March 16). In *Online Etymology Dictionary.* https://www.etymonline.com/word/architect

Architecture. (2024a, June 29). In *Wikipedia.* https://en.wikipedia.org/w/index.php?title=Architecture&oldid=1231602375

Architecture. (2024b, July 22). In *Britannica.* https://www.britannica.com/topic/architecture

Arens, Y., & Rosenbloom, P. (Eds.). (2002). *Responding to the unexpected: Report of the workshop held in New York City, February 27–March 1, 2002.*

Arkin, R. C., Ulam, P., & Duncan, B. A. (2009). *An ethical governor for constraining lethal action in an autonomous system.* Georgia Institute of Technology Mobile Robot Laboratory Technical Report GIT-GVU-09-02.

Arp, R., Smith, B., & Spear, A. D. (2015). *Building ontologies with basic formal ontology.* MIT Press.

Arrieta, A. B., Díaz-Rodríguez, N., Del Ser, J., Bennetot, A., Tabik, S., Barbado, A., Garcia, S., Gil-Lopez, S., Molina, D., Benjamins, R., Chatila, R., & Herrera, F. (2020). Explainable artificial intelligence (XAI): Concepts, taxonomies, opportunities and challenges toward responsible AI. *Information Fusion*, 58, 82–115.

Art. (2024, July 21). In *Wikipedia.* https://en.wikipedia.org/w/index.php?title=Art&oldid=1235812306

Artificial Intelligence (AI). (2024). In *Webopedia.* https://www.webopedia.com/definitions/ai/

Arvan, M. (2025). AI Is too unpredictable to behave according to human goals. *Scientific American*. https://www.scientificamerican.com/article/ai-is-too-unpredictable-to-behave-according-to-human-goals/

Augustine (400). *Confessiones*.

Aurelius (180). *Meditations*.

Baars, B. J. (2002). The conscious access hypothesis: Origins and recent evidence. *Trends in Cognitive Sciences*, *6*(1), 47–52.

Bader, S., & Hitzler, P. (2005). Dimensions of neural-symbolic integration—a structured survey. In S. N. Artëmov, H. Barringer, A. S. d'Avila Garcez, L. C. Lamb, & J. Woods (Eds.), *We Will show them! Essays in honour of Dov Gabbay* (pp. 167–194). College Publications.

Bahdanau, D., Cho, K., & Bengio, Y. (2015). Neural machine translation by jointly learning to align and translate. In *Proceedings of the International Conference on Learning Representations*.

Baker, A. (2016). Simplicity. In E. N. Zalta (Ed.), *The Stanford Encyclopedia of Philosophy* (Summer 2022 Edition). https://plato.stanford.edu/archives/sum2022/entries/simplicity/

Ball, P. (2022). *The book of minds: How to understand ourselves and other beings, from animals to AI to aliens*. Picador.

Barnes, E. C. (2008). *The paradox of predictivism*. Cambridge University Press.

Baum, L. E., & Petrie, T. (1966). Statistical inference for probabilistic functions of finite state Markov chains. *The Annals of Mathematical Statistics*, *37*(6), 1554–1563.

Baumeister, R. F., Sparks, E. A., Stillman, T. F., & Vohs, K. D. (2001). Free will in consumer behavior: Self-control, ego depletion, and choice. *Journal of Consumer Psychology*, *18*(1), 4–13.

Becker, K., Becker, M., & Schwarz, J. H. (2006). *String theory and M-theory: A modern introduction*. Cambridge University Press.

Bell, C. G., & Newell, A. (1971). *Computer structures: Readings and examples*. McGraw-Hill.

Bellman, R. (1957). A Markovian decision process. *Journal of Mathematics and Mechanics*, *6*(5), 679–684.

Bengio, Y. (2023). How rogue AIs may arise. https://yoshuabengio.org/2023/05/22/how-rogue-ais-may-arise/

Berliner, H. J., & Ebeling, C. (1989). Pattern knowledge and search: The SUPREM Architecture. *Artificial Intelligence*, *38*(2), 161–198.

Bigge, M. L. (1976). *Learning theories for teachers, third edition*. Harper & Row Publishing.

Bilalić, M., McLeod, P., & Gobet, F. (2008). Expert and "novice" problem solving strategies in chess: Sixty years of citing de Groot (1946). *Thinking & Reasoning, 14*(4), 395–408.

Binz, M., Akata, E., Bethge, M., Brändle, F., Callaway, F., Coda-Forno, J., Dayan, P., Demircan, C., Eckstein, M. K., Éltető, N., Griffiths, T. L., Haridi, S., Jagadish, A. K., Ji-An, L., Kipnis, A., Kumar, S., Ludwig, T., Mathony, M., Mattar, M., . . . Schulz, E. (2024). Centaur: A foundation model of human cognition. arXiv:2410.20268

Binz, M., & Schulz, E. (2023). Using cognitive psychology to understand GPT-3. *Proceedings of the National Academies of Science, 120*(6), e2218523120.

Bissiri, P. G., Holmes, C. C., & Walker, S. G. (2016). A general framework for updating belief distributions. *Journal of the Royal Statistical Society: Series B, 78*(5), 1103–1130.

Bledsoe, W.W. (1971). Splitting and reduction heuristics in automatic theorem proving. *Artificial Intelligence, 2*(1), 55–77.

Blum, L., & Blum, M. (2022). A theory of consciousness from a theoretical computer science perspective: Insights from the Conscious Turing Machine. *Proceedings of the National Academy of Sciences of the United States of America, 119*(21), e2115934119.

Blythe, J. (1999). Decision-theoretic planning. *AI Magazine, 20*(2), 37–54.

Bonasso, R. P., Firby, R. J., Gat, E., Kortenkamp, D., Miller, D., & Slack, M. (1997). Experiences with an architecture for intelligent reactive agents. *Journal of Experimental and Theoretical Artificial Intelligence, 9*(2–3), 237–256.

Bordini, R. H., Braubach, L., Dastani, M., El Fallah-Seghrouchni, A., Gómez-Sanz, J. J., Leite, J., O'Hare, G. M. P., Pokahr, A., & Ricci, A. (2006). A survey of programming languages and platforms for multi-agent systems. *Informatica (Slovenia), 30*, 33–44.

Bostrom, N. (2014). Superintelligence: Paths, dangers, strategies. Oxford University Press.

Bothell, D. (2017). *ACT-R 7 Reference Manual*. http://act-r.psy.cmu.edu/actr7/reference-manual.pdf

Botvinick, M., Ritter, S., Wang, J. X., Kurth-Nelson, Z., Blundell, C., & Hassabis, D. (2019). Reinforcement learning, fast and slow. *Trends in Cognitive Sciences, 23*(5), 408–422.

Bounded rationality. (2024, July 21). In *Wikipedia*. https://en.wikipedia.org/w/index.php?title=Bounded_rationality&oldid=1235772127

Boutilier, C., Dean, T., & Hanks, S. (1999). Decision-theoretic planning: Structural assumptions and computational leverage. *Journal of Artificial Intelligence Research, 11*(1), 1–94.

Bower, G. H., & Clapper, J. P. (1989). Experimental methods in cognitive science. In M. I. Posner (Ed.), *Foundations of cognitive science* (pp. 93–131). Bradford Books/MIT Press.

Boyle, A., Cooperrider, K., Cheke, L., Halina, M., & Cave, S. (2022). *The atlas of intelligences: A diverse intelligences resource.* Apollo--University of Cambridge Repository.

Broad, W. J. (2024). How hallucinatory A.I. helps science dream up big breakthroughs. *The New York Times.* https://www.nytimes.com/2024/12/23/science/ai-hallucinations-science.html

Brodkin, J. (2022). Google fires Blake Lemoine, the engineer who claimed AI chatbot is a person. *Ars Technica.*

Brooks, D. (2024). You might be a late bloomer. *The Atlantic.*

Brown, J. S., Collins, A., & Duguid, S. (1989). Situated cognition and the culture of learning. *Educational Researcher, 18*(1), 32–42.

Brown, T. B., Mann, B., Ryder, N., Subbiah, M., Kaplan, J., Dhariwal, P. Neelakantan, A., Shyam, P., Sastry, G., Askell, A., Agarwal, S., Herbert-Voss, A., Krueger, G., Henighan, T., Child, R., Ramesh, A., Ziegler, D. M., Wu, J., Winter, C., . . . Amodei, D. (2020). Language models are few-shot learners. In *Proceedings of the 34th Conference on Neural Information Processing Systems* (pp. 1877–1901).

Bryson, J. J. (2010). Robots should be slaves. In Y. Wilks (Ed.), *Close engagements with artificial companions: Key social, psychological, ethical and design issues.* John Benjamins.

Bubeck, S., Chandrasekaran, V., Eldan, R., Gehrke, J., Horvitz, E., Kamar, E., Lee, P., Lee, Y. T., Li, Y., Lundberg, S., Nori, H., Palangi, H., Ribeiro, M. T., & Zhang, Y. (2023). Sparks of artificial general intelligence: Early experiments with GPT-4. arXiv:2303.12712.

Cai, L. W., Dai, W. Z., Huang, Y. X., Li, Y. F., Muggleton, S. H., & Jiang, Y. (2021). Abductive learning with ground knowledge base. In *Proceedings of the 30th International Joint Conference on Artificial Intelligence* (pp. 1815–1821).

Card, S., Moran, T., & Newell, A. (1983). *The psychology of human computer interaction.* Lawrence Erlbaum Associates.

Chai, W. J., Abd Hamid, A. I., & Abdullah, J. M. (2018). Working memory from the psychological and neurosciences perspectives: A review. *Frontiers in Psychology, 9,* 401.

Chan, K. M. A. (2011). Ethical extensionism under uncertainty of sentience: Duties to non-human organisms without drawing a line. *Environmental Values, 20*(3), 323–346.

Chen, J., Demski, A., Han, T., Morency, L-P., Pynadath, D., Rafidi, N., & Rosenbloom, P. S. (2011). Fusing symbolic and decision-theoretic problem solving+perception in a graphical cognitive architecture. In *Proceedings of the 2nd International Conference on Biologically Inspired Cognitive Architectures* (pp. 64–72).

Chin, K. O., Gan, K. S., Alfred, R., Anthony, P., & Lukose, D. (2014). Agent architecture: An overview. *Transactions on Science and Technology, 1*(1), 18–35.

Cho, B., Rosenbloom, P. S., & Dolan, C. P. (1991). Neuro-Soar: A neural-network architecture for goal-oriented behavior. In *Proceedings of the 13th Annual Conference of the Cognitive Science Society* (pp. 673–677).

Cho, K., Merrienboer, B. V., Gülçehre, Ç., Bahdanau, D., Bougares, F., Schwenk, H., & Bengio, Y. (2014). Learning phrase representations using RNN encoder-decoder for statistical machine translation. In *Proceedings of the Conference on Empirical Methods in Natural Language Processing* (pp. 1724–1734).

Choi, A., & Darwiche, A. (2018). On the relative expressiveness of Bayesian and neural networks. *Proceedings of Machine Learning Research, 72*, 157–168.

Choi, D. (2020). ICARUS' implications for the Standard Model of Mind. *Common Model of Cognition Bulletin, 1*(2), 308–310.

Chomsky, N. (1965). *Aspects of the theory of syntax*. MIT Press.

Chong, H.-Q., Tan, A.-H., & Ng, G.-W. (2007). Integrated cognitive architectures: A survey. *The Artificial Intelligence Review, 28*(2), 103–130.

Chu, D., & Ho, W. K. (2006). A category theoretical argument against the possibility of artificial life: Robert Rosen's central proof revisited. *Artificial Life, 12*(1), 117–134.

Cibber, C. 1740. *An apology for the life of Mr. Colley Cibber*. John Watts.

Clements, P., Bachman, F., Bass, L., Garlan, D., Ivers, J., Little, R., Merson, P., Nord, R., & Stafford, J. (2010). *Documenting software architectures: Views and beyond* (2nd ed.). Addison-Wesley.

Cohen, N. J., Eichenbaum, H., Deacedo, B. S., & Corkin, S. (1985). Different memory systems underlying acquisition of procedural and declarative knowledge. *Annals of the New York Academy of Sciences, 444*, 54–71.

Cohen, P. R. (1995). *Empirical methods for artificial intelligence*. The MIT Press.

Cohen, W. W., & Singer, Y. (1999). A simple, fast, and effective rule learner. In *Proceedings of the Sixteenth National Conference on Artificial Intelligence* (pp. 335–342).

Cole, K. C. (1999). A world of difference between invention and discovery. *Los Angeles Times*. https://www.latimes.com/archives/la-xpm-1999-apr-29-me-32280-story.html

Colmerauer, A., & Roussel, P. (1993). The birth of Prolog. *ACM SIGPLAN Notices, 28*(3), 37–52.

The Committee on European Computing Education. (2017). *Informatics education in Europe: Are we all in the same boat?* New York: Association for Computing Machinery.

Comparative psychology. (2024, June 25). In *Wikipedia*. https://en.wikipedia.org/w/index.php?title=Comparative_psychology&oldid=1230956524

Contour Crafting. (2023, December 22). In *Wikipedia*. https://en.wikipedia.org/w/index.php?title=Contour_crafting&oldid=1191245309

Contour Crafting Corporation. (2024). https://www.contourcrafting.com/

Cooper, R., Fox, J., Farringdom, J., & Shallice, T. (1996). A systematic methodology for cognitive modelling. *Artificial Intelligence, 85*(1–2), 3–44.

Coulom, R. (2006). Efficient selectivity and backup operators in Monte-Carlo tree search. In *Proceedings of the 5th International Conference on Computers and Games* (pp. 72–83).

Cox, M. T. (2005). Metacognition in computation: A selected research review. *Artificial Intelligence, 169*(2), 104–141.

Crawford, K. (2021). *Atlas of AI: Power, politics, and the planetary costs of artificial intelligence*. Yale University Press.

Cristian, F. (1982). Exception handling and software fault tolerance. *IEEE Transactions on Computers, 31*(6), 531–540.

Customer Review. (2015). https://www.amazon.com/gp/customer-reviews/R3OSS4Q6YXWCZK/ref=cm_cr_dp_d_rvw_ttl?ie=UTF8&ASIN=0262528282

Dalrymple, D., Skalse, J., Bengio, Y., Russell, S., Tegmark, M., Seshia, S. A., Omohundro, S., Szegedy, C., Goldhaber, B., Ammann, N., Abate, A., Halpern, J., Barrett, C., Zhao, D., Tan, Z., Wing, J., & Tenenbaum, J. B. (2024). Towards guaranteed safe AI: A framework for ensuring robust and reliable AI systems. arXiv:2405.06624.

Dancy, C. L. (2022). Using a cognitive architecture to consider antiblackness in design and development of AI systems. In *Proceedings of the 20th International Conference on Cognitive Modeling* (pp. 65–72).

Darwiche, A. (2002). A logical approach to factoring belief networks. In *Proceedings of the 8th International Conference on Principles of Knowledge Representation and Reasoning* (pp. 409–420).

Darwiche, A. (2003). A differential approach to inference in Bayesian networks. *Journal of the ACM, 50*(3), 280–305.

Das, J. P. (1986). On definition of intelligence. In R. J. Sternberg & D. K. Detterman (Eds.), *What is intelligence?: Contemporary viewpoints on its nature and definition* (pp. 55–56). Ablex.

Dasgupta, S. (2016). *Computer science: A very short introduction.* Oxford University Press.

Davis, R. (1983). Reasoning from first principles in electronic troubleshooting. *International Journal of Man-Machine Studies, 19*(5), 403–423.

Davis, R., Shrobe, H., & Szolovits, P. (1993). What is a knowledge representation? *AI Magazine, 14*(1), 17–33.

Dechter, R. (2003). *Constraint processing.* Morgan Kaufmann.

de Groot, A. D. (1965). *Thought and choice in chess.* Mouton Publishers.

Dehaene, S. (2014). *Consciousness and the brain: Deciphering how the brain codes our thoughts.* Viking Press.

de Kleer, J. (1986). An assumption-based TMS. *Artificial Intelligence, 28*(2), 27–162.

Demski, A. (2022). *Cognitive equations: Fixed-point finding as an intermediate layer in cognitive architecture* [PhD dissertation, University of Southern California].

Dennett, D. C. (1991). *Consciousness explained.* Little, Brown & Company.

Dennett, D. C. (1996). *Kinds of minds: Toward an understanding of consciousness.* Basic Books.

Dennett, D. C. (2003). *Freedom evolves.* Penguin Books.

Denning, P. J. (2005). Is computer science science? *Communications of the ACM, 48*(4), 27–31.

Denning, P. J. (2007). Computing is a natural science. *Communications of the ACM, 50*(7), 13–18.

Denning, P. J. (2013). The science in computer science. *Communications of the ACM, 56*(5), 35–38.

Denning, P. J., & Rosenbloom, P. S. (2009). Computing: The fourth great domain of science. *Communications of the ACM, 52*(9), 27–29.

Derbinsky, N., Li, J., & Laird, J. E. (2012). Algorithms for scaling in a general episodic memory (Extended Abstract). In *Proceedings of the 11th International Conference on Autonomous Agents and Multiagent Systems* (pp. 1387–1388).

de Raedt, L., Kersting, K., Natarajan, S., & Poole, D. (2016). *Statistical relational artificial intelligence: Logic, probability, and computation.* Morgan & Claypool.

Desprez, F., Fox, G., Jeannot, E., Keahey, K., Kozuch, M., Margery, D., Neyron, P., Nussbaum, L., Perez, C., Richard, O., Smith, W., & von Laszewski. G. (2012). *Supporting experimental computer science.* Argonne National Laboratory, Rocquencourt, France, Technical Memo 362, 2012.

Determinism. (2024, July 12). In *Britannica.* https://www.britannica.com/topic/determinism

Deutsch, D. (2011). *The beginning of infinity: Explanations that transform the world.* Penguin Books.

Dietterich, T. G. (1986). Learning at the knowledge level. *Machine Learning, 1*(3), 287–316.

Digital physics. (2024, April 11). In *Wikipedia.* https://en.wikipedia.org/w/index.php?title=Digital_physics&oldid=1218342297

Dissect. (2024). In *Dictionary.com.* https://www.dictionary.com/browse/dissect

Domingos, P., & Lowd, D. (2009). *Markov logic: An interface layer for artificial intelligence.* Morgan & Claypool.

Douglas, H. E. (2013). The value of cognitive values. *Philosophy of Science, 80*(5), 796–806.

Douglass, S., Ball, J., & Rodgers, S. (2009). Large declarative memories in ACT-R. In *Proceedings of the 9th International Conference of Cognitive Modeling* (paper 234).

Dove, G. (2023). Rethinking the role of language in embodied cognition. *Philosophical Transactions of the Royal Society of London. Series B, Biological Sciences, 378,* 20210375(1870).

Doyle, J. A (1979). A truth maintenance system. *Artificial Intelligence, 12*(3), 251–272.

Dwivedi, V. P., Joshi, C. K., Luu, A. T., Laurent, T., Bengio, Y., & Bresson, X. (2022). Benchmarking graph neural networks. arXiv:2003.00982.

Education. (2024). In *Dictionary.com.* https://www.dictionary.com/browse/education

Ehrlinger, L., & Wöß, W. (2016). Towards a definition of knowledge graphs. In *Proceedings of Semantics* (pp. 13–16).

Eisner, J., & Filardo, N. W. (2011). Dyna: Extending Datalog for modern AI. In O. de Moor, G. Gottlob, T. Furche, & A. Sellers (Eds.), *Datalog reloaded.* Springer-Verlag.

Eliasmith, C. (2013). *How to build a brain: A neural architecture for biological cognition.* Oxford University Press.

Elman J. L. (1990). Finding structure in time. *Cognitive Science, 14*(2), 179–211.

Enderby, J. (2009). Lumpers and splitters: Darwin, Hooker, and the search for order. *Science, 326*(5959), 1496–1499.

Environmental, Social, and Governance. (2025, February 25). In *Wikipedia*. https://en.wikipedia.org/w/index.php?title=Environmental,_social,_and_governance&oldid=1277558879

Erman, L. D., Hayes-Roth, F., Lesser, V. R., & Reddy R. (1980). The Hearsay-II speech-understanding system: Integrating knowledge to resolve uncertainty. *Computing Surveys, 12*(2), 213–253.

Euclid. (c. 300 B.C.E.). *Elements*.

Eurisko. (2024, June 21). In *Wikipedia*. https://en.wikipedia.org/w/index.php?title=Eurisko&oldid=1230179089

Evidently AI Team. (2024). LLM guide: 20 LLM evaluation benchmarks and how they work. https://www.evidentlyai.com/llm-guide/llm-benchmarks

Explanatory power. (2023, December 28). In *Wikipedia*. https://en.wikipedia.org/w/index.php?title=Explanatory_power&oldid=1192309075

Falkenhainer, B., Forbus, K. D., & Gentner, D. (1989). The structure-mapping engine: Algorithm and examples. *Artificial Intelligence, 41*(1), 1–63.

Fann, K. T. (1970). *Pierce's theory of abduction*. Martinus Nijhoff.

Fedus, W., Zoph, B., & Shazeer, N. (2022). Switch transformers: Scaling to trillion parameter models with simple and efficient sparsity. *Journal of Machine Learning Research, 23*(1), 1–39.

Feigenbaum, E. (1992). *Expert systems: Principles and practice*. Stanford KSL Tech. Report 91–79.

Feldman, J. A., & Ballard, D. (1982). Connectionist models and their properties. *Cognitive Science, 6*(3), 205–254.

Ferguson, C., & Korf, R. E. (1988). Distributed tree search and its applications to alpha-beta pruning. In *Proceedings of the 7th National Conference on Artificial Intelligence* (pp. 128–132).

Feurer, M., & Hutter, F. (2019). Hyperparameter optimization. In F. Hutter, L. Kotthoff, & J. Vanschoren (Eds.), *Automated machine learning: Methods, systems, challenges* (pp. 3–38). Springer.

Fikes, R. E., & Kehler, T. (1985). The role of frame-based representation in knowledge representation and reasoning. *Communications of the ACM, 28*(9), 904–920.

Fine, T. L. (1999). *Feedforward neural network methodology*. Springer.

Fink, P. K., Lusth, J. C., & Duran, J. W. (1985). A general expert system design for diagnostic problem solving. *IEEE Transactions on Pattern Analysis and Machine Intelligence, 7(5)*, 553–560.

Fitts, P. M., & Seeger, C. (1953). S-R compatibility: Spatial characteristics of stimulus and response codes. *Journal of Experimental Psychology, 46*(3), 199–210.

Flenner, A., Fraune, M. R., Hiatt, L. M., Kendall, T., Laird, J. E., Lebiere, C., Rosenbloom, P. S., Stein, F., Topp, E. A., Unhelkar, V. V., & Zhao, Y. (2018). Reports of the AAAI 2017 fall symposium series. *AI Magazine, 39*(2), 81–86.

FLOPS. (2024, July 11). In *Wikipedia*. https://en.wikipedia.org/w/index.php?title=FLOPS&oldid=1233877946

Fluid and crystallized intelligence. (2024, July 21). In *Wikipedia*. https://en.wikipedia.org/w/index.php?title=Fluid_and_crystallized_intelligence&oldid=1235893893

Fodor, J. A. (1983). *Modularity of mind: An essay on faculty psychology*. MIT Press.

Fodor, J. A., & Pylyshyn, Z. W. (1988). Connectionism and cognitive architecture: A critical analysis. *Cognition, 28*(1–2), 3–71.

For All Mankind. (2024, July 17). In *Wikipedia*. https://en.wikipedia.org/w/index.php?title=For_All_Mankind_(TV_series)&oldid=1235129870

Forbus, K. D. (2019). *Qualitative representations: How people reason and learn about the continuous world*. MIT Press.

Forbus, K. D., & Hinrichs, T. R. (2006). Companion cognitive systems: A step towards human-level AI. *AI Magazine, 27*, 83–95.

Forgy, C. L. (1982). Rete: A fast algorithm for the many pattern/many object pattern match problem. *Artificial Intelligence, 19*(1), 17–37.

François, J. (1977). Evolution and tinkering. *Science, 196*(4295), 1161–1166.

Francois-Lavet, V., Henderson, P., Islam, R., Bellemare, M. G., & Pineau, J. (2018). An introduction to deep reinforcement learning. *Foundations and Trends in Machine Learning, 11*(3–4), 219–354.

Frasconi, P., Gori, M., & Sperduti, A. (1998). A general framework for adaptive processing of data structures. *IEEE Transactions on Neural Networks, 9*(5), 768–786.

Free will (2024, August 14). In *Dictionary.com*. https://www.dictionary.com/browse/free%20will

Frey, P. W. (1981). The Santa Cruz Open Othello tournament for computers. *BYTE, 6*(7), 26–37.

Friston, K. (2010). The free-energy principle: A unified brain theory? *Nature Reviews Neuroscience, 11*, 127–138.

FTC. (2023). FTC authorizes compulsory process for AI-related products and services. https://www.ftc.gov/news-events/news/press-releases/2023/11/ftc-authorizes-compulsory-process-ai-related-products-services

Gardner, H. (1983). *Frames of mind: The theory of multiple intelligences*. Basic Books.

Gardner, M. R. (1979). Realism and instrumentalism in 19th-ventury atomism. *Philosophy of Science, 46*(1), 1–34.

Gat, E. (1998). Three-layer architectures. In D. Kortenkamp, R. P. Bonasso, & R. Murphy (Eds.), *Artificial intelligence and mobile robots: Case studies of successful robot systems* (pp. 195–210). AAAI Press.

Gellers, J. C. (2021). *Rights for robots: Artificial intelligence, animal and environmental law*. Routledge.

Garnham, N. (2005). A personal intellectual memoir. *Media, Culture & Society, 27*(4), 469–493.

Genesereth, M. R., & Nilsson, N. J. (1987). *Logical foundations of artificial intelligence*. Morgan Kaufmann.

Gentner, D., & Forbus, K. D. (2011). Computational models of analogy. *Wiley Interdisciplinary Reviews: Cognitive Science, 2*(3), 266–276.

Gentner, D., Holyoak, K. J., & Kokinov, B. N. (Eds.). (2001). *The analogical mind: Perspectives from cognitive science*. Bradford Books.

Gershman, S. J., Horvitz, E. J., & Tenenbaum, J. B. (2015). Computational rationality: A converging paradigm for intelligence in brains, minds, and machines. *Science, 349*(6245), 273–278.

Getoor, L., & Taskar, B. (2007). *Introduction to statistical relational learning*. MIT Press.

Ghallab, M., Nau, D., & Traverso, P. (2004). *Automated planning: Theory and practice*. Morgan Kaufman.

Gilbey, J. (2012). Virtually there. *Nature, 491*, 331.

Ginsberg, M. L. (1989). Universal planning research: A good or bad idea? *AI Magazine, 10*(4), 40–44.

Girardi, R., & Leite, A. (2013). A survey on software agent architectures. *IEEE Intelligent Informatics Bulletin, 14*(1), 8–20.

Gluck, K. A., & Laird, J. E. (Eds.). (2018). *Interactive task learning: Humans, robots, and agents acquiring new tasks through natural interactions*. MIT Press.

Gobet, F., & Lane, P. C. R. (2020). Constructing a standard model: Lessons from CHREST. *Common Model of Cognition Bulletin, 1*(2), 338–343.

Goertzel, B. (2014). Artificial general intelligence: Concept, state of the art, and future prospects. *Journal of Artificial General Intelligence*, *5*(1), 1–46.

Goertzel, B., Pennachin, C., & Geisweiller, N. (2014). *Engineering general intelligence*. Atlantis Press.

Goller, C., & Kuchler, A. (1996). Learning task-dependent distributed representations by backpropagation through structure. In *Proceedings of International Conference on Neural Networks* (pp. 347–352).

Gonzalez, C., Lerch, F. J., & Lebiere, C. (2003). Instance-based learning in dynamic decision making. *Cognitive Science, 27*(4), 591–635.

Goodfellow, I., Bengio, Y., & Courville, A. (2016). *Deep learning*. MIT Press.

Goodfellow, I., Pouget-Abadie, J., Mirza, M., Xu, B., Warde-Farley, D., Ozair, S., Courville, A., & Bengio, Y. (2014). Generative adversarial networks. In *Proceedings of the International Conference on Neural Information Processing Systems* (pp. 2672–2680).

Google Maps. (2023). Paramount Studios Bronson Gate. https://www.google.com/maps/uv?pb=!1s0x80c2b986faea9da7%3A0xcf60649badeabf6a!3m1!7e115!4shttps%3A%2F%2Flh5.googleusercontent.com%2Fp%2FAF1QipPCXkwozHPkg2Y0ncY5_XCJqermNKQSSGGo9KK_%3Dw426-h320-k-no!5sparamount%20gates%20-%20Google%20Search!15sCgIgAQ&imagekey=!1e10!2sAF1QipPk9X7NHMehujI9eMQ_9IabR_Ic1SMxY-yTUwUY&hl=en&sa=X&ved=2ahUKEwjF7KSu867zAhVyMX0KHUsICrAQoip6BAhlEAM

Google Scholar. (2025, March 15). Paul Rosenbloom. https://scholar.google.com/citations?user=FgcThs0AAAAJ&hl=en

Gottfredson, L. S. (1997). Mainstream science on intelligence: An editorial with 52 signatories, history, and bibliography. *Intelligence, 24*(1), 1997.

Gracián, B. (1647). *Oráculo Manual y Arte de Prudencia*. Don Vincencio Juan de Lastanosa.

Graf, P., & Schacter, D. L. (1985). Implicit and explicit memory for new associations in normal and amnesic subjects. *Journal of Experimental Psychology: Learning, Memory, and Cognition, 11*(3), 501–518.

Grattafiori, A. et al. (2024). The Llama 3 herd of models. arXiv:2407.21783.

Graves, A., Wayne, G., & Danihelka, I. (2014). Neural Turing machines. arXiv:1410.5401.

Gray, H. M., Gray, K., & Wegner, D. M. (2007). Dimensions of mind perception. *Science, 315*(5812), 619.

Greengard, S. (2021). *The internet of things, revised and updated edition*. MIT Press.

Gunkel, D. J. (2018). The other question: Can and should robots have rights? *Ethics and Information Technology*, *20*(3), 87–99.

Haddon, M. (2003). *The curious incident of the dog in the night-time*. Doubleday.

Haddon, M. (2006). *A spot of bother*. Doubleday.

Hamilton. (2024, July 17). In *Wikipedia*. https://en.wikipedia.org/w/index.php?title=Hamilton_(musical)&oldid=1235021707

Hao, S., Gu, Y., Ma, H., Hong, J. J., Wang, Z., Wang, D. Z., & Hu, Z. (2023). Reasoning with language model is planning with world model. arXiv:2305.14992.

Harnad, S. (1990). The symbol grounding problem. *Physica D: Nonlinear Phenomena*, *42*(1–3), 335–346.

Hart, P. E., Nilsson, N. J., & Raphael, B. (1968). A formal basis for the heuristic determination of minimum cost paths. *IEEE Transactions on Systems Science and Cybernetics*, *4*(2), 100–107.

Hayes-Roth, F., Waterman, D. A., & Lenat, D. B. (1983). *Building expert systems*. Addison-Wesley.

Heaven, D. (2019). Why deep-learning AIs are so easy to fool: Artificial-intelligence researchers are trying to fix the flaws of neural networks. *Nature*, *574*(7777), 163–166.

Hebb, D.O. (1949). *The organization of behavior*. Wiley & Sons.

Hempel, C. G., & Oppenheim, P. (1948). Studies in the logic of explanation. *Philosophy of Science*, *15*(2), 135–175.

Hennessy, J. L., & Patterson, D. A. (2012). *Computer architecture: A quantitative approach* (5th ed.). Waltham, MA: Morgan Kaufmann

Hestness, J., Narang, S., Ardalani, N., Diamos, G., Jun, H., Kianinejad, H., Patwary, M. A., Yang, Y., & Zhou, Y. (2017). Deep learning scaling is predictable, empirically. arXiv:1712.00409.

Hill, R. W., Chen, J., Gratch, J., Rosenbloom, P. S., & Tambe, M. (1997). Intelligent agents for the synthetic battlefield: A company of rotary wing aircraft. In *Proceedings, 9th Conference on Innovative Applications of Artificial Intelligence* (pp. 1006–1012).

Hinton, G. E., Dean, J., & Vinyals, O. (2015). Distilling the knowledge in a neural network. arXiv:1503.02531

Hinton, G. E., McClelland, J. L., & Rumelhart, D. E. (1986). Distributed representations. In J. L. McClelland & D. E. Rumelhart (Eds.), *Parallel distributed processing: Explorations in the microstructure of cognition*, vol. 1 (pp. 77–109). MIT Press.

Hochreiter, S., & Schmidhuber, J. (1997). Long short-term memory. *Neural Computation*, *9*(8), 1735–1780.

Hohwy, J. (2013). *The predictive mind.* Oxford University Press.

Holland, J. H. (1992). *Adaptation in natural and artificial systems.* MIT Press.

Hopfield, J. J. (1982). Neural networks and physical systems with emergent collective computational abilities. *Proceedings of the National Academy of Sciences, 79*(8), 2554–2558.

Hossenfelder, S. (2018). *Lost in Math: How beauty leads physics astray.* Basic Books.

Howard, R. A., & Matheson, J. E. (1984). Influence diagrams. In R. A. Howard & J. E. Matheson (Eds.), *Readings on the principles and applications of decision analysis* (pp. 719–762). Strategic Decisions Group.

Hu, K., & Tong, A. (2024). OpenAI and others seek new path to smarter AI as current methods hit limitations. *Reuters.* https://www.reuters.com/technology/artificial-intelligence/openai-rivals-seek-new-path-smarter-ai-current-methods-hit-limitations-2024-11-11/

Hutter, M. (2005). *Universal artificial intelligence: Sequential decisions based on algorithmic probability.* Springer-Verlag.

IBM System/360. (2024, July 10). In *Wikipedia.* https://en.wikipedia.org/w/index.php?title=IBM_System/360&oldid=1233685943

Ingrand, F., & Ghallab, M. (2017). Deliberation for autonomous robots: A survey. *Artificial Intelligence, 247,* 10–44.

Intelligence. (2024, July 20). In *Wikipedia.* https://en.wikipedia.org/w/index.php?title=Intelligence&oldid=1235692514

Interregnum. (2024). In *Dictionary.com.* https://www.dictionary.com/browse/interregnum

Ji, J., Qiu, T., Chen, B., Zhang, B., Lou, H., Wang, K., Duan, Y., He, Z., Zhou, J., Zhang, Z., Zeng, F., Ng, K. Y., Dai, J., Pan, X., O'Gara, A., Lei, Y., Xu, H., Tse, B., Fu, J., . . . Gao, W. (2024). AI alignment: A comprehensive survey. arXiv:2310.19852.

Ji, Z., Lee, N., Frieske, R., Yu, T., Su, D., Xu, Y., Ishii, E., Bang, Y., Dai, W., Madotto, A., & Fung, P. (2022). Survey of hallucination in natural language generation. arXiv:2202.03629.

Jilk, D., Lebiere, C., O'Reilly, R., & Anderson, J. (2008). SAL: An explicitly pluralistic cognitive architecture. *Journal of Experimental and Theoretical Artificial Intelligence, 20*(3), 197–218.

John, B. E., & Newell, A. (1990). Toward an engineering model of stimulus-response compatibility. In R. W. Proctor & T. G. Reeve (Eds.), *Stimulus-response compatibility: An integrated perspective.* North-Holland.

John, B. E., Rosenbloom, P. S., & Newell, A. (1985). A theory of stimulus-response compatibility applied to human-computer interaction. In *Proceedings of the SIGCHI Conference on Human Factors in Computing Systems* (pp. 212–219).

Jones, C., & Bergen, B. (2024). People cannot distinguish GPT-4 from a human in a Turing test. arXiv:2405.08007.

Jones, G., & Ritter, F. E. (2003). Production systems and rule-based inference. In L. Nadel (Ed.), *Encyclopedia of cognitive science*. Wiley & Sons.

Jones, M. N., & Mewhort, D. J. (2007). Representing word meaning and order information in a composite holographic lexicon. *Psychological Review, 114*(1), 1–37.

Jones, R. M., Laird, J. E., Nielsen, P. E., Coulter, K. J., Kenny, P., & Koss, F. V. (1999). Automated intelligent pilots for combat flight simulation, *AI Magazine, 20*(1), 27–41.

Jordan, M. I., & Sejnowski, T. J. (2001). *Graphical models: Foundations of neural computation*. MIT Press.

Joshi, H., Rosenbloom, P. S., & Ustun, V. (2014). Isolated word recognition in the Sigma cognitive architecture. *Biologically Inspired Cognitive Architectures, 10*(6), 1–9.

Joshi, H., Rosenbloom, P. S., & Ustun, V. (2016). Continuous phone recognition in the Sigma cognitive architecture. *Biologically Inspired Cognitive Architectures, 18*(2), 23–32.

Joshi, H., Rosenbloom, P. S., & Ustun, V. (2018). Exact, tractable inference in the Sigma cognitive architecture via sum-product networks. *Advances in Cognitive Systems, 7*, 39–55.

Kahneman, D. (2011). *Thinking fast and slow*. Farrar, Straus and Giroux.

Kakkasageri, M. S., Hridya, C. P., Vibha, N. S., Manvi S. S., & Basarkod, P. I. (2012). A Survey on cognitive agent architectures. *IETE Journal of Education, 53*(1), 21–37.

Kaminka, G. A., & Burghart, C. R. (2007). *Evaluating architectures for intelligence: Papers from the 2007 AAAI workshop*. Technical Report WS-07-04. AAAI Press.

Kaplan, J., McCandlish, S., Henighan, T.J., Brown, T.B., Chess, B., Child, R., Gray, S., Radford, A., Wu, J., & Amodei, D. (2020). Scaling laws for neural language models. arXiv:2001.08361.

Karp, P. D., & Wilkins, D. C. (1989). An analysis of the distinction between deep and shallow expert systems. *International Journal of Expert Systems, 1*(2), 1–32.

Keas, M. N. (2018). Systematizing the theoretical virtues. *Synthese, 195*, 2761–2793.

Kejriwal, M., Knoblock, C. A., & Szekely, P. (2021). *Knowledge graphs: Fundamentals, techniques, and applications*. MIT Press.

Kelley, T. D., Thomson, R., & Milton, J. R. (2018). Standard model of mind: Episodic memory. *Common Model of Cognition Bulletin, 1*(3), 717–723.

Kelly, M. A., & Reitter, D. (2018). How language processing can shape a common model of cognition. *Procedia Computer Science, 145*, 724–729.

Ketola, M., Stocco, A., Thompson, S., Madhyastha, T. M., & Grabowski, T. J. (2020). Applying the common model of cognition to resting-state fMRI leads to the identification of abnormal functional connectivity in Parkinson's disease. In *Proceedings of the 42nd Annual Conference of the Cognitive Science Society* (pp. 2839–2845).

Khoshnevis (2012). *Contour crafting: Automated construction.* TedXOjai video at https://www.youtube.com/watch?v=JdbJP8Gxqog

Kieras, D. E. (2017). A summary of the EPIC cognitive architecture. In S. F. Chipman (Ed.), *The Oxford handbook of cognitive science* (pp. 37–48). Oxford University Press.

Kieras, D. E. (2020). EPIC lessons for the proposed Standard Model of the Mind. *Common Model of Cognition Bulletin, 1*(2), 388–393.

Kieras, D. E., & Meyer, D. E. (1997). An overview of the EPIC architecture for cognition and performance with application to human-computer interaction. *Human-Computer Interaction, 12*(4), 391–438.

Kim, T. W., & Strudler, A. (2023). Should robots have rights or rites? *Communications of the ACM, 66*(6), 78–85.

Kingma, D. P., & Welling, M. (2019). An Introduction to variational autoencoders. *Foundations and Trends in Machine Learning, 12*(4), 307–392.

Kivunja, C. (2018). Distinguishing between theory, theoretical framework, and conceptual framework: A systematic review of lessons from the field. *International Journal of Higher Education, 7*(6), 44–53.

Kleyko, D., Rachkovskij, D. A., Osipov, E., & Rahimi, A. (2022). A survey on hyperdimensional computing aka vector symbolic architectures, Part I: Models and data transformations. *ACM Computing Surveys, 55*(6), 1–40.

Knowledge graph. (2024, July 15). In *Wikipedia*. https://en.wikipedia.org/w/index.php?title=Knowledge_graph&oldid=1234644296

Koller, D., & Friedman, N. (2009). *Probabilistic graphical models: Principles and techniques.* MIT Press.

Kolodner, J. L. (1992). An introduction to case-based reasoning. *Artificial Intelligence Review, 6*(1), 3–34.

Korf, R. E. (1985). Macro-operators: A weak method for learning. *Artificial Intelligence, 26*(1), 35–77.

Kortenkamp, D., & Simmons, R. (2008). Robotic systems architectures and programming. In O. Khatib & B. Siciliano (Eds.), *Springer handbook of robotics* (pp. 187–206). Springer-Verlag.

Kotseruba, I., & Tsotsos, J. K. (2020). 40 years of cognitive architectures: Core cognitive abilities and practical applications. *Artificial Intelligence Review, 53*(1), 17–94.

Kowalski, R. (1979). Algorithm = logic + control. *Communications of the ACM, 22*(7), 424–436.

Kowalski, R. A. (1988). The early years of logic programming. *Communications of the ACM, 31*(1), 38–43.

Koza, J. R. (1992). *Genetic programming: On the programming of computers by means of natural selection.* MIT Press.

Kralik, J. D., Lee, J. H., Rosenbloom, P. S., Jackson, P. C., Epstein, S. L., Romero, O. J., Sanz, R., Larue, O., Schmidtke, H. R., Lee, S. W., & McGreggor, K. (2018). Metacognition for a common model of cognition. *Procedia Computer Science, 145*, 740–746.

Kschischang, F. R., Frey, B. J., & Loeliger, H.-A. (2001). Factor graphs and the sum-product algorithm. *IEEE Transactions on Information Theory, 47*(2), 498–519.

Kuhn, T. (1962). *The structure of scientific revolutions*. University of Chicago Press.

Kumova, B. İ., & Heye, S. B. (2017). A survey of robotic agent architectures. In *Proceedings of the International Artificial Intelligence and Data Processing Symposium* (pp. 1–8).

Kurfess, F. J. (2002). Integrating symbol-oriented and sub-symbolic reasoning methods into hybrid systems. In B. Apolloni & F. Kurfess (Eds.), *From synapses to rules: Discovering symbolic rules from neural processed data* (pp. 275–292). Kluwer.

Lacroix, G. L., & Cousineau, D. (2006). The introduction to the special issue on "RT(N) = a + b N-c: The Power Law of Learning 25 years later." *Tutorials in Quantitative Methods for Psychology, 2*(2), 77–83.

Laird, J. E. (1986). Universal subgoaling. In J. E. Laird, P. S. Rosenbloom, & A. Newell (Eds.), *Universal subgoaling and chunking: The automatic generation and learning of goal hierarchies* (pp. 1–131). Kluwer.

Laird, J. E. (2008). Extending the Soar cognitive architecture. In *Artificial General Intelligence: Proceedings of the 1st AGI Conference* (pp. 224–235).

Laird, J. E. (2012). *The Soar cognitive architecture*. MIT Press.

Laird, J. E. (2022). Introduction to Soar. arXiv:2205.03854.

Laird, J. E., Congdon, C. B. Assanie, M., Derbinsky, N., & Xu, J. (2024). *The Soar user's manual (Version 9.6.2).* https://soar.eecs.umich.edu/downloads/Documentation/SoarManual.pdf

Laird, J. E., Gluck, K., Anderson, J., Forbus, K. D., Jenkins, O. C., Lebiere, C., Salvucci, D., Scheutz, M., Thomaz, A., Trafton, G., Wray, R. E., Mohan, S., & Kirk, J. R. (2017). Interactive task learning. *IEEE Intelligent Systems, 32(4)*, 6–21.

Laird, J. E., Lebiere, C., & Rosenbloom, P. S. (2017). A standard model of the mind: Toward a common computational framework across artificial intelligence, cognitive science, neuroscience, and robotics. *AI Magazine, 38*(4), 13–26.

Laird, J. E., Lebiere, C., Rosenbloom, P. S., & Stocco, A. (2025). A proposal to extend the common model of cognition with metacognition. arXiv:2506.07807.

Laird, J. E., & Mohan, S. (2018). Learning fast and slow: Levels of learning in general autonomous intelligent agents. In *Proceedings of the 32nd AAAI Conference on Artificial Intelligence* (7983–7987).

Laird, J. E., & Newell, A. (1983). A universal weak method: Summary of results. In *Proceedings of the 8th International Joint Conference on Artificial Intelligence* (pp. 771–773).

Laird, J. E., Newell, A., & Rosenbloom, P. S. (1987). Soar: An architecture for general intelligence. *Artificial Intelligence, 33*(1), 1–64.

Laird, J. E., & Rosenbloom, P. S. (1990). Integrating execution, planning, and learning in Soar for external environments. In *Proceedings of the 8th National Conference on Artificial Intelligence* (pp. 1022–1029).

Laird, J. E., & Rosenbloom, P. S. (1992). In pursuit of mind: The research of Allen Newell. *AI Magazine, 13*(4), 17–45.

Laird, J. E., & Rosenbloom, P. S. (1996). The evolution of the Soar cognitive architecture. In D. M. Steier & T. M. Mitchell (Eds.), *Mind matters: A tribute to Allen Newell* (pp. 1–50). Lawrence Erlbaum Associates.

Laird, J. E., Rosenbloom, P. S., & Newell, A. (1984). Towards chunking as a general learning mechanism. In *Proceedings of the National Conference on Artificial Intelligence* (pp. 188–192).

Laird, J. E., Rosenbloom, P. S., & Newell, A. (1986a). Chunking in Soar: The anatomy of a general learning mechanism. *Machine Learning, 1*(1), 11–46.

Laird, J. E., Rosenbloom, P. S., & Newell, A. (1986b). *Universal subgoaling and chunking: The automatic generation and learning of goal hierarchies.* Kluwer Academic Publishers.

Lakatos, I. (1978). *The methodology of scientific research programmes: Philosophical papers,* Vol. 1. Cambridge University Press.

Lakatos, I. (2025, March 16). In *Stanford encyclopedia of philosophy.* https://plato.stanford.edu/entries/lakatos/

Landhuis, E. (2020). Deep learning takes on tumours. *Nature, 580*(7804), 551–553.

Langley, P., Laird, J. E., & Rogers, S. (2009). Cognitive architectures: Research issues and challenges. *Cognitive Systems Research, 10*(2), 141–160.

Langley, P., & Messina, E. (2004), Experimental studies of integrated cognitive systems. In *Proceedings of the Performance Metrics for Intelligent Systems Workshop*.

Langley, P., Thompson, K., Iba, W., Gennari, J., & Allen, A. J. (1989). *An integrated cognitive architecture for autonomous agents*. Technical Report 89–28 Department of Information & Computer Science, University of California, Irvine, CA.

La Rochefoucauld. (1665). *Maximes*.

Larue, O., West, R., Rosenbloom, P. S., Dancy, C. L., Samsonovich, A. V., Petters, D., & Juvina, I. (2018). Emotion in the common model of cognition. *Procedia Computer Science, 145*, 730–739.

Laskey, K. B. (2006). Quantum physical symbol systems. *Journal of Logic, Language and Information, 15*(1–2), 109–154.

Leake, D. B. (Ed.). (1996). *Case-based reasoning: Experiences, lessons, and future directions*. AAAI Press/MIT Press.

LeCun, Y. (2022). A path towards autonomous machine intelligence. https://openreview.net/pdf?id=BZ5a1r-kVsf

LeCun, Y., & Bengio, Y. (1995). Convolutional networks for images, speech, and time series. In M. Arbib (Ed.), *The handbook of brain theory and neural nets*. MIT Press.

LeCun, Y., Bengio, Y., & Hinton, G. E. (2015). Deep learning. *Nature, 521*(7553), 436–444.

LeCun, Y., Boser, B., Denker, J. S., Henderson, D., Howard, R. E., Hubbard, W., & Jackel, L. D. (1989). Backpropagation applied to handwritten zip code recognition. *Neural Computation, 1*(4), 541–551.

Lee, S. M. (2025). Teens are doing AI research now. Is that a good thing? *The Chronicle of Higher Education*. https://www.chronicle.com/article/teens-are-doing-ai-research-now-is-that-a-good-thing

Legg, S. (2023). 71 Definitions of Intelligence. https://www.calculemus.org/lect/08szt-intel/materialy/Definitions%20of%20Intelligence.html

Legg, S., & Hutter, M. (2006). A formal measure of machine intelligence. In *Proceedings of the Annual Machine Learning Conference of Belgium and The Netherlands* (pp. 73–80).

Legg, S., & Hutter, M. (2007). Universal intelligence: A definition of machine intelligence. *Minds and Machines, 17*(4), 391–444.

Lehman, J. F., Laird, J. E., & Rosenbloom, P. S. (1998). A gentle introduction to Soar, an architecture for human cognition. In D. N. Osherson, S. Sternberg, & D.

Scarborough (Eds.), *An invitation to cognitive science (Second edition), volume 4: Methods, models and conceptual issues* (pp. 211–253). MIT Press.

Lehman, J. F., Lewis, R. L., & Newell, A. (1991). Integrating knowledge sources in language comprehension. In *Proceedings of the 13th Annual Conferences of the Cognitive Science Society* (pp. 461–466).

Lehman, J. F., Newell, A., & Newell, P. (1994). Soar. Video at https://www.dropbox.com/s/k0uugyu7szo7pm6/SOAR_access.mov?dl=0

Lenat, D. B. (1983). Eurisko: Eurisko: A program that learns new heuristics and domain concepts. *Artificial Intelligence, 21*(1–2), 61–98.

Lenat, D. B., & Guha, R. V. (1989). *Building large knowledge-based systems; Representation and inference in the Cyc project*. Addison-Wesley Longman.

Lenat D. B., & Marcus, G. (2023). Getting from generative AI to trustworthy AI: What LLMs might learn from Cyc. arXiv:2308.04445.

Lenci, A. (2018). Distributional models of word meaning. *Annual Review of Linguistics, 4*, 151–171.

Lent, J. (2017). AI has already taken over. It's called the corporation. https://patternsofmeaning.com/2017/11/30/ai-has-already-taken-over-its-called-the-corporation/

Levin, M., & Yuste, R. (2022). Modular cognition: Powerful tricks from computer science and cybernetics show how evolution "hacked" its way to intelligence from the bottom up. *aeon*, https://aeon.co/essays/how-evolution-hacked-its-way-to-intelligence-from-the-bottom-up

Levy, D. N. L. (Ed.). (1988). *Computer games, Volumes I & II*. Springer-Verlag.

Li, L., Qi, X., Xie, T., & Li, B. (2020). SoK: Certified robustness for deep neural networks. arXiv:2009.04131.

Lieto, A. (2021). *Cognitive design for artificial minds*. Routledge.

Linnaeus, C. (1735). *Systema naturæ*.

Liu, L., Gurney, N., McCullough, K., & Ustun, V. (2021). Graph neural network based behavior prediction to support multi-agent reinforcement learning in military training simulations. In *Proceedings of the 2021 Winter Simulation Conference* (pp. 1–12).

Lloyd, J. W. (1994). Practical advantages of declarative programming. In *Proceedings of the 1994 Joint Conference on Declarative Programming* (pp. 3–17).

Long, R. Sebo, J., Butlin, P., Finlinson, K, Fish, K., Harding, J., Pfau, J., Sims, T., Birch, J., & Chalmers, D. (2024). Taking AI welfare seriously. arXiv:2411.00986v1.

Lotem, A., Nau, D., & Hendler, J. (1999). Using planning graphs for solving HTN planning problems. In *Proceedings of the 16th National Conference on Artificial Intelligence* (pp. 534–540).

Lucentini, D. F., & Gudwin, R. R. (2015). A Comparison among cognitive architectures: A theoretical analysis. *Procedia Computer Science, 71*, 56–61.

Lucretius. (1st Century BCE). *De Rerum Natura.*

Ma, Y. L. (2025). Tribe becomes 1st in US history to pass unprecedented constitutional amendment: "It is our responsibility to protect her." *The Cool Down.* https://www.thecooldown.com/outdoors/rappahannock-tribe-constitution-river-nature/

Maes, P., & Nardi, D. (1988). *Meta-level architectures and reflection.* North Holland.

Mai, J. K., Majtanik, M., & Paxinos, G. (2015). *Atlas of the human brain.* Elsevier.

Map. (2025, March 16). In *Dictionary.com.* https://www.dictionary.com/browse/map.

Marcus, G., & Davis, E. (2019). *Rebooting AI: Building artificial intelligence we can trust.* Pantheon Books.

Marino, M. C. (2020). *Critical code studies.* MIT Press.

McCarthy, J. (1963). *Situations, actions, and causal laws.* Stanford Artificial Intelligence Memo #2.

McCarthy, J. (1980). Circumscription—A form of non-monotonic reasoning. *Artificial Intelligence, 13*(1–2), 27–29.

McClelland, J. L. (2010). Emergence in cognitive science. *Topics in Cognitive Science, 2*(4), 751–70.

McClelland, J. L., & Rumelhart, D. E. (1981). An interactive activation model of the effect of context in perception: Part 1. An account of basic findings. *Psychological Review, 88*(5), 375–407.

McCrae, R. R., & John, O. P. (1992). An introduction to the five-factor model and its applications. *Journal of Personality, 60*(2), 175–215.

McCulloch, W. S., & Pitts, W. (1943). A logical calculus of the ideas immanent in nervous activity. *Bulletin of Mathematical Biophysics, 5*(4), 115–133.

McDermott, D. V. (2001). *Mind and mechanism.* MIT Press.

Medeiros, A. A. D. (1998). A survey of control architectures for autonomous mobile robots. *Journal of the Brazilian Computer Society, 4*(3).

Meireles, L., Alves, L., & Cruz, J. (2018). Conceptualization and measurement of individual Situation Awareness (SA) in expert populations across operational domains: A systematic review of the literature with a practical purpose on our minds. In *Proceedings of the Human Factors and Ergonomics Society Annual Meeting* (pp. 1093–1097).

Michalski, R. S. (1980). Knowledge acquisition through conceptual clustering: A theoretical framework and an algorithm for partitioning data into conjunctive concepts. *International Journal of Policy Analysis and Information Systems, 4*(3), 219–244.

Mikolov, T., Sutskever, I., Chen, K., Corrado, G. S., & Dean, J. (2013). Distributed representations of words and phrases and their compositionality. *Advances in Neural Information Processing Systems, 26*, 3111–3119.

Miller, G. A. (1956). The magical number seven, plus or minus two: Some limits on our capacity for processing information. *Psychological Review, 63*(2), 81–97.

Milnes, B. G., Pelton, G., Doorenbos, R., Hucka, M., Laird, J., Rosenbloom, P., & Newell, A. (1992). *A specification of the Soar cognitive architecture in Z*. Tech Report CMU-CS-92-169. Carnegie Mellon University.

Mind-body problem. (2024, July 22). In *Wikipedia*. https://en.wikipedia.org/w/index.php?title=Mind–body_problem&oldid=1235948383

Minsky, M. (1975). A framework for representing knowledge. In P. H. Winston (Ed.), *The psychology of computer vision*. McGraw-Hill.

Minsky, M. (1986). *The society of mind*. Simon & Schuster.

Minsky, M. (2012). Future models for mind-machines. https://www.cs.bham.ac.uk/research/projects/cogaff/dam00/papers/minsky.pdf

Minsky, M., & Papert, S. (1969). *Perceptrons*. MIT Press.

Minton, S. (1988). Quantitative results concerning the utility of explanation-based learning. In *Proceedings of the 7th National Conference on Artificial Intelligence* (pp. 564–569).

Mitchell, K. J. (2023). *Free agents: How evolution gave us free will*. Princeton University Press.

Mitchell, T. M., Keller, R. M., & Kedar-Cabelli, S. T. (1986). Explanation-based generalization: A unifying view. *Machine Learning, 1*(1), 47–80.

Mnih, V., Kavukcuoglu, K., Silver, D., Rusu, A. A., Veness, J., Bellemare, M. G., Graves, A., Riedmiller, M. A., Fidjeland, A. K., Ostrovski, G., Petersen, S., Beattie, C., Sadik, A., Antonoglou, I., King, H., Kumaran, D., Wierstra, D., Legg, S., & Hassabis, D. (2015). Human-level control through deep reinforcement learning. *Nature, 518*(7540), 529–533.

Mohan, S. (2021). Exploring the role of common model of cognition in designing adaptive coaching interactions for health behavior change. *ACM Transactions on Interactive Intelligent Systems, 11*(1), 1–30.

Monett, D., & Lewis, C. W. P. (2018). Getting clarity by defining artificial intelligence—A survey. In V. C. Müller (Ed.), *Philosophy and theory of artificial intelligence 2017* (pp. 212–214). Springer Nature Switzerland.

Mormann, T. (2007). The structure of scientific theories in logical empiricism. In A. Richardson & T. Uebel (Eds.), *The Cambridge companion to logical empiricism*. Cambridge University Press.

Morris, M. R., Sohl-Dickstein, J., Fiedel, N., Warkentin, T., Dafoe, A., Faust, A., Farabet, C., & Legg, S. (2023). Levels of AGI: Operationalizing progress on the path to AGI. arXiv:2311.02462.

Morsella, E., & Poehlman, T. A. (2013). The inevitable contrast: Conscious vs. unconscious processes in action control. *Frontiers in Psychology, 4*, 590.

Muggleton, S., & de Raedt, L. (1994). Inductive logic programming: Theory and methods. *The Journal of Logic Programming, 19–20*, 629–679.

Müller, J. (1998). Architectures and applications of intelligent agents: A survey. *The Knowledge Engineering Review, 13*(4), 353–380.

Murphy, K. P. (2012). *Machine learning: A probabilistic perspective*. MIT Press.

Myers–Briggs Type Indicator. (2024, July 16). In *Wikipedia*. https://en.wikipedia.org/w/index.php?title=Myers–Briggs_Type_Indicator&oldid=1234853473

Natural and legal rights. (2024, July 2). In *Wikipedia*. https://en.wikipedia.org/w/index.php?title=Natural_rights_and_legal_rights&oldid=1232247035

Neely, E. (2014). Machines and the moral community. *Philosophy & Technology, 27*(1), 97–111.

Nelson, T. O. (Ed.). (1992). *Metacognition: Core readings*. Allyn & Bacon.

Neves. D. M., & Anderson. J. R. (1981). Knowledge compilation: Mechanisms for the automatization of cognitive skills. In J. R. Anderson (Ed.), *Cognitive skills and their acquisition*. Lawrence Erlbaum Associates.

Newell, A. (1973a). You can't play 20 questions with nature and win: Projective comments on the papers of this symposium. In W. G. Chase (Ed.), *Visual information processing: Proc. of the 8th annual Carnegie symposium on cognition*. Academic Press.

Newell, A. (1973b). Production systems: Models of control structures. In W. G. Chase (Ed.), *Visual information processing* (pp. 283–308). Academic Press.

Newell, A. (1980). Physical symbol systems. *Cognitive Science, 4*(2), 135–183.

Newell, A. (1981). The knowledge level. *AI Magazine, 2*(2), 1–20.

Newell, A. (1983). Intellectual issues in the history of artificial intelligence. In F. Machlup & U. Mansfield (Eds.), *The study of information: Interdisciplinary messages*. John Wiley & Sons.

Newell, A. (1990). *Unified theories of cognition*. Harvard University Press.

Newell, A. (1991). *Desires and diversions*. Video at https://www.youtube.com/watch?v=hCe0ZPGap_k

Newell, A., & Rosenbloom, P. S. (1981). Mechanisms of skill acquisition and the law of practice. In J. R. Anderson (Ed.), *Cognitive skills and their acquisition* (pp. 1–55). Erlbaum.

Newell, A., Rosenbloom, P. S., & Laird, J. E. (1989). Symbolic architectures for cognition. In M. I. Posner (Ed.), *Foundations of cognitive science*. Bradford Books/MIT Press.

Newell, A., Shaw, J. C., & Simon, H.A. (1959). Report on a general problem-solving program. In *Proceedings of the International Conference on Information Processing* (pp. 256–264).

Newell, A., & Simon, H. A. (1956). The logic theory machine: A complex information processing system. *IRE Transactions on Information Theory, 2*(3), 61–79.

Newell, A., & Simon, H. A. (1972). *Human problem solving*. Prentice-Hall.

Newell, A., & Simon, H. A. (1976). Computer science as empirical inquiry: Symbols and search. *Communications of the ACM, 19*(3), 113–126.

Ng, A. Y., & Jordan, M. I. (2001). On discriminative vs. generative classifiers: A comparison of logistic regression and naive bayes. In *Proceedings of the 14th International Conference on Neural Information Processing Systems: Natural and Synthetic* (pp. 841–848).

Nilsson, N. J. (1971). *Problem-solving methods in artificial intelligence*. McGraw-Hill.

Nilsson, N. J. (1986). Probabilistic logic. *Artificial Intelligence, 28*(1), 71–87.

Nilsson, N. J. (2007). The Physical symbol system hypothesis: Status and prospects. In M. Lungarella, F. Iida, J. Bongard, & R. Pfeiffer (Eds.), *50 years of AI: Essays dedicated to the 50th anniversary of AI*. Springer.

Nussbaum, M. C. (2023). *Justice for animals: Our collective responsibility*. Simon & Schuster.

Nyhan, J., Terras, M. M., & Vanhoutte. E. (Eds.). (2013). *Defining digital humanities: A reader*. Ashgate.

Oberauer, K., & Lewandowky, S. (2019). Addressing the theory crisis in psychology. *Psychonomic Bulletin & Review, 26*(5), 1596–1618.

OECD. (2023). OECD AI Principles overview. https://oecd.ai/en/ai-principles https://oecd.ai/en/ai-principles.

Oishi, S., & Westgate, E. C. (2022). A psychologically rich life: Beyond happiness and meaning. *Psychological Review, 129*(4), 790–811.

Olson, R. S., La Cava, W., Orzechowski, P., Urbanowicz, R. J., & Moore, J. H. (2017). PMLB: A large benchmark suite for machine learning evaluation and comparison. *BioData Mining, 10*(1), 36.

Only in Your State. (2022). In 1977, Pittsburgh plunged into an arctic freeze that makes this year's winter look downright mild. https://www.onlyinyourstate.com/pennsylvania/pittsburgh/coldest-winter-1977-pittsburgh/

OpenAI. (2022). ChatGPT: Optimizing language models for dialogue. https://openai.com/blog/chatgpt/

O'Reilly, R. C., & Munakata, Y. (2000). *Computational explorations in cognitive neuroscience: Understanding the mind by simulating the brain.* MIT Press.

Oremus, W. (2022). Google's AI passed a famous test—and showed how the test is broken. *Washington Post.* https://www.washingtonpost.com/technology/2022/06/17/google-ai-lamda-turing-test/

Ortony, A., Norman, D. A., & Revelle, W. (2005). Affect and proto-affect in effective functioning. In J. M. Fellous & M. A. Arbib (Eds.), *Who needs emotions? The brain meets the machine.* Oxford University Press.

P versus NP problem. (2024, Jul 14). In *Wikipedia.* https://en.wikipedia.org/w/index.php?title=P_versus_NP_problem&oldid=1234373347

Park, J. S., O'Brien, J. C., Cai, C. J., Morris, M. R., Liang, P., & Bernstein, M. S. (2023). Generative agents: Interactive simulacra of human behavior. arXiv:2304.03442.

Parkinson, E. M. (1960). *Triple trouble for Rupert.* Scholastic Book Services.

Pausch, R. (2007). *Last lecture: Achieving your childhood dreams.* Video at https://www.youtube.com/watch?v=j7zzQpvoYcQ

Pausch, R. (2008). *The last lecture.* Hachette USA.

Pearl, J. (1984). *Heuristics: Intelligent search strategies for computer problem solving.* Addison-Wesley.

Pearl, J. (1988). *Probabilistic reasoning in intelligent systems: Networks of plausible inference.* Morgan Kaufman.

Pearl, J., & Mackenzie, D. (2018). *The book of why: The new science of cause and effect.* Basic Books.

Pearson, J., Naselaris, T., Holmes, E. A., & Kosslyn, S. M. (2015). Mental imagery: Functional mechanisms and clinical applications. *Trends in Cognitive Sciences, 19*(10), 590–602.

Perry, D. E., & Wolf, A. L. (1992). Foundations for the study of software architecture. *ACM SIGSOFT Software Engineering Notes, 17*(4), 40–52.

Pezzulo, G., Parr, T., Cisek, P., Clark, A., & Friston, K. (2024). Generating meaning: Active inference and the scope and limits of passive AI. *Trends in Cognitive Sciences, 28*(2), 97–112.

Pinchal, S., & Hassabis, D. (2023). Introducing Gemini: our largest and most capable AI model. https://blog.google/technology/ai/google-gemini-ai/#sundar-note

Pohl, K. R. (2021). *Random reflections on 18 years at the Genome Browser.* video at https://hgwdev.gi.ucsc.edu/~ann/talks/Kate/KateRosenbloomSwanSong6-21-2021.mp4

Polk, T. A., & Seifert, C. M. (Eds.). (2002). *Cognitive modeling.* MIT Press.

Pólya, G. (1945). *How to solve it.* Doubleday.

Poon, H., & Domingos, P. (2011). Sum-product networks: A new deep architecture. In *Proceedings of the IEEE International Conference on Computer Vision Workshops* (pp. 689–690).

Post, E. (1943). Formal reductions of the general combinatorial decision problem. *American Journal of Mathematics, 65*, 197–215.

Predictive power. (2024, June 10). In *Wikipedia.* https://en.wikipedia.org/w/index.php?title=Predictive_power&oldid=1228388892

Pynadath, D. V., Rosenbloom, P. S., & Marsella, S. C. (2014). Reinforcement learning for adaptive Theory of Mind in the Sigma cognitive architecture. In *Proceedings of the 7th Annual Conference on Artificial General Intelligence* (pp. 143–154).

Pynadath, D. V., Rosenbloom, P. S., Marsella, S. C., & Li, L. (2013). Modeling two-player games in the Sigma graphical cognitive architecture. In *Proceedings of the 6th Conference on Artificial General Intelligence* (pp. 98–108).

Quinlan, J. R. (1986). Induction of decision trees. *Machine Learning, 1*(1), 81–106.

Randy Pausch. (2024, July 21). In *Wikipedia.* https://en.wikipedia.org/w/index.php?title=Randy_Pausch&oldid=1235890996

Ray, S., & Tadepalli, P. (2011). Model-based reinforcement learning. In C. Sammut & G. I. Webb (Eds), *Encyclopedia of machine learning.* Springer.

Reiter, R. (1980). A logic for default reasoning. *Artificial Intelligence, 13*(1–2), 81–132.

Renn, J. (Ed.) (2005), *Albert Einstein—Chief engineer of the universe: Einstein's life and work in context.* Wiley-VCH.

Reversi. (2024, July 19). In *Wikipedia.* https://en.wikipedia.org/w/index.php?title=Reversi&oldid=1235422883

Ribeiro, D. N., & Forbus, K. (2021). Combining analogy with language models for knowledge extraction. In *Proceedings of the 3rd Conference on Automated Knowledge Base Construction.*

Rich, E., & Knight, K. (1991). *Artificial intelligence* (2nd ed.). McGraw Hill.

Rights of nature. (2024, July 15). In *Wikipedia*. https://en.wikipedia.org/w/index.php?title=Rights_of_nature&oldid=1234623077

Rigoni, D., & Brass, M. (2014). From intentions to neurons: Social and neural consequences of disbelieving in free will. *Topoi*, *33*(1), 5–12.

Ritz, B., & Vaidyanathan, B. (2025). Aha=wow. *aeon*. https://aeon.co/essays/how-the-search-for-beauty-drives-scientific-enquiry

Robison, K. (2024). OpenAI cofounder Ilya Sutskever says the way AI is built is about to change. *The Verge*. https://www.theverge.com/2024/12/13/24320811/what-ilya-sutskever-sees-openai-model-data-training

Romero, O. J., Zimmerman, J., Steinfeld, A., & Tomasic, A. (2023). Synergistic integration of large language models and cognitive architectures for robust AI: An exploratory analysis. arXiv:2308.09830

Rosch, E. H. (1973). Natural categories. *Cognitive Psychology*, *4*(3), 328–350.

Rosen, R. (1991). *Life itself: A comprehensive inquiry into the nature, origin, and fabrication of life*. Columbia University Press.

Rosenblatt, F. (1958). The Perceptron: A probabilistic model for information storage and organization in the brain. *Psychological Review*, *65*(6), 386–408.

Rosenbloom, P. S. (1982). A world-championship-level Othello program. *Artificial Intelligence*, *19*(3), 279–320.

Rosenbloom, P. S. (1983). *The chunking of goal hierarchies: A model of practice and stimulus-response compatibility*. [Ph.D. Thesis, Carnegie-Mellon University].

Rosenbloom, P. S. (1986). The chunking of goal hierarchies: A model of practice and stimulus-response compatibility. In J. E. Laird, P. S. Rosenbloom, & A. Newell (Eds.), *Universal subgoaling and chunking: The automatic generation and learning of goal hierarchies*. Kluwer.

Rosenbloom, P. S. (2004). A new framework for computer science and engineering. *IEEE Computer*, *37*(11), 23–28.

Rosenbloom, P. S. (2006). A cognitive odyssey: From the power law of practice to a general learning mechanism and beyond. *Tutorials in Quantitative Methods for Psychology*, *2*(2), 43–51.

Rosenbloom, P. S. (2009). Towards a new cognitive hourglass: Uniform implementation of cognitive architecture via factor graphs. In *Proceedings of the 9th International Conference on Cognitive Modeling* (pp. 116–121).

Rosenbloom, P. S. (2010a). Combining procedural and declarative knowledge in a graphical architecture. In *Proceedings of the 10th International Conference on Cognitive Modeling* (pp. 205–210).

Rosenbloom, P. S. (2010b). Computing and computation. *ACM Ubiquity Symposium: "What is Computation?"*

Rosenbloom, P. S. (2011a). Mental imagery in a graphical cognitive architecture. In *Proceedings of the 2nd International Conference on Biologically Inspired Cognitive Architectures* (pp. 314–323).

Rosenbloom, P. S. (2011b). Bridging dichotomies in cognitive architectures for virtual humans. In *Proceedings of the AAAI Fall Symposium on Advances in Cognitive Systems.*

Rosenbloom, P. S. (2012a). *On computing: The fourth great scientific domain.* MIT Press.

Rosenbloom, P. S. (2012b). Deconstructing reinforcement learning in Sigma. In *Proceedings of the 5th Conference on Artificial General Intelligence* (pp. 262–271).

Rosenbloom, P. S. (2012c). Extending mental imagery in Sigma. In *Proceedings of the 5th Conference on Artificial General Intelligence* (pp. 272–281).

Rosenbloom, P. S. (2012d). Towards a 50 msec cognitive cycle in a graphical architecture. In *Proceedings of the 11th International Conference on Cognitive Modeling* (pp. 305–310).

Rosenbloom, P. S. (2012e). Towards a conceptual framework for the digital humanities. *Digital Humanities Quarterly, 6*(2).

Rosenbloom, P. S. (2013). *AI and the future,* videos at https://www.youtube.com/watch?v=CRwMnngqQ4U and https://www.youtube.com/watch?v=Uov0i4BkndY

Rosenbloom, P. S. (2014). Deconstructing episodic learning and memory in Sigma. In *Proceedings of the 36th Annual Conference of the Cognitive Science Society* (pp. 1317–1322).

Rosenbloom, P. S. (2019a). (A)symmetry × (non)monotonicity: Towards a deeper understanding of key cognitive di/trichotomies and the common model of cognition. In *Proceedings of the 17th Annual Meeting of the International Conference on Cognitive Modeling.*

Rosenbloom, P. S. (2019b). (A)symmetry × (non)monotonicity: Towards a deeper understanding of key cognitive di/trichotomies and the common model of cognition. Submitted to the *41st Annual Meeting of the Cognitive Science Society.*

Rosenbloom, P. S. (2022a). Thoughts on architecture. In *Proceedings of the 15th International Conference on Artificial General Intelligence* (pp. 364–373).

Rosenbloom, P. S. (2022b). *From designing minds to mapping disciplines.* Video at https://www.youtube.com/watch?v=myC7oteFsWM

Rosenbloom, P. S. (2023a). In search of insight: My life as an architectural explorer (Author's cut). https://www.dropbox.com/scl/fi/pgy2xtoq6fymsd0oq1jun/Memoir-11-7-23-AC.pdf?rlkey=bmdf9t9l0xx4fpy7bnx53mtcz&dl=0

Rosenbloom, P. S. (2023b). Rethinking the physical symbol systems hypothesis. In *Proceedings of the 16th International Conference on Artificial General Intelligence* (pp. 207–216). Springer.

Rosenbloom. (2023c). Defining and exploring the intelligence space. arXiv: 2306.06499.

Rosenbloom, P. S. (2024). Writing the Original UARC Proposal. https://ict.usc.edu/news/writing-the-original-uarc-proposal/

Rosenbloom, P. S. (2025). In search of insight: My life as an architectural explorer. *Journal of Artificial General Intelligence, 15*, 3–61.

Rosenbloom, P. S., Demski, A., Han, T., & Ustun, V. (2013). Learning via gradient descent in Sigma. In *Proceedings of the 12th International Conference on Cognitive Modeling* (pp. 35–40).

Rosenbloom, P. S., Demski, A., & Ustun, V. (2016a). The Sigma cognitive architecture and system: Towards functionally elegant grand unification. *Journal of Artificial General Intelligence, 7*(1), 1–103.

Rosenbloom, P. S., Demski, A., & Ustun, V. (2016b). Rethinking Sigma's graphical architecture: An extension to neural networks. In *Proceedings of the 9th Conference on Artificial General Intelligence* (pp. 84–94).

Rosenbloom, P. S., Demski, A., & Ustun, V. (2017). Toward a neural-symbolic Sigma: Introducing neural network learning. In *Proceedings of the 15th Annual Meeting of the International Conference on Cognitive Modeling*.

Rosenbloom, P. S., & Forbus, K. D. (2019). Expanding and repositioning cognitive science. *Topics in Cognitive Science, 11*(4), 918–927.

Rosenbloom, P. S., Gratch, J., & Ustun, V. (2015). Towards emotion in Sigma: From appraisal to attention. In *Proceedings of the 8th Conference on Artificial General Intelligence* (pp. 142–151).

Rosenbloom, P. S., Joshi, H., & Ustun, V. (2019a). (Sub)symbolic × (a)symmetric × (non)combinatory: A map of AI approaches spanning symbolic/statistical to neural/ML. *Advances in Cognitive Systems, 8*, 113–131.

Rosenbloom, P. S., Joshi, H., & Ustun, V. (2019b). A/The language of AI? Submitted as a Blue Sky Idea to the *33rd AAAI Conference on Artificial Intelligence*.

Rosenbloom, P. S., Joshi, H., & Ustun, V. (2019c). (Sub)symbolic × (a)symmetric × (non)combinatoric: Connecting symbolic/statistical and neural/ML approaches to intelligence. Submitted to the *28th International Joint Conference on Artificial Intelligence*.

Rosenbloom, P. S., & Laird, J. E. (1993). On Unified theories of cognition: A response to the reviews. *Artificial Intelligence, 59*(1–2), 389–413.

Rosenbloom, P. S., Laird, J. E., Lebiere, C., Stocco, A., Granger, R. H., & Huyck, C. (2024). A proposal for extending the common model of cognition to emotion. In *Proceedings of the 22nd International Conference on Cognitive Modeling.*

Rosenbloom, P. S., Laird, J. E., McDermott, J., Newell; A., & Orciuch, E. (1985). R1-Soar: An experiment in knowledge-intensive programming in a problem-solving architecture. *IEEE Transactions on Pattern Analysis and Machine Intelligence, 7*(5), 561–569.

Rosenbloom, P. S., Laird, J. E., & Newell, A. (1987). Knowledge level learning in Soar. In *Proceedings of the 6th National Conference on Artificial Intelligence* (pp. 499–504).

Rosenbloom, P. S., Laird, J. E., & Newell, A. (1988). Meta-levels in Soar. In P. Maes & D. Nardi (Eds.), *Meta-level architectures and reflection* (pp. 227–240). North Holland.

Rosenbloom, P. S., Laird, J. E., & Newell, A. (1989). The chunking of skill and knowledge. In B. A. G. Elsendoorn & H. Bouma (Eds.), *Working models of human perception* (pp. 391–410). Academic Press.

Rosenbloom, P. S., Laird, J. E., & Newell, A. (Eds.). (1993). *The Soar papers: Research on integrated intelligence, Volumes One & Two.* MIT Press.

Rosenbloom, P. S., Lebiere, C., & Laird, J. E. (2022). Cross-pollination among neuroscience, psychology and AI research yields a foundational understanding of thinking. *The Conversation.*

Rosenbloom, P. S., Lee, S., & Unruh, A. (1990). Responding to impasses in memory-driven behavior: A framework for planning. In *Proceedings of the Workshop on Innovative Approaches to Planning, Scheduling, and Control* (pp. 181–191).

Rosenbloom, P. S., Lee, S., & Unruh, A. (1993). Bias in planning and explanation-based learning. In S. Minton (Ed.), *Machine learning methods for planning.* Morgan Kaufmann.

Rosenbloom, P. S., & Newell, A. (1986). The chunking of goal hierarchies: A generalized model of practice. In R. S. Michalski, J. G. Carbonell, & T. M. Mitchell (Eds.), *Machine learning: An artificial intelligence approach, Volume II.* Morgan Kaufmann.

Rosenbloom, P. S., & Newell, A. (1988). An integrated computational model of stimulus-response compatibility and practice. In G. H. Bower (Ed.), *The psychology of learning and motivation Vol. 21*(pp. 1–52). Academic Press.

Rosenbloom, P. S., & Ustun, V. (2016). *The Sigma Cognitive Architecture and System (IVA Tutorial).* https://cogarch.ict.usc.edu/iva-tutorial-2016/

Rosenbloom, P. S., & Ustun, V. (2019). An architectural integration of Temporal Motivation Theory for decision making. In *Proceedings of the 7th Annual Meeting of the International Conference on Cognitive Modeling.*

Rosenblueth, A., Wiener, N., & Bigelow, J. (1943). Behavior, purpose and teleology. *Philosophy of Science, 10*(1), 18–24.

Rosenfeld, J.S., Rosenfeld, A., Belinkov, Y., & Shavit, N. (2019). A constructive prediction of the generalization error across scales. arXiv:1909.12673.

Rossi, F., van Beek, P., & Walsh, T. (2006). *Handbook of constraint programming*. Elsevier.

Rousseau, J.-J. (1782). *Confessions*.

Rovelli, C. (2023). The secret to unlocking one of the universe's greatest mysteries. *New York Times*. https://www.nytimes.com/2023/10/16/opinion/black-holes-science-imagination.html

Rumelhart, D. E., Hinton, G. E., & Williams, R. J. (1986). Learning representations by back-propagating errors. *Nature, 323*(6088), 533–536.

Russell, B. (2009). *The philosophy of logical atomism*. Routledge Classics.

Russell, S. (2019). *Human compatible artificial intelligence and the problem of control*. Viking.

Russell, S., Binder, J., Koller, D., & Kanazawa, K. (1995). Local learning in probabilistic networks with hidden variables. In *Proceedings of the 14th International Joint Conference on AI* (pp. 1146–1152).

Russell, S. J., & Norvig, P. (2010). *Artificial intelligence: A modern approach* (3rd ed.). Pearson.

Rychener, M. D. (1983). The Instructable Production System: A retrospective analysis. In R. S. Michalski, J. G. Carbonell, & T. M. Mitchell (Eds.), *Machine learning: An artificial intelligence approach*. Morgan Kaufmann.

Rychener, M. D., & Newell, A. (1978). An instructable production system: Basic design issues. In D. A. Waterman & F. Hayes-Roth (Eds.), *Pattern-directed inference systems*. Academic Press.

Rytting, A. (2000). NL-Soar and WordNet: Fitting the lexicon into a cognitive architecture. *Deseret Language and Linguistic Society Symposium, 26*, 23–35.

Sacerdoti, E. D. (1974). Planning in a hierarchy of abstraction spaces. *Artificial Intelligence, 5*(2), 115–135.

Samek, W., Montavon, G., Lapuschkin, S., Anders, C. J., & Müller, K.-R. (2021). Explaining deep neural networks and beyond: A review of methods and applications. *Proceedings of the IEEE, 109*(3), 247–278.

Samsonovich, A.V. (2010). Towards a unified catalog of implemented cognitive architectures. In A. V. Samsonovich, K. R. Jóhansdóttir, A. Chella, & B. Goertzel (Eds.), *Biologically inspired cognitive architectures 2010* (pp. 195–244). Amsterdam: IOS Press.

Samsonovich, A.V. (2011). Comparative analysis of implemented cognitive architectures. In A. V. Samsonovich & K. R. Jóhansdóttir (Eds.), *Biologically inspired cognitive architectures 2011* (pp. 469–479). IOS Press.

Santoro, A., Raposo, D., Barrett, D. G., Malinowski, M., Pascanu, R., Battaglia, P., & Lillicrap, T. (2017). A simple neural network module for relational reasoning. In *Proceedings of the 31st Conference on Neural Information Processing Systems* (pp. 4967–4976).

Sapolsky, R. M. (2023). *Determined: A science of life without free will.* Penguin Press.

Savage, C. W. (1990). Preface. In C. W. Savage (Ed.), *Minnesota studies in the philosophy of science (Volume XIV: Scientific Theories).* University of Minnesota Press.

Scarselli, F., Gori, M., Tsoi, A. C., Hagenbuchner, M., & Monfardini, G. (2009). The graph neural network model. *IEEE Transactions on Neural Networks, 20*(1), 61–80.

Scerri, P., Pynadath, D. V., Johnson, L., Rosenbloom, P., Schurr, N., & Tambe, M. (2003). A prototype infrastructure for distributed robot-agent-person teams. In *Proceedings of the 2nd International Joint Conference on Autonomous Agents & Multiagent Systems* (pp. 433–440).

Schneider, W., & Chien, J. M. (2003). Controlled & automatic processing: behavior, theory, and biological mechanisms. *Cognitive Science, 27*(3), 525–559.

Schneider, W., & Shiffrin, R. M. (1977). Controlled and automatic human information processing: I. Detection, search, and attention. *Psychological Review, 84*(1), 1–66.

Schoppers, M. J. (1987). Universal plans for reactive robots in unpredictable domains. In *Proceedings of the 10th International Joint Conference on Artificial Intelligence* (pp. 1039–1046).

Schwitzgebel, E., & Garza, M. (2015). A defense of the rights of artificial intelligences. *Midwest Studies in Philosophy, 39*(1), 98–119.

Science Daily. (2018). Dissecting artificial intelligence to better understand the human brain. https://www.sciencedaily.com/releases/2018/03/180325115759.htm.

Searle, J. (1980). Minds, brains and programs. *Behavioral and Brain Sciences, 3*(3), 417–457.

Seetharaman, D. (2024). The next great leap in AI is behind schedule and crazy expensive. *The Wall Street Journal.* https://www.wsj.com/tech/ai/openai-gpt5-orion-delays-639e7693

Semantics. (2024, July 13). In *Wikipedia.* https://en.wikipedia.org/w/index.php?title=Semantics&oldid=1234212174

Shannon, C. E. (1948). A mathematical theory of communication. *The Bell System Technical Journal, 27*(3), 379–423, 623–656.

Shapiro, E. Y. (1981). The model inference system. In *Proceedings of the 7th International Joint Conference on Artificial intelligence* (pp. 1064).

Shapiro, S. C. (1992). Artificial intelligence. In S. C. Shapiro (Ed.), *Encyclopedia of artificial intelligence* (2nd ed.) (pp. 54–57). John Wiley & Sons.

Shastri, L. (1988). A connectionist approach to knowledge representation and limited interference. *Cognitive Science, 12*(3), 331–392.

Shmueli, G. (2010). To explain or to predict? *Statistical Science, 25*(3), 289–310.

Shoshan, V., Hazan, T., & Plonsky, O. (2024). Using machine learning to create an adaptable, scalable, and interpretable behavioral model. *Decision, 11*(4), 652–667.

Sica, A., & Sætra, H. (2024). Artificial emotions and the evolving moral status of social robots. In *Proceedings of the 19th Annual ACM/IEEE International Conference on Human Robot Interaction* (pp. 649–657).

Siegelmann, H. T., & Sontag, E. D. (1991). Turing computability with neural nets. *Applied Mathematics Letters, 4*(6), 77–80.

Silva, L. de, Meneguzzi, F., & Logan, B. (2020). BDI agent architectures: A survey. In *Proceedings of the 29th International Joint Conference on Artificial Intelligence* (pp. 4914–4921).

Silver, D., Hubert, J., Anonoglou, I., Lai, M., Guez, A., Lanctot, M., Sifre, L., Kumaran, D., Graepel, T., Lillicrap, T., Simonyan, K., & Hassabis, D. (2018). A general reinforcement learning algorithm that masters chess, shogi, and Go through self-play. *Science, 362*(6419), 1140–1144.

Silver, D., Singh, S., Precup, D., & Sutton, R. S. (2021). Reward is enough. *Artificial Intelligence, 299*, 103535.

Simon, H. A. (1969). *The sciences of the artificial*. MIT Press.

Simon, H. A. (1974). How big is a chunk? *Science, 183*(4124), 482–488.

Simon, H. A., & Chase W. G. (1973). Skill in chess. *American Scientist, 61*(4), 394–403.

Simon, H. A., & Kaplan, C. A. (1993). Foundations of cognitive science. In M. I. Posner (Ed.), *Foundations of cognitive science* (pp. 154–170). MIT Press.

Singer, P. (2023). *Animal liberation now: The definitive classic renewed.* Diversion Books.

Sipser, M. (2012). *Introduction to the theory of computation* (3rd ed.). Cengage Learning.

Situ, X., Zukerman, I., Paris, C., Maruf, S., & Haffari, G. (2021). Learning to explain: Generating stable explanations fast. In *Proceedings of the 59th Annual Meeting of the Association for Computational Linguistics and the 11th International Joint Conference on Natural Language Processing* (pp. 5340–5355).

Slagle, J. R., & Bursky, P. (1968). Experiments with a multipurpose, theorem-proving heuristic program. *Journal of the ACM, 15*(1), 85–99.

Sloman, A. (1984). The structure of the space of possible minds. In S. Torrance (Ed.), *The mind and the machine: Philosophical aspects of artificial intelligence*. Ellis Horwood.

Sloman, S. A. (1996). The empirical case for two systems of reasoning. *Psychological Bulletin, 119*(1), 3–22.

Smolensky, S., & Legendre, G. (2006). *The harmonic mind: From neural computation to optimality-theoretic grammar*. MIT Press.

Smolensky, P., McCoy, R. T., Fernandez, R., Goldrick, M., & Gao, J. (2022). Neurocompositional computing: From the Central Paradox of Cognition to a new generation of AI systems. *AI Magazine, 43*(3), 308–322.

Soar. (2017). *The Soar 9 tutorial*. https://soar.eecs.umich.edu/tutorials/soar_tutorial/

Solomonoff, R. J. (1964). A formal theory of inductive inference: Parts 1 and 2. *Information and Control, 7*(1–2), 1–22 & 224–254.

Spinoza, B. (1677). *Ethica, Ordine Geometrico Demonstrata*.

Squire L. R. (1987). *Memory and brain*. Oxford University Press.

Steels, L. (1989). The deepening of expert systems. In J. A. Campbell (Ed.), *Perspectives in artificial intelligence. Vol. 1: Expert systems applications and technical foundations* (pp. 17–29). Halsted Press.

Steier, D. M., Laird, J. E., Newell, A., Rosenbloom, P. S., Flynn, R., Golding, A., Polk, T. A., Shivers, O. G., Unruh, A., & Yost, G. R. (1987). Varieties of learning in Soar: 1987. In *Proceedings of the 4th International Workshop on Machine Learning* (pp. 300–311).

Sternberg, R. J. (1985). *Beyond IQ: A triarchic theory of intelligence*. Cambridge University Press.

Sternberg, R. J. (Ed.). (2000). *Handbook of intelligence*. Cambridge University Press.

Steunebrink, B. R., & Schmidhuber, J. (2012). Towards an actual Gödel machine implementation: A lesson in self-reflective systems. In P. Wang & B. Goertzel (Eds), *Theoretical foundations of artificial general intelligence*. Atlantis Press.

Stocco, A., Lebiere, C., & Samsonovich, A. (2010). The B-I-C-A of biologically inspired cognitive architectures. *International Journal of Machine Consciousness, 2*(2), 171–192.

Stocco, A., Steine-Hanson, Z., Koh, N., Laird, J. E., Lebiere, C., & Rosenbloom, P. S. (2021). Analysis of the human connectome data supports the notion of a "common model of cognition" for human and humanlike intelligence. *Neuroimage, 235*, 118035.

Stockholm syndrome. (2024, July 6). In *Wikipedia*. https://en.wikipedia.org/w/index.php?title=Stockholm_syndrome&oldid=1232942681

Stone, C. D. (1972). Should trees have legal standing?—Towards legal rights for natural objects. *Southern California Law Review*, *45*(2), 450–501.

Stross, C. (2017). Dude, you broke the future! https://media.ccc.de/v/34c3-9270-dude_you_broke_the_future

Sullivan, D. (2020). A reintroduction to our Knowledge Graph and knowledge panels. https://blog.google/products/search/about-knowledge-graph-and-knowledge-panels/

Sumers, T., Yao, S., Narasimhan, K., & Griffiths, T. L. (2023). Cognitive architectures for language agents. arXiv:2309.02427.

Sun, R. (2004) Desiderata for cognitive architectures. *Philosophical Psychology*, *17*, 341–373.

Sun, R. (Ed.). (2008). *The Cambridge handbook of computational psychology*. Cambridge University Press.

Sun, R. (2016). *Anatomy of the mind: Exploring psychological mechanisms and processes with the Clarion cognitive architecture*. Oxford University Press.

Sun, R. (2017). The CLARION cognitive architecture: Towards a comprehensive theory of the mind. In S. Chipman (Ed.), *The Oxford handbook of cognitive science* (pp. 117–133). Oxford University Press.

Sun, R. (2024). Roles of LLMs in the overall mental architecture. arXiv:2419.19937.

Sunkin, S. M., Ng, L., Lau, C., Dolbeare, T., Gilbert, T. L., Thompson, C. L., Hawrylycz, M., & Dang, C. (2013). Allen Brain Atlas: An integrated spatio-temporal portal for exploring the central nervous system. *Nucleic Acids Research*, *41*, D996-D1008.

Surma, K. (2024). A river in Washington state now has enforceable legal rights. *Inside Climate News*. https://insideclimatenews.org/news/05122024/washington-state-snohomish-river-watershed-legal-rights/

Surma, K. (2025). A court says coastal marine ecosystems have intrinsic value—and legal rights. *Inside Climate News*. https://insideclimatenews.org/news/17012025/ecuador-constitutional-court-marine-ecosystem-rights/#:~:text=The%20Constitutional%20Court%20of%20Ecuador,human%20activities%20like%20industrial%20fishing

Sutton, R. S., & Barto, A. G. (2018). *Reinforcement learning: An introduction* (2nd ed.). MIT Press.

Sutskever, I., Vinyals, O., & Le, Q. V. (2014). Sequence to sequence learning with neural networks. In *Proceedings of the 28th Conference on Neural Information Processing Systems* (pp. 3104–3112).

Swartout, W. R., Gratch, J., Hill, R. W., Hovy, E. H., Marsella, S., Rickel, J., & Traum, D. R. (2006). Toward virtual humans, *AI Magazine, 27*(2), 96–108.

Symbol. (2024). In *Dictionary.com.* https://www.dictionary.com/browse/symbol

Szalavitz, M. (2023). I don't mind if you say you have "a little OCD." *New York Times.* https://www.nytimes.com/2023/01/22/opinion/neurodiverse-ocd-mental-health.html

Szegedy, C., Zaremba, W., Sutskever, I., Bruna, J., Erhan, D., Goodfellow, I., & Fergus, R. (2014). Intriguing properties of neural networks. In *Proceedings of the 2nd International Conference on Learning Representations.*

Taatgen, N. A., & van Rijn, H. (2007). ACT-R tutorial. In *Proceedings of the 29th Annual Meeting of the Cognitive Science Society* (pp. 49).

Tabula rasa. (2024, July 3). In *Wikipedia.* https://en.wikipedia.org/w/index.php?title=Tabula_rasa&oldid=1232310177

Tallant, J., & Andow, J. (2020). English language and philosophy. In S. Adolphs & D. Knight (Eds.), *The Routledge handbook of English language and digital humanities.* Routledge.

Tambe, M. (1997). Agent architectures for flexible, practical teamwork. In *Proceedings of the 14th National Conference on Artificial Intelligence* (pp. 22–28).

Tambe, M., Johnson, W. L., Jones, R. M., Koss, F., Laird, J. E., Rosenbloom, P. S., & Schwamb, K. B. (1995a). Intelligent agents for interactive simulation environments. *AI Magazine, 16*(1), 15–39.

Tambe, M., Newell, A., & Rosenbloom, P. S. (1990). The problem of expensive chunks and its solution by restricting expressiveness. *Machine Learning, 5,* 299–348.

Tambe, M., Schwamb, K. B., & Rosenbloom, P. S. (1995b). Building intelligent pilots for simulated rotary wing aircraft. In *Proceedings of the 5th Conference on Computer Generated Forces and Behavioral Representation* (pp. 39–44).

Taunk, K., De, S., Verma, S., & Swetapadma, A. (2019). A brief review of nearest neighbor algorithm for learning and classification. In *2019 International Conference on Intelligent Computing and Control Systems* (pp. 1255–1260).

Tedre, M. (2015). *The science of computing: Shaping a discipline.* CRC Press.

Tesauro, G. (1995). Temporal difference learning and TD-Gammon. *Communications of the ACM, 38*(3), 58–68.

Thagard, P. (1993). *Computational philosophy of science.* MIT Press.

The Big Benchmarks Collection. (2024, November 18). *Hugging Face.* https://huggingface.co/collections/open-llm-leaderboard/the-big-benchmarks-collection-64faca6335a7fc7d4ffe974a

Theory. (2024a, July 10). In *Dictionary.com*. https://www.dictionary.com/browse/theory

Theory. (2024b, July 10). In *Wikipedia*. https://en.wikipedia.org/w/index.php?title=Theory&oldid=1233692127

Theory of Everything. (2024, December 23). In *Wikipedia*. https://apastyle.apa.org/style-grammar-guidelines/references/examples/wikipedia-references

Theory of Mind. (2025, February 8). In *Wikipedia*. https://en.wikipedia.org/w/index.php?title=Theory_of_mind&oldid=1274659155

Thiebes, S., Lins, S., & Sunyaev, A. (2020). Trustworthy artificial intelligence. *Electronic Markets*, *31*(2), 447–464.

Thomsen, K. (2018). It Is time to dissolve old dichotomies in order to grasp the whole picture of cognition. In D. Fagan, C. Martín-Vide, M. O'Neill, & M. A. Vega-Rodríguez (Eds), *Theory and practice of natural computing*. Springer.

Thórisson, K., & Helgasson, H. (2012). Cognitive architectures and autonomy: A comparative review. *Journal of Artificial General Intelligence*, *3*(2), 1–30.

Three-body problem. (2024, July 17). In *Wikipedia*. https://en.wikipedia.org/w/index.php?title=Three-body_problem&oldid=1234982567

Titus, L. M. (2024). Does ChatGPT have semantic understanding? A problem with the statistics-of-occurrence strategy. *Cognitive Systems Research*, *83*, 101174.

Toews, R. (2020). Deep learning's carbon emissions problem. *Forbes*. https://www.forbes.com/sites/robtoews/2020/06/17/deep-learnings-climate-change-problem/?sh=47079a796b43

Townsend, J. T. (1972). Some results concerning the identifiability of parallel and serial processes. *British Journal of Mathematical and Statistical Psychology, 25*(2), 168–199.

Townsend J., Chaton T., & Monteiro J. M. (2020). Extracting relational explanations from deep neural networks: A survey from a neural-symbolic perspective. *IEEE Transaction on Neural Networks and Learning Systems*, *31*(9), 3456–3470.

Treisman, A. M., & Gelade, G. (1980). A feature-integration theory of attention. *Cognitive Psychology*, *12*(1), 97–136.

Tulving, E. (1972). Episodic and semantic memory. In E. Tulving & W. Donaldson (Eds.), *Organization of memory*. Academic Press.

Turing, A. M. (1937). On computable numbers, with an application to the entscheidungsproblem. *Proceedings of the London Mathematical Society Series 2*, *42*(1), 230–265.

Turing, A. M. (1950). Computing machinery and intelligence, *Mind*, *59*, 433–460.

UCSC Genome Browser. (2023). *UCSC*. https://genome.ucsc.edu/

Understand. (2024). In *Dictionary.com*. https://www.dictionary.com/browse/understand

UNESCO. (2021). *Recommendations on the ethics of artificial intelligence*. UNESCO.

U.S. Copyright Office. (2025). *Copyright and artificial intelligence, Part 2: Copyrightability*. https://www.copyright.gov/ai/Copyright-and-Artificial-Intelligence-Part-2-Copyrightability-Report.pdf

Ustun, V., Kumar, R., Reilly, A., Sajjadi, S., & Miller, A. (2021). Adaptive synthetic characters for military training. In *Proceedings of the Interservice/Industry Training, Simulation, and Education Conference (I/ITSEC)*.

Ustun, V., & Rosenbloom, P. S. (2015). Towards adaptive, interactive virtual humans in Sigma. In *Proceedings of the 15th International Conference on Intelligent Virtual Agents* (pp. 98–108).

Ustun, V., Rosenbloom, P. S., Kim, J., & Li, L. (2015). Building high fidelity human behavior models in the Sigma cognitive architecture. In *Proceedings of the 2015 Winter Simulation Conference* (pp. 3124–3125).

Ustun, V., Rosenbloom, P. S., Sagae, K., & Demski, A. (2014). Distributed vector representations of words in the Sigma cognitive architecture. In *Proceedings of the 7th Annual Conference on Artificial General Intelligence* (pp. 196–207).

Ustun, V., Rosenbloom, P. S., Sajjadi, S., & Nuttall, J. (2018). Controlling synthetic characters in simulations: A case for cognitive architectures and Sigma. In *Proceedings of the Interservice/Industry Training, Simulation, and Education Conference (I/ITSEC)*.

Varma, S. (2011). Criteria for the design and evaluation of cognitive architectures. *Cognitive Science, 35*(7), 1329–1351.

Vaswani, A., Shazeer, N., Parmar, N., Uszkoreit, J., Jones, L., Gomez, A. N., Kaiser, Ł., & Polosukhin, I. (2017). Attention is all you need. In *Proceedings of the 31st Annual Conference on Neural Information Processing Systems* (pp. 5998–6008).

Vera, A. H., & Simon, H. A. (1993). Situated action: A symbolic interpretation. *Cognitive Science, 17*(1), 7–48.

Vinge, V. (1993). The coming technological singularity: How to survive in the post-human era. *Vision-21 Symposium*.

Voegelin, E. (1989). *Autobiographical reflections*. Louisiana State University Press.

Vohs, K. D., & Schooler, J. W. (2008). The value of believing in free will: Encouraging a belief in determinism increases cheating. *Psychological Science, 19*(1), 49–54.

von Goethe, J. W. (1892). *Maxims and reflections*.

Waibel, A., Hanazawa, T., Hinton, G. E., Shikano, J., & Lang, K. J. (1989). Phoneme recognition using time-delay neural networks. *IEEE Transactions on Acoustics, Speech, and Signal Processing, 37*(3), 328–339.

Waldrop, M. M. (1988a). Toward a unified theory of cognition. *Science, 241*(4861), 27–29.

Waldrop, M. M. (1988b). Soar: A unified theory of cognition? *Science, 241*(4863), 296–298.

Wang, H-n., Liu, N., Zhang, Y-y., Feng, D-w., Huang, F., Li, D-s., & Zhang, Y-m. (2020). Deep reinforcement learning: A survey. *Frontiers of Information Technology & Electronic Engineering, 21*(12), 1726–1744.

Wang, L., Ma, C., Feng, X., Zhang, Z., Yang, H. Zhang, J., Chen, Z., Tang, J., Chen, X., Lin, Y., Zhao, W. X., Wei, Z., & Wen, J.-R. (2023). A survey on large language model based autonomous agents. arXiv:2308.11432.

Wang, P. (2019). On defining artificial intelligence. *Journal of Artificial General Intelligence, 10*(2), 1–37.

Warren, M. A. (1998). *Moral status: Obligations to persons and other living things.* Oxford University Press.

Wei, J., Wang, X., Schuurmans, D., Bosma, M., Chi, E. H., Xia, F., Le, Q., & Zhou, D. (2022). Chain of thought prompting elicits reasoning in large language models. arXiv:2201.11903

What is the difference between a model and theory. (2023). In *ResearchGate.* https://www.researchgate.net/post/What_is_the_difference_between_a_model_and_a_theory.

Whewell, W. (1840). *The philosophy of the inductive sciences, founded upon their history.* London: John W. Parker.

Wickelgren, W. A. (1977). Speed-accuracy tradeoff and information processing dynamics. *Acta Psychologica, 41*(1), 67–85.

Wiggers, K. (2024). Microsoft claims its new tool can correct AI hallucinations, but experts advise caution. *TechCrunch.* https://techcrunch.com/2024/09/24/microsoft-claims-its-new-tool-can-correct-ai-hallucinations-but-experts-caution-it-has-shortcomings/.

Wilczek, F. (2015). Why is physics beautiful? *World Economic Forum.*

Wilczek, F. (2016). In A. Nevala-Lee (Ed.), *Quote of the Day.* https://nevalalee.wordpress.com/2016/01/21/quote-of-the-day-1285/

William James Lectures. (2024, April 17). In *Wikipedia.* https://en.wikipedia.org/w/index.php?title=William_James_Lectures&oldid=1219328787.

Winograd, T. (1975). Frame representations and the declarative/procedural controversy. In D. G. Bobrow & A. M. Collins (Eds.), *Representation and understanding: Studies in cognitive science* (pp. 185–210). Academic Press.

Wirth, N. (1976). *Algorithms + Data Structures = Programs*. Prentice-Hall.

Wolff, J. G. (2013). The SP theory of intelligence: An overview. *Information, 4*, 283–341.

Word ladder (2024, July 9). In *Wikipedia*. https://en.wikipedia.org/w/index.php?title=Word_ladder&oldid=1233521999

Xi, Z., Chen, W., Guo, X., He, W., Ding, Y., Hong, B., Zhang, M., Wang, J., Jin, S., Zhou, E., Zheng, R., Fan, X., Wang, X., Xiong, L., Zhou, Y., Wang, W., Jiang, C., Zou, Y., Liu, X., . . . Gui, T. (2023). The rise and potential of large language model based agents: A Survey. arXiv:2309.07864v3.

Ye, P., Wang, T., & Wang F.-Y. (2018). A survey of cognitive architectures in the past 20 years. *IEEE Transactions on Cybernetics, 48*(12), 3280–3290.

Young, R. M., Ritter, F. E., & Jones, G. (1996). Introduction to the Soar cognitive architecture. http://acs.ist.psu.edu/misc/nottingham/pst2/hungry-thirsty/full.html

Yu, P, Xu, J., Weston, J., & Kulikov, I. (2024). Distilling system 2 into system 1. arXiv:2407.06023

Zhang, Z., Beck, M. W., Winkler, D. A., Huang, B., Sibanda, W., & Goyal, H. (2018). Opening the black box of neural networks: Methods for interpreting neural network models in clinical applications. *Annals of Translational Medicine, 6*(11), 216.

Zhou, J., Cui, G., Zhang, Z., Yang, C., Liu, Z., & Sun, M. (2019). Graph neural networks: A review of methods and applications. ArXiv, abs/1812.08434.

Zhou, J., & Ustun, V. (2021). PySigma: Towards enhanced grand unification for the Sigma cognitive architecture. In *Proceedings of the 14th Conference on Artificial General Intelligence* (pp. 355–366).

Ziegler, D. M., Stiennon, N., Wu, J., Brown, T. B., Radford, A., Amodei, D., Christiano, P., & Irving, G. (2019). Fine-tuning language models from human preferences. ArXiv:1909.08593.

Zilberstein, S. (1996). Using anytime algorithms in intelligent systems. *AI Magazine, 17*(3), 73–83.

Index

Publisher contact:
The MIT Press
Massachusetts Institute of Technology
77 Massachusetts Avenue, Cambridge, MA 02139
mitpress.mit.edu

EU Authorised Representative:
Easy Access System Europe, Mustamäe tee 50,
10621 Tallinn, Estonia
gpsr.requests@easproject.com

Printed by Integrated Books International,
United States of America